VOLUME EIGHTY EIGHT

CURRENT TOPICS IN DEVELOPMENTAL BIOLOGY

Hox Genes

VOLUME EIGHTY EIGHT

Current Topics in DEVELOPMENTAL BIOLOGY

Hox Genes

Edited by

OLIVIER POURQUIÉ
Howard Hughes Medical Institute
Stowers Institute for Medical Research
Pourquie Laboratory
Kansas City, Missouri, USA

ELSEVIER

AMSTERDAM • BOSTON • HEIDELBERG • LONDON
NEW YORK • OXFORD • PARIS • SAN DIEGO
SAN FRANCISCO • SINGAPORE • SYDNEY • TOKYO
Academic Press is an imprint of Elsevier

Academic Press is an imprint of Elsevier
525 B Street, Suite 1900, San Diego, CA 92101-4495, USA
30 Corporate Drive, Suite 400, Burlington, MA 01803, USA
32, Jamestown Road, London NW1 7BY, UK
Linacre House, Jordan Hill, Oxford OX2 8DP, UK

First edition 2009

ISBN: 978-0-12-374529-3
ISSN: 0070-2153

For information on all Academic Press publications
visit our website at elsevierdirect.com

Printed and bound in USA
09 10 11 12 13 9 8 7 6 5 4 3 2 1

Contents

Contributors

Jeremy S. Dasen
Smilow Neuroscience Program, Department of Physiology and Neuroscience, New York University School of Medicine, New York, NY, USA

Nicolas Denans
Stowers Institute for Medical Research, Kansas City, Missouri, USA

Jacqueline Deschamps
Hubrecht Institute, Developmental Biology and Stem Cell Research, Uppsalalaan, Utrecht, The Netherlands

Walter J. Gehring
Department of Cell Biology, Biozentrum, University of Basel, Klingelbergstrasse Basel, Switzerland

Tadahiro Iimura
Tokyo Medical and Dental University, Global Center of Excellence (GCOE) Program, International Research Center for Molecular Science in Tooth and Bone Diseases, Tokyo, Japan

Thomas M. Jessell
Departments of Neuroscience, and Biochemistry and Molecular Biophysics, Howard Hughes Medical Institute, Kavli Institute for Brain Science, Columbia University, New York, NY, USA

Rohit Joshi
Department of Biochemistry and Molecular Biophysics, Columbia University, New York, NY, USA

François Karch
Department of Zoology and Animal Biology and NCCR Frontiers in Genetics, University of Geneva, Geneva, Switzerland

Urs Kloter
Department of Cell Biology, Biozentrum, University of Basel, Klingelbergstrasse Basel, Switzerland

Robb Krumlauf
Stowers Institute for Medical Research, Kansas City, Missouri, USA, and Department of Anatomy and Cell Biology, Kansas University Medical School, Kansas City, Kansas, USA

Katherine M. Lelli
Department of Genetics and Development, Columbia University, New York, NY, USA

Robert K. Maeda
Department of Zoology and Animal Biology and NCCR Frontiers in Genetics, University of Geneva, Geneva, Switzerland

Richard S. Mann
Department of Biochemistry and Molecular Biophysics, Columbia University, New York, NY, USA

Yuichi Narita
Friedrich Miescher Institute for Biomedical Research, Basel, Switzerland, and Institut de Génétique et de Biologie Moléculaire et Cellulaire, Illkirch Cedex, C.U. de Strasbourg, France

Olivier Pourquié
Stowers Institute for Medical Research, Howard Hughes Medical Institute, Kansas City, Missouri, USA, and Department of Anatomy and Cell Biology, The University of Kansas School of Medicine, Kansas City, Kansas, USA

Filippo M. Rijli
Friedrich Miescher Institute for Biomedical Research, Basel, Switzerland, and Institut de Génétique et de Biologie Moléculaire et Cellulaire, Illkirch Cedex, C.U. de Strasbourg, France

Hiroshi Suga
Department of Cell Biology, Biozentrum, University of Basel, Klingelbergstrasse Basel, Switzerland, and Present address: Barcelona Science Park, Universitat de Barcelona, Barcelona, Spain

Stefan Tümpel
Stowers Institute for Medical Research, Kansas City, Missouri, USA, and Present address: Institut für Molekulare Medizin und Max-Planck-Forschungsgruppe Stammzellalterung, Universität Ulm, Ulm, Germany

Deneen M. Wellik
Department of Internal Medicine, Division of Molecular Medicine & Genetics, and Department of Cell and Developmental Biology, University of Michigan Medical Center, Ann Arbor, Michigan, USA

Leanne M. Wiedemann
Stowers Institute for Medical Research, Kansas City, Missouri, USA, and Department of Pathology and Laboratory Medicine, Kansas University Medical School, Kansas City, Kansas, USA

Teddy Young
Hubrecht Institute, Developmental Biology and Stem Cell Research, Uppsalalaan, Utrecht, The Netherlands

Preface

Hox genes were discovered almost 30 years ago as one of the very first unifying principles in development of Bilaterian species. These genes code for a family of conserved transcription factors which, in most species, are organized in clusters along chromosomal territories. Their action and distribution along the anteroposterior axis of the embryo, as well as their deployment in time, exhibit a striking order which reflects their linear organization on the chromatin. This peculiar arrangement, termed colinearity, was first recognized by Ed Lewis in the fly embryo. Despite very active research during the past three decades, *Hox* gene regulation and function remain extremely mysterious. Particularly, the molecular mechanism underlying the fundamental property of colinearity found in organisms ranging from flies to humans remains unknown.

Hox mutations exhibit spectacular homeotic properties, whereby the identity of a body segment can be changed in that of a different segment. Whereas these effects are now well characterized in vertebrates and invertebrates at the phenotypic level, the molecular details of the targets and functions of Hox proteins underlying these identity changes are poorly understood. Thus, despite the wealth of research focused on *Hox* genes, major questions are still to be answered.

A large body of literature on *Hox* genes has been published since the seminal paper from Ed Lewis in 1978, but strikingly, very few monographs have been devoted to this fascinating topic. Thus, the goal of this book is to provide a comprehensive and up-to-date summary of recent developments in the field of Hox biology. This is a large field and due to space limitations, some areas might be covered more extensively than others. In this book, we cover some history of the characterization of the Hox complexes in the fly, as well as discussions of the organization, regulation, and function of *Hox* genes in patterning the body axis in invertebrates (essentially *Drosophila*) and in vertebrates.

The book begins with a chapter by Robert K. Maeda and François Karch who recapitulates the history of the discovery of the BX-C complex in the fly and describes the striking colinear organization of the *cis*-regulatory elements controlling expression of the *Ubx, AbdA*, and *AbdB* genes initially recognized by Ed Lewis. This organization is strikingly different from that of vertebrate *Hox* clusters where no such colinear distribution of the *cis*-regulatory sequences is observed. Genes involved in the early patterning of the embryo, such as the gap and pair rule genes, control the initiation of *Hox* gene expression, whereas their maintenance requires the genes of the Polycomb and Trithorax complexes that act on chromatin to respectively maintain the repressed or activated configurations of *Hox* genes. The chapter also describes our current understanding of the role of specific chromatin domains and regulators in the colinear regulation of the BX-D *cis*-regulatory elements. The next chapter by Walter Gehring, Urs Kloter, and Hiroshi Suga presents genetic and phylogenetic arguments, supporting the notion that the second thoracic segment in the fly (T2), which is specified by Antenapedia, corresponds to a developmental and evolutionary ground state in Bilaterians. In vertebrates, this ground state would correspond to the thoracic level patterned by the *Hox6* group. Based on these arguments, the authors further argue about the origin of the *Hox* clusters by duplication and unequal crossing-over, leading to the progressive addition of genes in between the two extremities of the cluster. In various circumstances in flies and vertebrates, the posterior *Hox* genes have been shown to be functionally dominant over the anterior ones, a property called posterior prevalence or phenotypic suppression. The arguments developed and the model proposed by Gehring and colleagues in this chapter clearly challenge this notion for the genes expressed anterior to T2 in flies.

Strikingly, very little is known about the mode of action and the targets regulated by Hox proteins. A longstanding paradox in the field is the relative lack of specificity of the Hox binding sequences compared to the exquisite developmental functions assumed by individual Hox proteins. In the third chapter, Richard Mann, Katherine Lelli, and Rohit Joshi discuss this question and argue in favor of different modes of Hox regulation. They survey the different kinds of Hox targets characterized, and distinguish very specific targets recognized by a single paralog, from targets showing less specificity and, hence, recognized by several Hox factors. Hox proteins can act with cofactors, such as the TALE proteins, that help cooperative binding to DNA. This cooperative binding induces conformational changes revealing novel, specific binding properties of the complexes to their DNA targets. Finally, they discuss how Hox proteins also interact with collaborators, forming what they call a Hoxasome, which influences the transcriptional outcome of the Hox-based regulation.

In the fly and vertebrates, *Hox* genes are not expressed in the anterior-most part of the brain (telencephalon, diencephalon, and mesencephalon), which is patterned by other Homeobox-containing genes, such as *Otx* in vertebrates or *orthodenticle* in the fly. In the vertebrate central nervous system, *Hox* genes are involved in the patterning of the hindbrain and the spinal cord. The hindbrain or rhombencephalon corresponds to the posterior part of the brain, which is transiently segmented into seven rhombomeres. These segments define compartments that acquire distinct functional identities during development, and which control a variety of physiological functions such as respiration. In the hindbrain, *Hox* genes are expressed segmentally, with their expression boundaries respecting the rhombomeric frontiers. Mutations in the mouse demonstrated that *Hox* genes play a key role in the control of the identity of the rhombomeres. The regulation and role of *Hox* genes in patterning the vertebrate nervous system has been extensively studied in the hindbrain and more recently in motoneurons. The fourth chapter by Stefan Tumpel, Leanne Wiedemann, and Robb Krumlauf summarizes our current understanding of the role of anterior *Hox* genes in early patterning of the hindbrain in vertebrates. It describes the *cis*-regulatory codes and regulatory networks established during hindbrain differentiation by the *Hox1* to *Hox4* paralog genes which are involved in initiating and maintaining *Hox* gene expression at the appropriate segmental level. Chapter 5, by Yuichi Narita and Filippo Rijli, focuses on the later functions of *Hox* genes in hindbrain development and, more specifically, on their role in the establishment of the complex neuronal connectivity that underlies various important physiological functions and behaviors.

In chapter 6, Jeremy Dasen and Tom Jessell discuss our understanding of the role of *Hox* transcription factors in the patterning of the spinal cord motoneurons. Recent elucidation of the role of these factors in the establishment of the various levels of motoneuron organization such as columns and pools is detailed, as well as the recognition of key cofactors such as FoxP1 in this process.

In vertebrates, aside from the nervous system, the role of *Hox* genes in axial patterning has been examined in great detail at the level of the vertebral column. The spine is progressively formed in a head-to-tail direction during embryogenesis by the rhythmic addition of vertebral precursors, termed somites. The somitic columns formed during embryogenesis become subsequently patterned into different anatomical regions when somitic derivatives differentiate to form the vertebrae and associated muscles. Mouse knock-out experiments have shown that somite regional identity is largely controlled by *Hox* genes. In chapter 7, Tadahiro Iimura, Nicolas Denans, and Olivier Pourquié describe the early colinear activation of *Hox* gene expression in the precursors of the vertebrate spine (the paraxial mesoderm), which roughly positions the Hox expression domains at the appropriate axial level. They discuss how this temporal expression sequence is translated into the characteristic colinear expression domains along the forming axial skeleton. A second repositioning phase involving the segmentation machinery is also required for the definitive positioning and subsequent maintenance of the Hox expression domains in the somites. Interestingly, such a two-step regulation of Hox expression in the developing embryo is also described for fly embryos in chapter 1 and for the hindbrain in chapter 4. Strikingly, in these different systems, it largely relies on different mechanisms.

Related issues are also discussed in chapter 8 by Teddy Young and Jacqueline Deschamps, which is furthermore concerned with the role of the *Cdx* genes in the regulation of *Hox* genes in the embryo. *Cdx* genes belong to the ParaHox cluster, which was proposed to share a common evolutionary origin with the Hox cluster. In chapter 9, Deneen Wellik details the later function of *Hox* genes in patterning the vertebrate axis once the complex nested expression patterns are established in the embryo. The regional patterning of vertebrae was originally proposed to be dependent on the combinatorial action of all Hox proteins in the vertebral precursors—the Hox code. However, this idea was subsequently challenged by the concept of posterior prevalence, which assumed that only the posterior-most genes expressed in a given segment are involved in patterning this segment. The knockout of entire paralog groups in the mouse somehow reconciles these two ideas, demonstrating that whereas posterior genes are clearly dominant over anterior ones, some level of combinatorial functions of adjacent paralog groups are required for the appropriate patterning of vertebrae.

While this book is expected to meet the expectations of Hox aficionados, it is also intended to provide a survey of the field to newcomers. We hope that this book will take its place as a useful tool for those working in the ever growing field of Hox biology.

I am indebted to all the authors for their excellent contributions. I also thank Tara Hoey at Elsevier for her continuous help and support. I am also grateful to Joanne Chatfield for her most valuable editorial assistance and to Silvia Esteban for the cover illustration.

CHAPTER ONE

The Bithorax Complex of *Drosophila*: An Exceptional *Hox* Cluster

Robert K. Maeda *and* François Karch

Contents

Abstract

In his 1978 seminal paper, Ed Lewis described a series of mutations that affect the segmental identities of the segments forming the posterior two-thirds of the *Drosophila* body plan. In each class of mutations, particular segments developed like copies of a more-anterior segment. Genetic mapping of the different classes of mutations led to the discovery that their arrangement along the chromosome paralleled the body segments they affect along the anteroposterior axis of the fly. As all these mutations mapped to the same cytological location, he named this chromosomal locus after its founding mutation. Thus the first homeotic gene (*Hox*) cluster became known as the *bithorax* complex (BX-C).

Even before the sequencing of the BX-C, the fact that these similar mutations grouped together in a cluster, lead Ed Lewis to propose that the homeotic genes arose through a gene duplication mechanism and that these clusters would be

Department of Zoology and Animal Biology and NCCR Frontiers in Genetics, University of Geneva, Geneva, Switzerland

Current Topics in Developmental Biology, Volume 88
ISSN 0070-2153, DOI: 10.1016/S0070-2153(09)88001-0

conserved through evolution. With the identification of the homeobox in the early 1980s, Lewis' first prediction was confirmed. The two cloned *Drosophila* homeotic genes, *Antennapedia* and *Ultrabithorax*, were indeed related genes. Using the homeobox as an entry point, homologous genes have since been cloned in many other species. Today, *Hox* clusters have been discovered in almost all metazoan phyla, confirming Lewis' second prediction. Remarkably, these homologous *Hox* genes are also arranged in clusters with their order within each cluster reflecting the anterior boundary of their domain of expression along the anterior-posterior axis of the animal. This correlation between the genomic organization and the activity along the anteroposterior body axis is known as the principle of "colinearity."

The description of the BX-C inspired decades of developmental and evolutionary biology. And although this first *Hox* cluster led to the identification of many important features common to all *Hox* gene clusters, it now turns out that the fly *Hox* clusters are rather exceptional when compared with the *Hox* clusters of other animals. In this chapter, we will review the history and salient features of bithorax molecular genetics, in part, emphasizing its unique features relative to the other *Hox* clusters.

1. Pseudoallelism and the History of the BX-C

The term "homeotic" was first introduced by William Bateson more than a century ago (1894) to describe phenotypic variations in which "something is changed into the likeness of something else" (Bateson, 1894). The first isolated homeotic mutation was described in 1915 by Calvin Bridges (in Bridges and Morgan, 1923). Like all insects, *Drosophilae* have three thoracic segments (T1, T2, and T3). The landmarks of these thoracic segments are pairs of legs emanating from each of thoracic segments, a pair of wings that develop from the dorsal part of T2, and a pair of flight organs, called halteres, that develop from T3. In Bridges' mutant, the anterior part of T3 develops like a copy of the anterior part of T2. This is visible on the fly as a transformation of the anterior haltere to a structure resembling the anterior part of the wing. As T2 is the most prominent thoracic segment, Bridges named his mutant *bithorax* (*bx*).

In 1919, Bridges isolated a second homeotic mutation showing a somewhat similar homeotic transformation of posterior haltere toward posterior wing. This mutation, which he named *bithoraxoid* (*bxd*), maps to approximately the same region of the *Drosophila* third chromosome as *bx*. Because of the similarity in phenotypes and map location of *bx* and *bxd*, Bridges and Morgan (1923) were surprised to observe that the two mutations complemented. In 1934, a third mutation affecting the identity of T3 was discovered by Hollander (1937). In this case, the effect of the mutation is dominant, with heterozygous flies harboring swollen halteres, a sign of a weak transformation toward wings. Although this mutation has been given many names, it acquired its definitive name, *Ultrabithorax* (*Ubx*), in 1950. Unlike the *bx* or *bxd* mutations, *Ubx* homozygotes die as first instar larvae. Interestingly, although *bx* and *bxd* mutations complement each other, *Ubx* mutations fail to complement both the *bx* and the *bxd* mutations (*bx/Ubx* animals look similar to *bx* homozygous flies and *bxd/Ubx* animals are similar to *bxd* homozygous flies). These complex genetic interactions, where two or more genes appear to occupy the same locus under certain conditions, and different loci under other conditions is called pseudoallelism. Lewis began his undergraduate training in 1937, with the conviction that understanding pseudoallelism was crucial in defining the nature of genes. For more than 30 years, he devoted his research to understanding pseudoallelism using the BX-C as model system (for details, see two excellent perspectives written by Duncan and Montgomery, 2002a,b in *Genetics*). During this time, he identified hundreds of mutations in the BX-C, among which was a deletion of the entire region around *bx*. This deletion displayed an astonishing phenotype that changed the way people thought about the BX-C. Larvae homozygous for this deficiency die at the first instar stage with T3 and all eight abdominal segments (A1-A8) developing like a copy of T2. This phenotype indicated that the *bithorax* locus contained not only "genes" specifying T3, but also other "genes" responsible for the identities of all the abdominal segments. In his 1978 paper, Lewis describes the series of mutations that affect each of these segments. The actual names of these mutations are *abx/bx*, *bxd/pbx*, and *iab-2* through *iab-8* (Lewis, 1978).

Phenotypic analysis indicated that each class of mutation defined an element that was required for the identity of a single segment. Remarkably enough, these elements mapped to the chromosomes in an order that corresponded to the body segment in which they acted. This correspondence between body axis and genomic organization is referred to as "colinearity" (see Figs 1.1 and 1.2). Because these different mutations formed a series of pseudoalleles, it was not entirely clear if they defined individual genes. Thus the term "segment-specific function" was commonly used to refer to the elements of this allelic series.

2. The Ed Lewis Model

Because embryos deficient for the whole *BX-C* have all their segments posterior to T2 developing as copies of T2, Lewis proposed that T2 represents the ground state of development (i.e., the default state) and that each class of mutation represents a segment-specific function that allows a more-posterior segment to differentiate away from the ground state. Furthermore, the fact that mutations in individual segment-specific functions

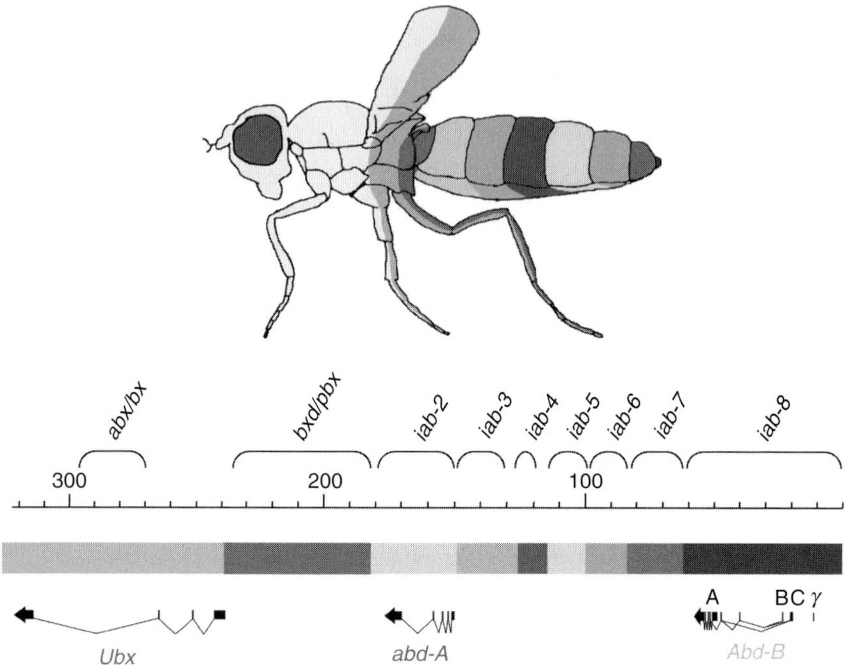

Figure 1.1 Organization of the BX-C. The 300-kb-long genomic DNA of the BX-C is displayed as the multicolored horizontal line. Map coordinate corresponds to the numbering of the original *Drosophila* genome project sequence of Martin *et al.* (1995). The structures of the three transcription units *Ubx*, *abd-A*, and *Abd-B* are depicted below the genomic bar with the arrows indicating the polarity of transcription. The extents of each of the nine segment-specific *cis*-regulatory domains are indicated by the different colors of the genomic DNA. The orange and red regions (*abx/bx* and *bxd/pbx*) regulate the expression of *Ubx*. The regions shaded in blue regulate *and-A* and correspond to *iab-2*, *iab-3*, and *iab-4*. Finally, the regions in green (*iab-5* through *iab-8*) regulate *Abd-B*. The corresponding adult segments affected by the mutations in each *cis*-regulatory domain are indicated in the same color on the drawing of the fly. Reproduced with permission of the Company of Biologists. (See Color Insert.)

always caused homeotic transformations toward the last unaffected more-anterior segment (and not always to T2), meant that everything required for more-anterior segment development had to be present in more-posterior segments. For example, *iab-3* homozygous flies have their A3 segment developing like a copy of A2. Thus, the role of *iab-3*$^+$ function must be to assign segmental identity to A3. However, because A3 is transformed into a copy of A2 instead of T2 (as in the BX-C deficiency), the *abx/bx*$^+$, *bxd/pbx*$^+$, and the *iab-2*$^+$ segment-specific functions required for A2 specification must normally be present in the developing A3 segment. Lewis summarized these findings into two rules: "…a [segment-specific function]

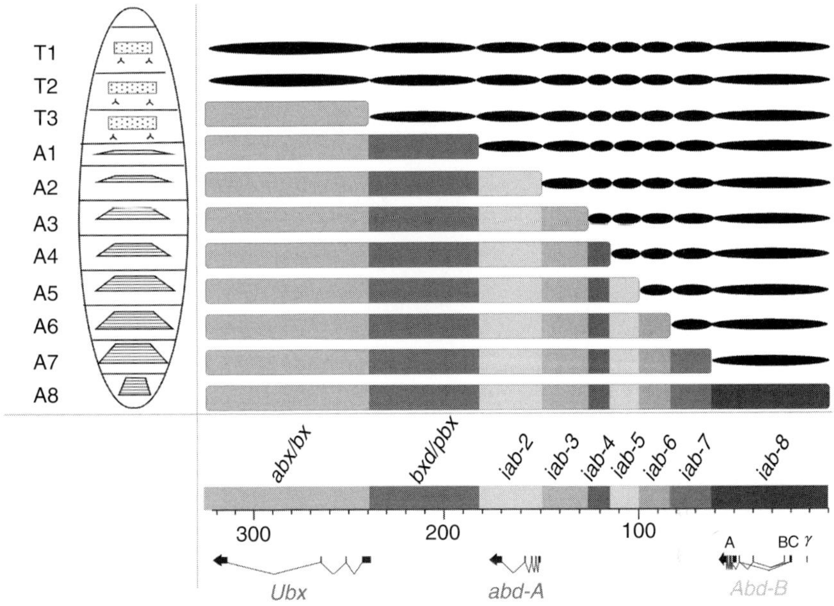

Figure 1.2 The model of Ed Lewis revisited. The diagram of a *Drosophila* larva is depicted on the left (the *y*-axis). The three thoracic segments (T1-T3) and eight abdominal segments (A1-A8) are indicated (as well as the correspondence with the parasegments). The genomic organization of the BX-C is represented in the *X*-axis. In his original version, the model of Ed Lewis was represented as a matrix with more and more dots in the posterior segments, symbolizing the on or off status of the segment-specific functions. In this most updated version, activation along the chromosome of the segment-specific functions (*abx/bx* through *iab-8*) is envisioned as a sequential opening of chromosomal domains (also referred as to the "open for business model"; Peifer *et al.*, 1987). In PS2/T1 and PS3/T2, none of the BX-C function is active, as represented by the black ovals, which symbolize inactive chromosomal domains. The junctions between the adjacent ovals represent the chromosomal boundaries. The domains are maintained inactive by the products of the *Pc-G* genes. In PS5/pT2aT3, the *abx/bx* domain opens, allowing the enhancers residing in the domain to regulate *Ubx* expression in a pattern specific for that parasegment. In parasegment 6/pT3aA1, the adjacent *bxd/pbx* domain opens up to regulate *Ubx* in a pattern specific for that parasegment. Like in its original version, this model envisages that the more posterior a segment is along the anteroposterior axis, the more segment-specific functions are active in it. The *Abd-B* transcription unit positioned 5′ from the *iab-5* through *iab-7* regulatory domains poses a problem to the chromosomal domain hypothesis. Indeed, it is unclear how *iab-5*, *iab-6*, or *iab-7* can regulate *Abd-B* in their respective segments, while their *Abd-B* target promoter still resides in a "closed" domain. While there is no answer to this apparent discrepancy, recent evidences regarding Pc-G regulation suggest that, contrary to the classical picture of their role, Pc-G complexes do not set a repressed chromatin state that is maintained throughout development but have a much more dynamic role. *Pc-G* target genes can become repressed or be reactivated or exist in intermediate states (see for instance Schwartz and Pirrotta, 2008). (See Color Insert.)

derepressed in one segment is derepressed in all segments posterior thereto. . .," and ". . .the more posterior a segment. . .the greater the number of *BX-C* [segment-specific functions] that are in a derepressed state" (Lewis, 1978). Lewis believed that the segment-specific functions acted in an additive fashion to progressively differentiate segments away from T2, an idea supported by the fact that some mutations in anterior segment-specific functions also caused slight changes in more-posterior segments (Fig. 1.2; see below). He synthesized all these findings, along with the peculiar colinearity of the BX-C segment-specific functions into model of where genes "opened" along the chromosome in a segmentally regulated fashion from anterior to posterior (see Fig. 1.2 for a modern visualization of the Lewis model).

It is important to note that all the mutations affecting individual segment-specific functions are viable as homozygotes. Lewis also reported the existence of lethal mutations within the BX-C. The *Ubx* class mentioned above that failed to complement the *bx*, *bxd*, and *pbx* mutations being one of them. In 1985, the groups of Gines Morata in Madrid and Robert Whittle in Sussex independently published two papers describing a different route to isolate mutations in the BX-C (Sanchez-Herrero *et al.*, 1985; Tiong *et al.*, 1985). Using the whole BX-C deficiency chromosome mentioned above, they performed a screen aimed at identifying mutations that fail to complement the lethality of the BX-C deficiency. Three independent complementation groups (each giving rise to homozygous lethality) were identified. One of these corresponded to the *Ubx* mutation. However, the two other complementation groups were new. The first, *abdominal-A* (*abd-A*), affected segments A2–A4, while the second, *Abdominal-B* (*Abd-B*), affected abdominal segments A5–A8.[1]

3. Molecular Genetics of the BX-C

During the course of these genetic screens, the cloning of the whole BX-C was reported in two successive papers (Bender *et al.*, 1983; Karch *et al.*, 1985). The cloning provided the molecular basis to explain much of the genetic data gathered by over the preceding decades. Overall, the BX-C was found to cover ~300 kb of DNA. All the mutations affecting the segment-specific functions were found to be associated with rearrangement breakpoints (such as translocations, inversions, deficiencies, or insertions of transposable elements). The lesions associated with a given class of mutations always clustered in a relatively small part of the *BX-C*, and different

[1] It should be noted that *abd-A* mutations are truly recessive lethal, while *Abd-B* heterozygous flies are sterile (this dominant sterility explains why *Abd-B* is written with a capital A, unlike *abd-A* alleles).

classes of mutations never overlapped. The collinear arrangement of the segment-specific functions along the chromosome and the body segments they specify along the AP axis was also confirmed (Fig. 1.1).

The observation that all the mutations in each class are associated with rearrangement breakpoints not only helped to localize them on the DNA map (more than hundred mutations have been localized) but also suggested that the segment-specific functions were probably not simple protein-coding regions (otherwise point mutations would have been recovered during the numerous screens performed). Further evidence to support this hypothesis came from the Hogness laboratory. With the help of Arthur Kornberg's laboratory, the Hogness lab used overlapping probes to scan developmental Northern blots to identify transcripts. From this analysis, they determined that only about 12 kb of the 300 kb of DNA from the BX-C are present as mature poly(A)$^+$ transcripts (Hogness *et al.*, 1985). By mid-1983, a cDNA spread across a 70 kb span of the DNA was isolated. This 70 kb span of DNA corresponded to the genomic region associated with *Ubx* mutations, and thus, the cDNA was soon identified as the *Ubx* gene product. At the same time, cloning of the *Antennapedia* complex (*Antp-C*) by Rick Garber and Matthew Scott in the Gehring and Kaufmann laboratories, respectively, led to the identification of the *Antp* transcription unit (covering 100 kb of DNA; Garber *et al.*, 1983; Scott *et al.*, 1983). It was not long before sequence comparisons between the two genes revealed a region of DNA similar in both genes. This sequence became known as the homeobox (McGinnis *et al.*, 1984; Scott and Weiner, 1984). The homeobox accelerated the identification of the remaining *Drosophila Hox* genes. Very quickly, two other homeobox genes were identified within the BX-C in the regions where the *abd-A* and *Abd-B* mutations had been mapped (Regulski *et al.*, 1985).

These molecular studies suggested that the whole BX-C encodes only three homeotic genes: *Ubx*, *abd-A*, and *Abd-B*. The first genetic confirmation of this was published in 1987 by Casanova *et al.* (1987), who showed that a *Ubx;abd-A;Abd-B* triple mutant embryo harbored the same phenotype as an embryo carrying a complete deletion of the entire *BX-C*. This hypothesis was later confirmed when the whole region was sequenced (Martin *et al.*, 1995).

But what are the nine segment-specific functions identified by Ed Lewis if genetic and molecular analysis indicates that the *BX-C* only encodes three homeotic proteins? The description of the expression patterns of *Ubx*, *abd-A*, and *Abd-B* brought an answer to this apparent paradox (Beachy *et al.*, 1985; Celniker *et al.*, 1990; Karch *et al.*, 1990; Macias *et al.*, 1990; Sanchez-Herrero, 1991; White and Wicox, 1985). Figure 1.3 shows the central nerve cord of wild-type and various mutant embryos (see below) stained with an antibody directed against *Abd-B*. Like *Ubx* and *abd-A*, though in a more-posterior area, *Abd-B* is expressed in an intricate pattern

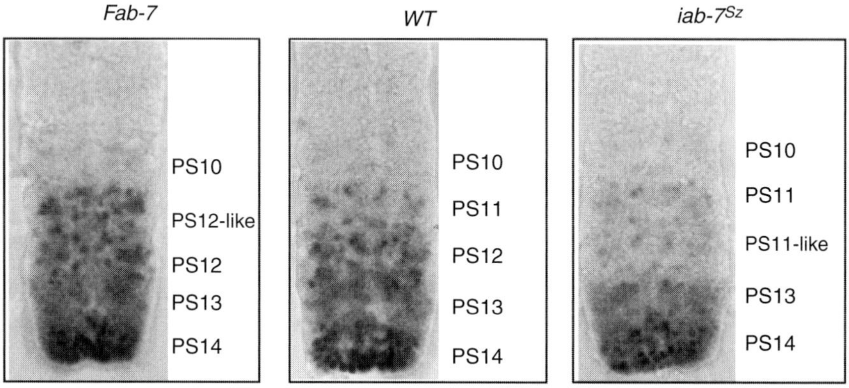

Figure 1.3 *Abd-B* expression in the embryonic central nervous system of WT and mutant embryos. After staining, the central nervous systems were dissected out from 12-h-old embryos. In wild type, the typical *Abd-B* expression pattern is characterized by an anterior-to-posterior gradient from PS10 to 14 in the number of expressing nuclei per parasegment, as well as by the intensity in each nucleus. Note that the Abd-B protein expressed in PS14 is an isoform derived from alternatively spliced transcripts initiating from the B, C, and γ promoters (see Fig. 1.1). In *Fab-7* mutant embryos, the PS11-specific expression pattern is replaced by the pattern expressed in PS12, resulting into the homeotic transformation of PS11/A6 into PS12/A7. In *iab-7*^{Sz}, the whole *iab-7* regulatory domain is deleted. As a consequence, the PS12-specific expression pattern is replaced by the pattern specific for PS11. (See Color Insert.)

that is finely tuned from one parasegment (PS) to the next.[2] By staining various mutant embryos, it was finally understood that the segment-specific functions corresponded to *cis*-regulatory regions that regulate the expression of *Ubx*, *abd-A*, or *Abd-B* in a segment-specific fashion. Mutations in any of the segment-specific regulatory regions alter the expression of its relevant target gene. For example, flies homozygous for the *iab-7*^{Sz} mutation have their seventh abdominal segment transformed into a copy of the sixth. Consistent with this, the embryonic *Abd-B* expression pattern characteristic for PS12/A7 is replaced by the pattern normally present in PS11/A6 (Fig. 1.3; Galloni *et al.*, 1993).

[2] Homeotic gene expression in *Drosophila* does not exactly respect segmental borders. They are shifted, being composed of the posterior part of one segment and the anterior part of the next segment. This unit of expression is called the parasegment (PS) and explains why the adult phenotypes observed in homeotic mutations often affect the posterior part of one segment and the anterior part of the next. For example, the *bxd* mutation mentioned above actually transforms the posterior part of T3 and the anterior part of A1 into copies of the posterior part of T2 and the anterior part of T3, respectively. This is less visible in the abdominal segments, where the anterior portion of each segment is hidden underneath the posterior part of the preceding segment.

The finding that the segment-specific functions correspond to *cis*-regulatory domains helped to explain the phenomenon of pseudoallelism in the BX-C. In Fig. 1.1 the *cis*-regulatory region of the *BX-C* is schematically detailed. The regulatory regions interacting with the *Ubx* gene are shown in red and orange. They include the *abx*/*bx* and *bxd*/*pbx* regions that regulate *Ubx* expression in PS5 and PS6, respectively (Beachy *et al.*, 1985; Little *et al.*, 1990; White and Wicox, 1985).[3] As explained above, *bx* and *bxd* mutations fully complement, but mutations in *Ubx* fail to complement both the *bx* and *bxd* mutations. This can now be explained by the fact that these segment-specific functions require *Ubx* function for their activity. For example, if we look at the contribution of each chromosome to *Ubx* expression independently, a chromosome carrying a *bx* mutation fails to produce Ubx protein in PS5 (where the *bx cis*-regulatory element is normally active), but produces the normal amount of Ubx protein in PS6 (where the *bxd*/*pbx cis*-regulatory element is active). The *Ubx* mutant chromosome in *trans*, however, does not produce a functional Ubx product in PS5 or PS6. The resulting *trans*-heterozygote is therefore $Ubx^{-/-}$ in PS5 but $Ubx^{+/-}$ in PS6. Because segment-specific functions behave as recessive mutations, *bx*/*Ubx* mutants resemble *bx* mutant flies.

4. Initiation and Maintenance Phases in *BX-C* Regulation

How the *cis*-regulatory elements control *BX-C* gene expression has been the focus of much research for the past 20 years. Through this work, it now seems clear that the regulation of homeotic gene expression can be divided into two phases: initiation and maintenance.

The initial determination of the AP axis during *Drosophila* embryogenesis is under the control of three classes of transcription factors that are deployed in a cascade and lead to the subdivision of the embryo into 14 parasegments (the maternal, gap, and pair-rule genes; for reviews see, e.g., DiNardo *et al.*, 1994; Hoch and Jackle, 1993; Ingham, 1988; Kornberg and Tabata, 1993). It is now known that these proteins interact with elements in each of the *cis*-regulatory regions of the *BX-C*, to determine the ultimate homeotic gene pattern (Casares and Sanchez Herrero, 1995; Irish *et al.*, 1989). For example,

[3] *abx* allele stands for *anterobithorax*. Ed Lewis distinguished these alleles from *bx* because they primarily affect the dorsal part of anterior T3. *bx* mutations, on the other hand, affect anterior part of T3 without affecting the dorsal region. However, both types of enhancers are part of the same regulatory region that is active in PS5 (mostly, anterior T3). A similar distinction can be made for the *bxd* and *pbx* elements that are both active in PS6 but in different regions. *pbx* is mostly active in the anterior part of PS6 (mostly corresponding to posterior T3) while *bxd* is mostly active in the posterior part of PS6 (corresponding to anterior A1 in the adult; see Fig. 1.5).

the combination of the gap and pair-rule gene products present in PS12 allows *iab-7*, but not *iab-8*, to control *Abd-B* expression in PS12/A7. However, because the gap and pair-rule genes are only transiently expressed in the early embryo, and the activity states of the segment-specific *cis*-regulatory regions is fixed for the life of the fly, a system to maintain homeotic gene expression is required in each *cis*-regulatory domain (Struhl and Akam, 1985). This maintenance system has been shown to require the products of the *Polycomb group* (*Pc-G*) and *trithorax group* (*trx-G*) genes. While the *Pc-G* products function as negative regulators, maintaining the inactive state of the *cis*-regulatory regions not in use, the *trx-G* products function as positive regulators, maintaining the active state of the active regulatory regions (Kennison, 1993; Paro, 1990; Pirrotta, 1997; Simon, 1995). Both the *Pc-G* and *trx-G* products are thought to maintain the active or inactive state of each parasegment-specific *cis*-regulatory region by modifying the chromatin structure of each region. Indeed, both Pc-G and trx-G proteins contain members that bind, modify or move histones. Thus, the current model suggests that Pc-G proteins compact chromatin to prevent activators from binding to the *cis*-regulatory regions, while trx-G proteins open the chromatin to keep the *cis*-regulatory domains accessible to activators. It should be noticed that recent evidences regarding Pc-G regulation suggest that, contrary to the classical picture of their role, Pc-G complexes do not set a repressed chromatin state that is maintained throughout development but have a much more dynamic role. *Pc-G* target genes can become repressed or be reactivated or exist in intermediate states (see for instance Schwartz and Pirrotta, 2007, 2008). The distinction between the initiation of expression and the maintenance of expression has led to the identification of DNA elements that mediate these distinct phases, cleverly called initiators and maintenance elements (also known as Pc-G or trx-G response elements—PREs/TREs: see next).

5. Initiators, Maintenance Elements, and Segment-Specific Enhancers

Confirmation of the segment-specific and biphasic nature of *BX-C* gene regulation has been done using reporter gene constructs. In these experiments, DNA fragments from the various regulatory regions were cloned upstream of a *lacZ* gene reporter. By making transgenic flies carrying these reporter constructs and studying their resulting patterns of expression, scientists have been able to identify specific DNA fragments from the BX-C that are required for initiating segment-specific expression, maintaining a restricted pattern of expression, and allowing segment-independent,

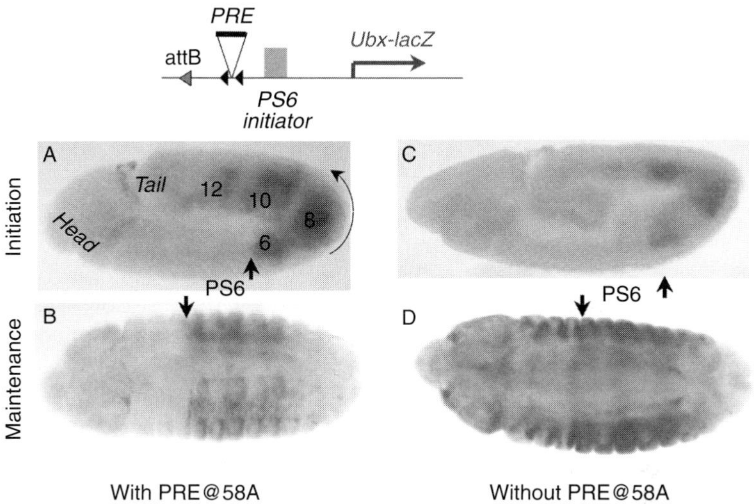

Figure 1.4 Identification of initiation and maintenance elements with lacZ reporter constructs. The structure of the lacZ reporter constructs used in these studies is shown on top of the figure. The plasmids have been transformed into flies using the new ΦC31 recombinase system, and are located in the 58A platform (Bischof *et al.*, 2007). All embryos contain the plasmid with a PS6 initiator element that activates the *lacZ* reporter gene in PS6 and more-posterior parasegments, following a pair-rule pattern (Qian *et al.*, 1991). Early embryos at initiation phase are shown in panels (A) and (C). At this particular stage called "germband extension," parasegments 8–14 curve around toward the dorsal side (as indicated by the curved arrow). At later stages of development (panels B and D), the germband has retracted such as the posterior parasegments are at the posterior pole of the embryo. In panel (B), the presence of the maintenance element (PRE) maintains the initial anterior lacZ expression pattern. In absence of the PRE (panel D) the initial expression pattern degenerates and β-galactosidase is detected anteriorly to the initial PS6 anterior border. The use of the ΦC31 recombinase system allows refuting any positional effect as both constructs are inserted within the same platform. The PRE used in these studies is derived from the *bxd/pbx* regulatory domain (Horard *et al.*, 2000; Sipos *et al.*, 2007). (See Color Insert.)

cell-type-specific expression. Figure 1.4 provides examples to highlight the differences between each type of element.

BX-C initiator elements can be defined as specific types of enhancers that confer a parasegmentally restricted pattern of expression to a reporter gene during early embryogenesis (Barges *et al.*, 2000; Busturia and Bienz, 1993; Mihaly *et al.*, 2006; Muller and Bienz, 1992; Qian *et al.*, 1991; Shimell *et al.*, 2000; Simon *et al.*, 1990; Zhou *et al.*, 1999). For example, Fig. 1.4A and C shows the lacZ expression pattern of an early embryo where lacZ expression is driven by a DNA element derived from the *bxd/pbx* region is normally responsible for *Abd-B* expression in PS6. Likewise, this element drives lacZ expression in PS6 and more-posterior parasegments.[4] This DNA element, therefore, displays the properties expected of an initiator element. It is able to read an early positional address and transmit that information to a promoter. At later stages of embryogenesis, however, the strict anterior border of expression derived from this construct is lost, and lacZ becomes expressed in all the parasegments along the AP axis (Fig. 1.4D). This degeneration of the initial pattern is probably due to the loss of positional information provided in the early embryo by the gap and pair-rule gene products. Supporting this hypothesis, some initiator elements have been mapped precisely enough to show a direct correlation with binding sites for the gap and pair-rule gene products (Qian *et al.*, 1991; Shimell *et al.*, 1994; Zhang *et al.*, 1991; Zhou *et al.*, 1999).

In most cases, the anterior border of expression of a reporter gene controlled by an initiator element is lost when the products of the gap and pair-rule genes decay (at the end of the initiation phase). However, a few larger fragments are able to maintain the initial anterior border of expression of a lacZ reporter. For example, the construct shown in Fig. 1.4A and B contains the same initiator element from *bxd/pbx* but in addition also includes a nearby maintenance element (the so-called BXD PRE; see Horard *et al.*, 2000) that can maintain the appropriate PS6-specific anterior border of expression (compare Fig. 1.4B and D). The ability to maintain the initial pattern of expression has been mapped to a fragment distinct from the initiator, called a maintenance element (Brock and van Lohuizen, 2001). These maintenance elements have since been shown to be binding sites for the Pc-G proteins, and consistent with this, the initial expression pattern is lost in *Pc-G* mutant backgrounds. Because of this, MEs are often referred to as *Polycomb* response elements (PREs). When associated with an initiator element, a maintenance element maintains the anterior limit of expression of a reporter gene throughout late embryogenesis and larval life (Busturia *et al.*, 2001; Chan *et al.*, 1994; Fritsch *et al.*, 1999; Muller and Bienz, 1991; Simon *et al.*, 1993). These maintenance elements do not have an intrinsic segmental address and can maintain different segmental expression patterns when combined with different initiator elements (Chiang *et al.*, 1995).

Cell-type or tissue-specific enhancers constitute a third type of regulatory element that has been identified within the segment-specific *cis*-regulatory regions of the *BX-C* (Barges *et al.*, 2000; Busturia and Bienz, 1993; Mihaly *et al.*, 2006; Pirrotta *et al.*, 1995; Simon *et al.*, 1990). In most cases, these elements confer a cell/tissue-specific expression pattern to a reporter gene that is reiterated in all the parasegments along the AP axis of the embryo. It must be noted, however, that within the *BX-C*, the cell/tissue-specific expression of the homeotic genes is restricted

[4] Note the pair-rule gene pattern.

parasegmentally, even though the enhancers examined outside of the complex are not. This apparent discrepancy between the true expression pattern and the transgenic reporter genes can easily be explained if the enhancers are coordinately regulated by initiators and maintenance elements (see below).

6. Organization of the *Cis*-Regulatory Regions into Chromosomal Domains

How can the various enhancers in a *cis*-regulatory region be coordinately regulated? Three kinds of observations provide compelling evidence that the *cis*-regulatory regions of the *BX-C* are organized into parasegmentally regulated chromosomal domains. All three are genetic in nature. The first came through discussion of the Cbx^{1} mutation in a 1987 review article by Peifer *et al.* (1987). Thus far, we have restricted our description of BX-C mutations to the loss-of-function (LOF) alleles that transform a given parasegment into a copy of the parasegment immediately anterior to it. However, there are also dominant gain-of-function (GOF) mutations that cause the opposite homeotic transformation, where a given parasegment is transformed into a copy of the parasegment immediately posterior to it. In other words, these alleles seem to cause the ectopic activation of a given function, one parasegment anterior to where it should normally be active. These GOF alleles have been instrumental in shaping the logic of the Lewis model.

The Cbx^{1} mutation, standing for *Contrabithorax*, is in some ways the most astonishing BX-C allele. Recovered by Ed Lewis in 1949, this mutation is dominant, with heterozygous flies having small wings (Lewis, 1954). Fine structure mapping by recombination revealed that the mutation is actually associated with two DNA lesions within the BX-C. Separately, the first mutation is associated with a dominant phenotype, mapping just to the right of *bx*. Meanwhile, the second mutation is associated with a recessive LOF phenotype, mapping just to the right of *bxd*. Remarkably, the two mutations cause opposite homeotic transformations. The small wing phenotype, associated with the dominant mutation, is caused by a transformation of the posterior part of the wing into the posterior part of the halter (transformation toward a more-posterior segment). Conversely, the recessive mutation (corresponding to the pbx^{1} mutation; see footnote 3)[5] causes a strong transformation of the posterior part of the halter into posterior wing (transformation toward a more-anterior segment).

[5]It is interesting to note that the initial double mutation Cbx^{1}, pbx^{1} complements for the loss of function of pbx^{1}.

In 1983, cloning of the *Cbx*[1] DNA lesions confirmed the genetic mapping data (Bender *et al.*, 1983). In the initial double mutant, a 17 kb piece of DNA had been deleted from the *bxd/pbx* regulatory region and reinserted 44 kb away, in the *abx/bx* region of the BX-C. Based on this mutation, Peifer *et al.* (1987) developed an ingenious model to explain not only how the *Cbx*[1] mutation works, but also how the whole BX-C might work. According to this model, in the original *Cbx*[1] double mutant, the DNA deleted in the *pbx* region contains enhancers regulating *Ubx* expression in specific cells in PS6 to create the posterior haltere (the D cells in Fig. 1.5). The homozygous deletion of these enhancers leads to the loss of *Ubx* expression in these cells, and the posterior halter is transformed toward the posterior wing (the default, T2 state). This is the same model that had been used to explain most of the LOF mutation in the BX-C. However, the transposition of these enhancers somehow leads to the more drastic, dominant *Cbx* phenotype. This was surprising since enhancers seem to be able to act on their target promoters independent of their position. Therefore, moving them from an upstream position (*bxd/pbx*) to downstream position (*abx/bx*) relative to the *Ubx* promoter should not affect their function. To explain this finding, Peifer *et al.* (1987) postulated that a parasegmental address might be conferred by the DNA domain in which the enhancers reside. In this model, each regulatory element would be imbedded in a DNA domain that would be coordinately activated in a parasegment-specific manner. Therefore, in the *Cbx*[1] mutation, the *pbx* enhancers placed in the *abx/bx* domain would be activated in the cells equivalent to the cells they activate in PS6, but in PS5, transforming the posterior wing into posterior haltere.

Since this early hypothesis, other evidences have supported the parasegment-specific domain model. The first comes from enhancers trap lines within the BX-C. In *Drosophila*, transgenic animals are generally made using P-element transposons. These transposons insert throughout the genome in a fairly random fashion. If these P-elements contain a basal promoter and a reporter construct, they can often be used to visualize the enhancer elements around an insertion site. This technique is called enhancer trapping (O'Kane and Gehring, 1987). A number of enhancer trap lines have been isolated in the BX-C. For example, the anterior border of lacZ expression for three transposons inserted within the ~75 kb region corresponding to the orange domain in Fig. 1.1 is PS5. Within this region, lies the *abx/bx* *cis*-regulatory region that regulates *Ubx* expression in PS5. Although the promoters of these three P-elements are obviously trapping different enhancer activities from the 75 kb of DNA, they are all active in PS5 and more-posterior parasegments, regardless of their position in this region. Meanwhile the anterior parasegmental boundary of expression for three inserts within the red regulatory region is shifted one parasegment toward the posterior, to PS6. This domain corresponds to the region that contains the *bxd/pbx* *cis*-regulatory region that drives *Ubx* expression in PS6.

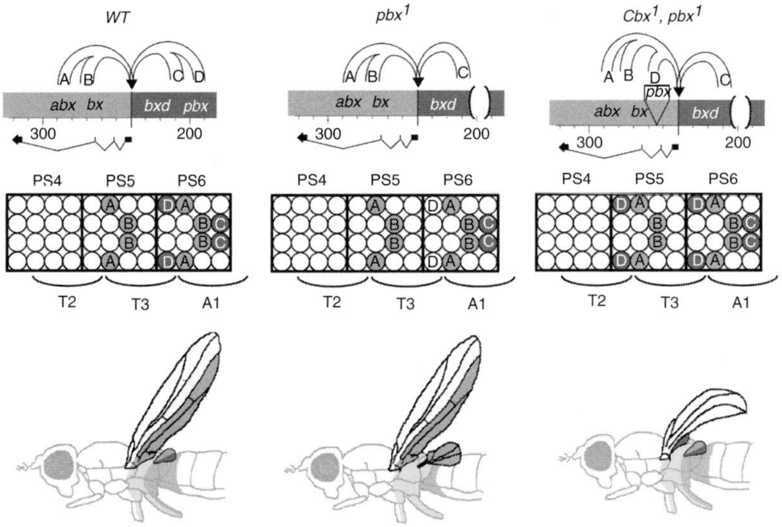

Figure 1.5 Organization of the *cis*-regulatory regions in parasegment-specific chromosomal domains. The first row depicts the *abx/bx* and *bxd/pbx* regions regulating *Ubx* expression in PS5 and PS6, respectively, in WT, in pbx^1 and in Cbx^1pbx^1. The second row represents a cartoon of an embryonic central nervous system with cells expressing *Ubx* (the correspondence between parasegmental and segmental boundaries is indicated). In *WT*, *Ubx* appears from PS5 at a relatively low level in a few cells. We envision that *Ubx* expression in A cells is regulated by the *abx* enhancers while in B cells, *Ubx* is activated by the *bx* enhancer. These expression patterns are reiterated in the posterior parasegments. In PS6, more cells express *Ubx* at a higher level under the control of *bxd* enhancers in the C cells and under the control of *pbx* enhancers in D cells. In pbx^1 mutants, the lack of the *pbx* enhancers result in the loss of *Ubx* expression in the D cells of PS6. Note that these D cells belong to the posterior part of T3 giving rise the posterior part of the halter in the adult (bottom row). In the absence of *Ubx* in D cells, the posterior part of the halter develops like the posterior wing. In the initial Cbx^1, pbx^1 double mutant the 17 kb fragment deleted in pbx^1 has transposed in the opposite orientation 44 kb away, in the second intron of the *Ubx* transcription unit. This dominant gain-of-function mutation results in the transformation of the posterior part of the wing into posterior halter; the *pbx* enhancers functions in the D cells, but one parasegment ahead. This observation led to the idea that cell specificity is provided by the enhancers and parasegmental address by the DNA domain (see text).

Examining the large number of enhancer trap lines isolated in the *BX-C* (Bender and Hudson, 2000) made two striking observations. First, lines spread out over quite great distances often have the same parasegmental anterior border of expression, while other lines, located just a few kbs away display a different anterior border of expression. Second, the anterior border of lacZ expression always progresses toward the posterior by increments of one parasegment. Although these lines are trapping different enhancers, lines can be grouped by the parasegment in which they start to express.

These observations are in perfect agreement with the initial proposal made by Peifer *et al.* (1987) (see above) that the *BX-C* enhancers reside in chromosomal domains that are coordinately regulated in parasegment-specific fashion (see also Maeda and Karch, 2006).

7. Chromatin Boundaries Flank the Parasegment-Specific Domains

The third line of evidence supporting the domain hypothesis is the presence of specialized elements called boundary elements. One prediction made by the domain hypothesis is the existence of elements to limit the extent of each domain. Indeed, the sharp parasegmental transition in expression of the enhancer trap lines supports this hypothesis. In Figs 1.1 and 1.2, the boundaries are symbolized by the sharp color transition between the adjacent domains symbolized by the colored rectangles. The presence of a boundary is postulated between each of the regulatory domains. Thus far, three boundaries, *Mcp*, *Fab-7*, and *Fab-8*, have been conclusively identified through molecular and mutational analysis (Barges *et al.*, 2000; Gyurkovics *et al.*, 1990; Karch *et al.*, 1994; Mihaly *et al.*, 1997, 1998). All three boundary mutations are associated with a dominant gain-of-function phenotype. The best characterized of them is *Fab-7* which separates the *iab-6 cis*-regulatory domain from the *iab-7 cis*-regulatory domain. In *Fab-7* mutants, PS11/A6 is transformed toward PS12/A7 identity (posterior-oriented transformation). In effect, *iab-7*, which is normally active only in PS12/A7, is activated one parasegment ahead, within PS11/A6. In agreement with this, *Abd-B* expression is regulated in a PS12-like pattern,[6] transforming cell identity from PS11 to PS12 (Fig. 1.3; Galloni *et al.*, 1993; Mihaly *et al.*, 1996). How can a deletion in a regulatory region result in a dominant gain-of-function phenotype? Perhaps the simplest explanation is to assume that the deletion removes binding sites for a repressor or silencer that normally keeps *iab-7* inactive in parasegments anterior to PS12. However, this does not seem to be the case. To understand the dominant *Fab-7* mutation, second-site mutations that revert the dominant phenotype were isolated (see Gyurkovics *et al.*, 1990).[7] Three classes of mutants were recovered that suppress the *Fab-7* phenotype. The first class carried mutations in the structural part of the *Abd-B* gene

[6] versus expression in PS11, *Abd-B* expression in PS12 is higher and expressed in more cells.

[7] In this experiment, *Fab-7*^{1} homozygous males were treated with X-rays and crossed to wild-type females. Nearly all progeny of this cross are heterozygotes (*Fab-7*^{1}/+) and should show the dominant transformation of PS11/A6 into PS12/A7. If the X-ray treatment hit a region necessary for the *Fab-7* phenotype, a fly will be easily recognized because of its wild-type appearance of PS11/A6.

itself. This is not surprising in as much as *Fab-7* is a mutation that causes the overexpression of *Abd-B* in PS11 (see footnote 5). By knocking down *Abd-B* in *cis*, the overexpression of *Abd-B* is prevented. The second class of revertants carried chromosomal rearrangements that disrupt the *iab-7* domain so that the *Fab-7* deletion, as well as the *iab-6* and *iab-5* regions, become separated from the *Abd-B* gene. Such rearrangements cause the loss of *Abd-B* expression in PS10 to PS12 and flies homozygous for these mutations have PS10–12 transformed into PS9(A4). This class confirms that to obtain the dominant *Fab-7* phenotype, *iab-7* and the *Fab-7* deletion must be in *cis* with the *Abd-B* gene. The third class of mutants, however, is the most instructive. All members of this class of mutations carry lesions that inactivate the *iab-6* domain. This result implies that the *Fab-7* transformation depends upon the presence of an intact *iab-6* in *cis*. This finding is critical because it rules out the hypothesis that the *Fab-7* region is simply a silencer that negatively regulates *iab-7* in PS11. Indeed, if this hypothesis was true, it would not be possible to revert the *Fab-7* phenotype by inactivating *iab-6*. Moreover, the fact that in the *iab-6* revertants, *Abd-B* expression appears normal in PS12 indicates that the *Fab-7* deletion does not remove sequences essential for *iab-7* activity. Based on these reversion experiment, it was proposed that in *Fab-7* mutants, the *iab-6* and *iab-7* domains become fused into a single functional unit with mixed characteristics: parasegment *specificity* being provided by *iab-6* (initiation), while parasegment *identity* is provided by *iab-7* (the enhancers driving *Abd-B* expression) (Gyurkovics *et al.*, 1990). In a case very similar to the Cbx^1 example, enhancers from one domain become controlled by the initiator of another domain. *Fab-7*, therefore, behaves as a domain boundary between the *iab-6* and *iab-7* *cis*-regulatory domains. Similar findings have been found for the *Mcp* boundary that separates *iab-4* from *iab-5*, and the *Fab-8* boundary that separates *iab-7* from *iab-8* (see Barges *et al.*, 2000; Karch *et al.*, 1985; 1994). Recently, the existence of a *Fab-6* boundary was also described by inference from the phenotypes of two relatively large deletions that fuse the *iab-5* and *iab-7* domains (Mihaly *et al.*, 2006).

8. Boundaries Versus Insulators and Long-Distance Interactions

Nearly all chromatin domain boundaries have been identified as chromatin insulators (such as the scs/scs′, gypsy, and β-globin 5′HS4 insulators; for review see, e.g., Valenzuela and Kamakaka, 2006). In flies, insulator activity is determined by a transgenic assay. DNA fragments suspected of insulator activity are placed between an enhancer and the promoter of a reporter gene. If the DNA fragment is able to suppress the reporter gene when placed in between the enhancer and the promoter (but not when placed elsewhere), the fragment is considered an insulator. Because the activity of boundary elements is reminiscent of the activity of chromatin insulators, separating enhancers, each boundary element has been tested for insulator activity. In the transgenic assay described above, each of the three molecularly isolated BX-C boundary elements (*Mcp*, *Fab-7*, and *Fab-8*) has been proven to have insulator activity (Barges *et al.*, 2000; Gruzdeva *et al.*, 2005; Hagstrom *et al.*, 1996; Zhou *et al.*, 1996, 1999). However, this finding leads to a paradox. Insulators, by definition, block enhancers from interacting with a target promoter when placed in between these two elements. In the BX-C, the boundary elements, like *Fab-7* and *Fab-8*, are located in between *BX-C* enhancers and their target promoter. How then can these enhancers ever reach their target promoter over so many intervening insulators? The answer to this paradox is still a mystery. However, two sets of experiments have suggested possible answers.

In 1999, Zhou and Levine asked for specific DNA fragments that could aid distal enhancers in bypassing intervening boundaries. The result of these experiments was the identification of an element that they called the promoter targeting sequence (PTS). This element, normally located in the *iab-7* domain, just adjacent to the *Fab-8* boundary is able to allow distal enhancers to bypass the *Fab-8* boundary in transgenic assays. Later, it was shown that this PTS element is also able to aid an enhancer in bypassing other insulators (like the gypsy insulator), suggesting that PTS function is independent of the insulator itself (Zhou and Levine, 1999). Recently, a new PTS element has been found in the *iab-6* domain (Chen *et al.*, 2005). Based on these results, it seems possible that each boundary element may be flanked by a PTS element to aid in insulator bypass. Although this is an attractive hypothesis, studies from our lab have complicated this issue. We recently recovered simple deletions of both known PTS element (in *iab-6* and *iab-7*, respectively). These deletions are not associated with visible phenotype (Mihaly *et al.*, 2006) indicating that if the PTS hypothesis is correct, each domain must contain multiple, redundant PTS elements.

The second possible resolution for this paradox is to simply disprove the existence of the paradox. What do we really know about the phenomenon of insulation, and can we resolve a mechanism for insulation that is consistent with the placement of boundary elements in the BX-C? To gain insights into the possible mechanism of insulation, it is best to first examine what is perhaps the best studied insulator in *Drosophila*, the gypsy insulator.

The gypsy retrotransposon contains 12 reiterated suppressor of Hairy-wing (Su(Hw)) protein-binding sites that are absolutely required for its ability to insulate (Geyer and Corces, 1992; Geyer *et al.*, 1988;

Parkhurst *et al.*, 1988; Peifer and Bender, 1988).[8] The Su(Hw) protein, however, is not specific to the gypsy retrotransposon. On polytene chromosomes, the Su(Hw) protein is bound to hundreds of native sites throughout the *Drosophila* genome. Because of the insulator activity of gypsy elements, it was proposed that the native *Drosophila* Su(Hw)-binding sites might normally insulate chromosomal regions from the influence of neighboring regions. The first clue to how this could be accomplished was discovered by Gerasimova *et al.* (2000). As mentioned above, on *Drosophila* polytene chromosomes, there are hundreds of Su(Hw)-binding sites throughout the genome. But in the preparation of polytene chromosomes, the cells of the larval salivary gland are flattened on a slide, causing the nuclei to explode, and the chromosomes to spread flat on the surface of the slide (hence the term, "polytene chromosome squash"). However, when Gerasimova and Corces stained normal, intact diploid cells for the Su (Hw) protein, instead of finding the diffuse chromosomal staining they expected to see from a protein that bind to so many sites, they found only about 20–25 spots of intense staining along the nuclear periphery.[9] Based on this staining and other experiments, they hypothesized that Su(Hw) insulators might interact with each other to organize the chromatin domains. Insulators, then, could be explained as simple chromatin-chromatin interaction motifs, creating chromosomal domains by forming chromatin loops. Consequently, insulation would be explained by having the enhancer in one domain physically blocked from interacting with a promoter in another domain by an intervening insulator.

This model was supported by the finding of gypsy insulator bypass. It turns out that if one places two gypsy insulators in between an enhancers and a promoter, instead of one, the insulator activity of the each gypsy is nullified (Cai and Shen, 2001; Muravyova *et al.*, 2001). Although the exact physical mechanism behind this bypass is unclear, when combined with previous findings, insulator bypass adds credence to the idea that insulators are chromatin-chromatin interaction motifs. In the *BX-C*, where there are often many boundary elements, this is a very attractive model.

Work from our lab provided support for the idea that BX-C boundaries are involved in mediating long-distance chromatin interactions. In this case, however, the interactions were found to take place between a boundary element and a target promoter. Using a modified DamID method, we were able to document an association between the *Fab-7* boundary element and a region near the *Abd-B* promoter (Cleard *et al.*, 2006). Interestingly, this

[8] Interestingly, the gypsy retrotransposon was first cloned from insertions in the BX-C (mutant alleles that all shared the characteristic of being suppressible by a second-site mutation in the *Suppressor of Hairy-wing* gene; Modolell *et al.*, 1983).

[9] Recently, these finding have been called into question by Golovnin *et al.* (2008) who claim that the aggregates found by Gerasimova and Corces are protein aggregates that do not contain the actual insulators.

interaction was only found in anterior tissues (where *Abd-B* is silenced), and not in more-posterior tissues (where *Abd-B* is expressed), suggesting that this interaction is regulated along the AP axis. Furthermore, we showed that this interaction is absolutely dependant on the presence of the *Fab-7* boundary element. Based on these observations and boundary element genetics, we have proposed a model for the regulation of *Abd-B* in which boundaries play an active role in targeting the enhancer regions to their promoters (Fig. 1.6). This model suggests that boundary elements are required to tether the inactive *cis*-regulatory domains to a region near the *Abd-B* promoter. In doing so, boundaries would form chromatin domains, keeping each domain autonomous and in part preventing enclosed enhancer from interacting with the *Abd-B* promoter. If the domain remains inactive, then these tethered domains would become silenced in a Pc-G-dependent manner. If the domain becomes activated, the boundary element

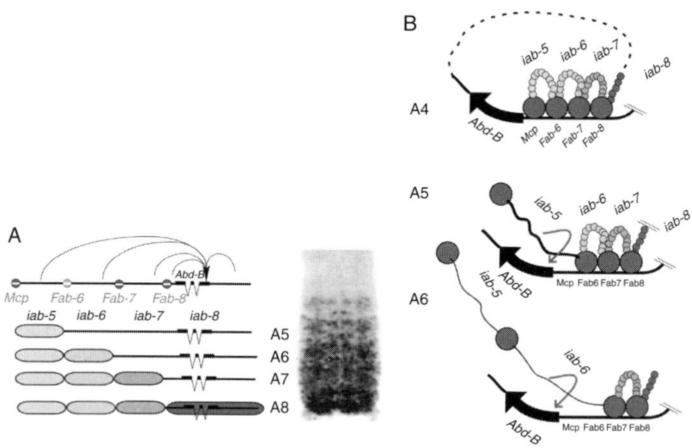

Figure 1.6 Panel (A) shows the state of each of the *iab-5* through *iab-8* domains that regulate *Abd-B* in the fifth, sixth, seventh, and eight abdominal segments, respectively (A5–A8). The boundaries are depicted as one-way signs (note that *Fab-6* is inferred from indirect evidences described in Mihaly *et al.* (2006)). Panel (B): boundary tethering model. In this model, boundaries are represented by red circles, the inactive regulatory regions are covered by green circles (representing Polycomb silencing), and active regulatory regions are depicted by black lines. Based on the DamID results (Cleard *et al.*, 2006), we believe that the boundaries tether the inactive *cis*-regulatory domains to a region near the *Abd-B* promoter. In doing so, boundaries form chromatin domains, keeping each domain autonomous and preventing the imbedded enhancers from interacting with the *Abd-B* promoter. Once a domain is activated, the boundary element would release from the tethering region and allow the formerly enclosed enhancers to interact with *Abd-B* promoter. For example, in A5, *Mcp* is released allowing the enhancers contained in *iab-5* to activate *Abd-B*. Since the next downstream regulatory domain (*iab-6*) is still tethered by the next boundary (*Fab-6*), only the appropriate regulatory *iab-5* domain is able to regulate *Abd-B* in A5. The elements are not drawn at scales. (See Color Insert.)

would release from the tethering region, prior to Pc-G-dependent silencing, and allow the formerly enclosed enhancers to interact with the nearby *Abd-B* promoter. Because the next downstream element would still be tethered by the next boundary, this would place the appropriate *cis*-regulatory elements in close proximity to the *Abd-B* promoter. This model is summarized in Fig. 1.6.

Of the nine postulated boundaries, only four have been identified genetically (*Mcp*, *Fab-6*, *Fab-7*, and *Fab-8*). Sequence comparisons of the *Drosophila* insulators have not allow highlighting a particular factor common to all of them (for review, see Maeda and Karch, 2007). Recently, new evidence has hinted at the exact locations of the other boundary elements in the BX-C. CTCF is a well-conserved zinc finger DNA-binding factor present in species from flies to humans. In vertebrates, most characterized insulators contain binding sites recognized by the CTCF factor (for review, see Wallace and Felsenfeld, 2007). In 2005 it was found that the fly homologue of CTCF binds to the *Fab-8* boundary element and is required for *Fab-8* insulator function (Moon *et al.*, 2005). A genome-wide study of the distribution of dCTCF-binding sites has been recently been performed by two groups (Holohan *et al.*, 2007; Mohan *et al.*, 2007). Strikingly, although the sequence comparisons did not originally predict CTCF-binding sites, the distribution of dCTCF protein within the BX-C coincides almost perfectly to the regions where boundary elements are positioned by the enhancer trap data. CTCF binding is also found near the *Abd-B* promoter. This finding hints at a possible association between CTCF-type boundaries and the *Abd-B* promoter, as predicted by our model above. In light of this, it is interesting to note that Kyrchanova *et al.* (2008) recently showed that the *Fab-8* boundary is able to mediate insulator bypass when combined with the *Abd-B* promoter.

9. Mixing the Old and the New

The Lewis model implies that once a segment-specific function is activated in the proper segment, it remains active and contributes to the identity of the more-posterior segments (Fig. 1.2). In this view, segmental identity reflects the buildup of active segment-specific functions that progressively differentiate segments toward more-posterior identity (a cumulative effect). As a result, the more posterior a segment is, the more differentiated away from the T2 ground state. This model is very attractive because it makes sense of the expression patterns of *Ubx*, *abd-A*, and *Abd-B*. In each of their realms of action, the intensity of expression of the three *Hox* genes increase in the more-posterior parasegments, as if each regulatory domain was adding a new layer of pattern on the patterns established by more-anterior domains. This is particularly visible for the *Abd-B* gene. Expression of *Abd-B* in the central nervous system seems to increase toward the posterior in a parasegmental step gradient from PS10/A5 to PS13/A8 (Fig. 1.3).

The cumulative model is supported by many observations. Perhaps the most convincing is the fact that some mutations in anterior segment-specific functions also cause slight changes in more-posterior segments. For example, *bxd* mutant flies not only have A1 developing like a copy of T3, but also have ventral pits, a thoracic feature, forming in A2–A7. Another piece of evidence in support of the cumulative model is the expression pattern of the BX-C enhancer trap (see Maeda and Karch, 2006) and initiator transgenic lines (see Fig. 1.4), which remain active in the segments posterior to the segment they specify.

While there are many pieces of information indicating that the segment-specific regulatory domains can be active in more than one segment, the available data suggesting that they are active in driving homeotic gene expression in more-posterior segments is not so clear. At least for the *Abd-B* portion of the BX-C, the data quite clearly suggests that *cis*-regulatory domains function in only one parasegment. As mentioned above, most of the early mutations in segment-specific functions were associated with chromosome rearrangement breaks that affect more than one segment-specific function. In the *Abd-B* region of the BX-C, many smaller internal deletions have been generated, allowing a more precise correlation between phenotype and lesion (Mihaly *et al.*, 2006). One such deletion is the *iab-7*^{Sz} deficiency, which deletes the entire *iab-7* domain. As expected, this deletion causes A7/PS12 to transform into A6/PS11 (Fig. 1.3). If *iab-7* was also required to determine the identity of the more-posterior segment, A8, then one would expect to find defects in A8/PS13. This is not what is found. In *iab-7*^{Sz} mutant flies, all abdominal segments outside of A7 are perfectly normal. This phenotype indicates that while *iab-7* is absolutely required in A7/PS12, it is dispensable for the identity of all other segments. It is important to note, however, that A7/PS12 is still transformed into a perfect copy of A6/PS11, which means that *iab-6* is capable of functioning in A7/PS12. This is also visible by the PS11-like expression pattern that is present in PS12 (see Fig. 1.3). Therefore, we believe that more-anterior domains remain capable of functioning in more-posterior domains, but only in the absence of an active posterior domain.

Internal deletions that affect more than one *cis*-regulatory region confirm these conclusions. In the *iab-6,7*^{IH} deletion, for example, where both the *iab-6* and the *iab-7* domains are deleted, both A6 and A7 acquire an A5 identity. However, A8/PS13 identity is again not affected by the deletion even though, in this case, two more-anterior domains are missing. Once again, the more-anterior *iab-5* domain remains capable of acting posterior to A5/PS10, but only in the absence of *iab-6* and *iab-7*. It is worthwhile noting these observations are compatible with the enhancer trap expression patterns that are active in more than one segment. Indeed, once a regulatory domain is released from regulating *Abd-B* in the more-posterior segments (i.e., *iab-5* in PS11/A6, PS12/A7, and PS13/A8), it can still regulate the nearby *lacZ* reporter gene. This does not mean, however, that it will regulate its target *Hox* gene promoter.

While the available data clearly demonstrate that the regulatory domains can be active in segments posterior to the segment they specify, it is not entirely understood how they are activated in those posterior segments. As mentioned above, experiments with lacZ reporter constructs have revealed the existence of initiator elements. In nearly all cases, the initiators activate a *lacZ* reporter gene in a pair-rule fashion, that is to say, in every other parasegment (Barges *et al.*, 2000; Busturia and Bienz, 1993; Mihaly *et al.*, 2006; Muller and Bienz, 1992; Qian *et al.*, 1991; Shimell *et al.*, 2000; Simon *et al.*, 1990; Zhou *et al.*, 1999). For example, this is the case for the PS6 initiator, which turns on a reporter gene in A1/PS6, A3/PS8, and A15/PS10 (Fig. 1.4). Yet, we know from mutational analysis that initiators can turn on in all segments posterior to the first segment activated. *iab-6*, for example, is active in A7/PS11 in the *iab-*7^{Sz} deficiency (as A7 is transformed into A6, see above and figure). Given that the activity state of each regulatory domain is thought to rely on the binding of appropriate gap and pair-rule gene products to the initiator elements during early embryogenesis, we still do not understand how initiators activate domains outside of the pair-rule pattern. One possibility is that there may be more than one initiator per regulatory domain. Thus far, however, all data have pointed toward having only one initiator (or one type of initiator) in each of the regulatory domains.[10]

Although many initiator elements have been discovered, the mechanism of initiation remains elusive. As mentioned above, initiators read a parasegmental address, probably through the binding of gap and pair-rule genes. But they must then transmit this information to the rest of the domain. How is this communication accomplished? One hypothesis currently being looked at is intergenic transcription. This hypothesis suggests that initiators bind gap and pair-rule genes to activate internal promoters to drive transcription across a *cis*-regulatory domain. The act of transcription would then spread a signal to all of the *cis*-regulatory elements in the domain. There is some evidence to support this model. First, intergenic transcription has been known in the BX-C for some time now (Cumberledge *et al.*, 1990; Lipshitz *et al.*, 1987; Sanchez-Herrero and Akam, 1989). Moreover, some of these transcripts precede the activation of the homeotic genes and are approximately expressed in the parasegments corresponding to the activity of the

[10] The *abx/bx* domain being the primary exceptions (Qian *et al.*, 1991; Simon *et al.*, 1990).

domain in which they reside (Bae *et al.*, 2002; Drewell *et al.*, 2002). And lastly, three recent works have shown that forced early transcription through a PRE maintenance element will permanently inactivate its ability to silence (Bender and Fitzgerald, 2002; Hogga and Karch, 2002; Schmitt *et al.*, 2005). Inactivation of the PREs would then leave the enhancers in the domain open for activator binding. Yet even with this data, it is unclear if the transcription is the cause or result of initiation. This is obviously a difficult question to answer. The current approach of studying initiators in transgenes is now limiting. Perhaps with the development of homologous recombination in *Drosophila*, these questions will finally be answered.

10. Colinearity in the BX-C

The evolutionary conservation of *Hox* gene clustering and colinearity suggests a conserved mechanism to *Hox* gene regulation that requires clustering. Surprisingly, although colinearity and clustering were first discovered in *Drosophila*, it turns out that the *Drosophila Hox* clusters are quite different from the textbook picture of a highly ordered *Hox* cluster. The *Drosophila melanogaster Hox* clusters are actually only half clusters. In many other animals, the homologues of the *Ant-C* and the BX-C are clustered together. This begs the question of whether or not *Hox* clustering and colinearity are actually important in flies. Although an intense discussion of this topic would be beyond the scope of this review (and the knowledge of the authors), we would like to point out a few areas where *Drosophila* genetics may provide insight. First, we must define what we mean by clustering and colinearity. With regards to the *Drosophila Hox* complexes, there is twofold meaning to clustering and colinearity. First relates to the way clustering and colinearity were first discovered in the BX-C and deals with the *cis*-regulatory domains. The second has to do location of the *Hox* genes (Hox protein-coding sequences) themselves.

With regards to the *cis*-regulatory domains, it seems quite clear that at least clustering is important. Mutations that translocate parts of the *cis*-regulatory domains to other areas of the chromosome can have drastic effects on gene expression. For example, in the *iab-7*^{770} mutation, a whole region of the third chromosome including the BX-C until *iab-7* is translocated to another area (T68;89E;91;94; Celniker *et al.*, 1990). This leaves only the *iab-8* and *iab-9* regions next to the *Abd-B* gene. As expected, in the *iab-7*^{770} mutation, A6-A7 are transformed into A5.[11] In this example,

[11] Note that when *iab-5* is moved away from the *Abd-B* promoter, A5 is normal. This is thought to be due to *iab-5*'s ability to regulate *abd-A* in the absence of a functional *Abd-B* promoter (Celniker *et al.*, 1990; Hendrickson and Sakonju, 1995; Karch *et al.*, 1985).

A7 is transformed in A5 because the lesion destroys the *cis*-regulatory domain. Presumably, A6 is transformed into A5 due to the displacement of *iab-6* from the vicinity of the *Abd-B*. Therefore, perhaps unsurprisingly, clustering of the *cis*-regulatory domains is important. Unfortunately, the importance of colinearity in the *D. melanogaster cis*-regulatory domains is still not known, as no small inversions have been found that do not affect multiple genes.

As for the clustering and colinearity of the genes themselves, the answer is more complex. As you recall, the *D. melanogaster Hox* genes are located in two separate complex: the *Antp-C* and the BX-C. These two complexes are both located on the *D. melanogaster* third chromosome, separated by about 10,000 kb. The break between the two complexes separates the *Antp* gene and the *Ubx* gene. However, if we disregard the separation, the genes themselves are indeed arranged along the chromosome in the order that they work from anterior to posterior. Based on these findings, it appears that in *Drosophila*, clustering is not absolutely required for *Hox* gene function. Many other evidences support this hypothesis. For example, in other species of *Drosophila* as well as in the silk moth *Bombyx mori*, there have been other breaks and rearrangements that not only cause a splitting of the complex, but also a break up in the colinearity (Lewis *et al.*, 2003; Negre *et al.*, 2003; Von Allmen *et al.*, 1996; Yasukochi *et al.*, 2004). In *Drosophila buzzatii*, the complex has been slit and rearranged to give two complexes with the order: *labial*, *abd-A*, *Abd-B* (complex 1) and *Ubx*, *Antp*, *Scr*, *Dfd*, *pb* (complex 2; Negre and Ruiz, 2007). And yet, all evidences suggest that *D. buzzatii* expresses the *Hox* genes in the same relative pattern as its cousin *D. melanogaster*. Results from mutational analysis in *D. melanogaster* also support the dispensability of *Hox* gene clustering and ordering, as most of the phenotypes associated with rearrangement breaks affecting the *cis*-regulatory domains can be attributed solely to the lesions themselves. For example, with the *iab-7*^{770} inversion mentioned above, the flies display only a phenotype consistent with *Abd-B* misexpression, as only *Abd-B cis*-regulatory domains have been affected. *Ubx* and *abd-A* expression remained unchanged, even though most of the BX-C is moved to another location and is inverted.

Based on the data from other *Drosophila* species and mutational analysis in *D. melanogaster*, it seems that *Hox* clustering and colinearity in the fly are simply remnants of a past mechanism which required clustering and colinearity. This is definitely a possibility. But if this was true, what kind of mechanism could have existed to account for the clustering and colinearity? One hypothesis centers on the idea of temporal colinearity. In many organisms, like mice, there is a second kind of colinearity called temporal colinearity. Temporal colinearity refers to the order of the homeotic genes along the chromosome corresponding to the timing of the initial expression the *Hox* genes (Duboule, 1992). Therefore, in many organisms the more spatially anterior genes are placed on one end of the complex and begin to be expressed earlier than more-posterior genes. This is particularly easy to see in the vertebrate somites, where new somites form sequentially from anterior to posterior, and progressively express more-posterior *Hox* genes. Because the genes start off in a silenced state and then sequentially become expressed, a model has been put forth suggesting that the *Hox* complex might start off grouped together in a silenced complex and overtime have genes activated by separating from this complex (or becoming separated from the complex by gene activation). Based on this model, the need for temporal colinearity could have been the driving force behind chromosomal clustering and chromosomal colinearity, as organized clusters would simplify the formation of the silenced complex and organize its sequential release.

So, what can *Drosophila* add to this model? *Drosophila* do not, for the most part, have temporal colinearity. As a long-germband insect, *Drosophila* develop all their segments simultaneously, with the *Hox* genes turning on, more or less at the same time. But *Drosophila* is only one type of insect. Other insects develop by sequentially budding off new segments, much like vertebrate somitic development. In these so-called short or intermediate germband insects, temporal colinearity does exist to some extent, giving credence to the idea that temporal colinearity could have driven clustering and chromosomal colinearity. In this view, it is worthwhile mentioning that *Tribolium*, an intermediate germband insect contains an intact cluster *Hox* cluster (Shippy *et al.*, 2008). While this model may be correct, we would like to argue that the *Drosophila Hox* clusters still require clustering and that this requirement could be part of the evolutionary driving force to keep the homeotic genes together.

Earlier in this section, we have pointed out pieces of evidence suggesting that *Hox* genes in *Drosophila* do not have to be clustered. Although this evidence is substantial, there is an intriguing phenomenon, called the *cis*-overexpression effect (or COE effect), that complicates the issue (a so-called "fly in the ointment"). The phenomenon, once again discovered by Ed Lewis, is that breaks in certain *cis*-regulatory regions not only caused loss-of-function phenotypes in the *cis*-regulatory domain in which the lesion took place (as above), but also caused a dominant overexpression phenotype in the *cis*-regulatory domain immediately anterior to it (Lewis, 1985). For example, breaks in *iab-4* not only cause an A4-A3 transformation, but also show a transformation of A2 toward A3. This finding suggests that *iab-3* (responsible for A3 development) must become activated one segment too early. Although not all BX-C mutants display this effect, many do. The exceptions mostly turn out to be cases where the homeotic genes themselves are destroyed or where the *cis*-regulatory domain that would be hyperactivated is separated form the gene it is to hyperactivate (like *iab-6* in the *iab-7*^{770} mutation). What the COE effect shows is that breaks in the complex often result in a lack of appropriate silencing in other parts of the complex, suggesting that, even in flies, clustering could be important to keep genes silenced (see also discussion in Chiang *et al.*, 1995). What we know about homeotic gene silencing supports this interpretation. Within the *Drosophila Hox* clusters, silencing is thought to be accomplished primarily by the Pc-G proteins. Many observations on the Pc-G proteins suggest that the local concentration of Pc-G proteins is important for gene silencing. First, many Pc-G proteins display dominant loss-of-function phenotypes, suggesting that Pc-G silencing is extremely concentration dependent (for review, see Paro, 1990). Secondly, Pc-G response elements often work better in pairs. Due to a phenomenon called pairing sensitive repression, we know that heterozygous PRE reporter constructs often showing leaky expression of a reporter gene, become more completely silenced when homozygous (Bantignies *et al.*, 2003; Chan *et al.*, 1994; Gindhart and Kaufman, 1995; Hagstrom *et al.*, 1997; Kassis *et al.*, 1991; Muller *et al.*, 1999; Vazquez *et al.*, 2006). And third, Pc-G proteins seem to cluster in the nucleus in elements called Polycomb bodies (Buchenau *et al.*, 1998). Polycomb bodies are far less numerous than the number of PREs in the genome and therefore, it is believed that Pc silencing occurs only at distinct hubs. Because of the concentration effect of Pc-G protein, we believe that the homeotic genes may need clustering to attract enough Pc-G proteins to properly silence inactive domains. Without clustering, the *cis*-regulatory domains become incompletely silenced, resulting in visible phenotypes when the transitions zones between active and inactive domains are near the translocation breakpoints. But then, why do breaks seem to occur so often in the *Drosophila* genus? One possible explanation is that stable breaks could be permitted in the *Hox* complexes if breaks relocated the misplaced genes near another source of Pc. This type of break might not be as uncommon as one might believe given the number of PREs and Polycomb bodies in the nucleus. So, if this is true, then perhaps flies and vertebrates do share a somewhat common mechanism in the control of *Hox* gene expression that could be the driving force behind *Hox* gene clustering and colinearity.

REFERENCES

Bae, E., Calhoun, V. C., Levine, M., Lewis, E. B., and Drewell, R. A. (2002). Characterization of the intergenic RNA profile at *abdominal-A* and *abdominal-B* in the *Drosophila* bithorax complex. *Proc. Natl. Acad. Sci. USA* **99**, 16847–16852.

Bantignies, F., Grimaud, C., Lavrov, S., Gabut, M., and Cavalli, G. (2003). Inheritance of Polycomb-dependent chromosomal interactions in *Drosophila*. *Genes Dev.* **17,** 2406–2420.

Barges, S., Mihaly, J., Galloni, M., Hagstrom, K., Müller, M., Shanower, G., Schedl, P., Gyurkovics, H., and Karch, F. (2000). The *Fab-8* boundary defines the distal limit of the bithorax complex *iab-7* domain and insulates *iab-7* from initiation elements and a PRE in the adjacent *iab-8* domain. *Development* **127,** 779–790.

Bateson, W. (1894). "Materials for the Study of Variation Treated with Especial Regards to Discontinuity in the Origins of Species." Macmillan, New York.

Beachy, P. A., Helfand, S. L., and Hogness, D. S. (1985). Segmental distribution of bithorax complex proteins during *Drosophila* development. *Nature* **313,** 545–551.

Bender, W., and Fitzgerald, D. P. (2002). Transcription activates repressed domains in the *Drosophila* bithorax complex. *Development* **129,** 4923–4930.

Bender, W., and Hudson, A. (2000). P element homing to the *Drosophila* bithorax complex. *Development* **127,** 3981–3992.

Bender, W., Akam, M., Karch, F., Beachy, P. A., Peifer, M., Spierer, P., Lewis, E. B., and Hogness, D. S. (1983). Molecular genetics of the bithorax complex in *Drosophila melanogaster. Science* **221,** 23–29.

Bischof, J., Maeda, R. K., Hediger, M., Karch, F., and Basler, K. (2007). An optimized transgenesis system for *Drosophila* using germ-line-specific phiC31 integrases. *Proc. Natl. Acad. Sci. USA* **104,** 3312–3317.

Bridges, C. B., and Morgan, T. H. (1923). "The Third-Chromosome Group of Mutant Characters of *Drosophila melanogaster.*" The Carnegie Institution, Washington, DC.

Brock, H. W., and van Lohuizen, M. (2001). The Polycomb group—No longer an exclusive club?. *Curr. Opin. Genet. Dev.* **11,** 175–181.

Buchenau, P., Hodgson, J., Strutt, H., and Arndt-Jovin, D. J. (1998). The distribution of Polycomb-group proteins during cell division and development in *Drosophila* embryos: Impact on models for silencing. *J. Cell Biol.* **141,** 469–481.

Busturia, A., and Bienz, M. (1993). Silencers in *abdominal-B*, a homeotic *Drosophila* gene. *EMBO J.* **12,** 1415–1425 (ISSN: 0261-4189).

Busturia, A., Lloyd, A., Bejarano, F., Zavortink, M., Xin, H., and Sakonju, S. (2001). The MCP silencer of the *Drosophila Abd-B* gene requires both Pleiohomeotic and GAGA factor for the maintenance of repression. *Development* **128,** 2163–2173.

Cai, H. N., and Shen, P. (2001). Effects of *cis* arrangement of chromatin insulators on enhancer-blocking activity. *Science* **291,** 493–495.

Casanova, J., Sanchez-Herrero, E., Busturia, A., and Morata, G. (1987). Double and triple mutant combinations of the bithorax complex of *Drosophila. EMBO J.* **6,** 3103–3109.

Casares, F., and Sanchez Herrero, E. (1995). Regulation of the infraabdominal regions of the bithorax complex of *Drosophila* by gap genes. *Development* **121,** 1855–1866 (ISSN: 0950-1991).

Celniker, S. E., Sharma, S., Keelan, D. J., and Lewis, E. B. (1990). The molecular genetics of the bithorax complex of *Drosophila*: *Cis*-regulation in the *Abdominal-B* domain. *EMBO J.* **9,** 4277–4286.

Chan, C. S., Rastelli, L., and Pirrotta, V. (1994). A Polycomb response element in the *Ubx* gene that determines an epigenetically inherited state of repression. *EMBO J.* **13,** 2553–2564 (ISSN: 0261-4189).

Chen, Q., Lin, L., Smith, S., Lin, Q., and Zhou, J. (2005). Multiple promoter targeting sequences exist in *Abdominal-B* to regulate long-range gene activation. *Dev. Biol.* **286,** 629–636.

Chiang, A., O'Connor, M. B., Paro, R., Simon, J., and Bender, W. (1995). Discrete Polycomb-binding sites in each parasegmental domain of the bithorax complex. *Development* **121,** 1681–1689 (ISSN: 0950-1991).

Cleard, F., Moshkin, Y., Karch, F., and Maeda, R. K. (2006). Probing long-distance regulatory interactions in the *Drosophila melanogaster* bithorax complex using Dam identification. *Nat. Genet.* **38,** 931–935.

Cumberledge, S., Zaratzian, A., and Sakonju, S. (1990). Characterization of two RNAs transcribed from the *cis*-regulatory region of the abd-A domain within the *Drosophila* bithorax complex. *Proc. Natl. Acad. Sci. USA* **87,** 3259–3263.

DiNardo, S., Heemskerk, J., Dougan, S., and O'Farrell, P. H. (1994). The making of a maggot: Patterning the *Drosophila* embryonic epidermis. *Curr. Opin. Genet. Dev.* **4,** 529–534.

Drewell, R. A., Bae, E., Burr, J., and Lewis, E. B. (2002). Transcription defines the embryonic domains of *cis*-regulatory activity at the *Drosophila* bithorax complex. *Proc. Natl. Acad. Sci. USA* **99,** 16853–16858.

Duboule, D. (1992). The vertebrate limb: A model system to study the *Hox/HOM* gene network during development and evolution. *Bioessays* **14,** 375–384.

Duncan, I., and Montgomery, G. (2002a). E. B. Lewis and the bithorax complex: Part I. *Genetics* **160,** 1265–1272.

Duncan, I., and Montgomery, G. (2002b). E. B. Lewis and the bithorax complex: Part II. From *cis-trans* test to the genetic control of development. *Genetics* **161,** 1–10.

Fritsch, C., Brown, J. L., Kassis, J. A., and Muller, J. (1999). The DNA-binding Polycomb group protein Pleiohomeotic mediates silencing of a *Drosophila* homeotic gene. *Development* **126,** 3905–3913.

Galloni, M., Gyurkovics, H., Schedl, P., and Karch, F. (1993). The bluetail transposon: Evidence for independent *cis*-regulatory domains and domain boundaries in the bithorax complex. *EMBO J.* **12,** 1087–1097 (ISSN: 0261-4189).

Garber, R. L., Kuroiwa, A., and Gehring, W. J. (1983). Genomic and cDNA clones of the homeotic locus *Antennapedia* in *Drosophila. EMBO J.* **2,** 2027–2036.

Gerasimova, T. I., Byrd, K., and Corces, V. G. (2000). A chromatin insulator determines the nuclear localization of DNA. *Mol. Cell* **6,** 1025–1035.

Geyer, P. K., and Corces, V. G. (1992). DNA position-specific repression of transcription by a *Drosophila* zinc finger protein. *Genes Dev.* **6,** 1865–1873.

Geyer, P. K., Green, M. M., and Corces, V. G. (1988). Mutant gene phenotypes mediated by a *Drosophila melanogaster* retrotransposon require sequences homologous to mammalian enhancers. *Proc. Natl. Acad. Sci. USA* **85,** 8593–8597.

Gindhart, J. G. Jr., and Kaufman, T. C. (1995). Identification of Polycomb and trithorax group responsive elements in the regulatory region of the *Drosophila* homeotic gene Sex combs reduced. *Genetics* **139,** 797–814 (ISSN: 0016-6731).

Golovnin, A., Melnikova, L., Volkov, I., Kostuchenko, M., Galkin, A. V., and Georgiev, P. (2008). 'Insulator bodies' are aggregates of proteins but not of insulators. *EMBO Rep.* **9,** 440–445.

Gruzdeva, N., Kyrchanova, O., Parshikov, A., Kullyev, A., and Georgiev, P. (2005). The *Mcp* element from the bithorax complex contains an insulator that is capable of pairwise interactions and can facilitate enhancer-promoter communication. *Mol. Cell. Biol.* **25,** 3682–3689.

Gyurkovics, H., Gausz, J., Kummer, J., and Karch, F. (1990). A new homeotic mutation in the *Drosophila* bithorax complex removes a boundary separating two domains of regulation. *EMBO J.* **9,** 2579–2585 (ISSN: 0261-4189).

Hagstrom, K., Muller, M., and Schedl, P. (1996). *Fab*-7 functions as a chromatin domain boundary to ensure proper segment specification by the *Drosophila* bithorax complex. *Genes Dev.* **10,** 3202–3215.

Hagstrom, K., Muller, M., and Schedl, P. (1997). A *Polycomb* and GAGA dependent silencer adjoins the *Fab*-7 boundary in the *Drosophila* bithorax complex. *Genetics* **146,** 1365–1380.

Hendrickson, J. E., and Sakonju, S. (1995). *Cis* and *trans* interactions between the iab regulatory regions and *abdominal-A* and *abdominal-B* in *Drosophila melanogaster. Genetics* **139,** 835–848 (ISSN: 0016-6731).

Hoch, M., and Jackle, H. (1993). Transcriptional regulation and spatial patterning in *Drosophila*. *Curr. Opin. Genet. Dev.* **3,** 566–573.

Hogga, I., and Karch, F. (2002). Transcription through the iab-7 *cis*-regulatory domain of the bithorax complex interferes with maintenance of Polycomb-mediated silencing. *Development* **129,** 4915–4922.

Hogness, D. S., Lipshitz, H. D., Beachy, P. A., Peattie, D. A., Saint, R. B., Goldschmidt-Clermont, M., Harte, P. J., Gavis, E. R., and Helfand, S. L. (1985). Regulation and products of the Ubx domain of the bithorax complex. *Cold Spring Harb. Symp. Quant. Biol.* **50,** 181–194.

Hollander, W. F. (1937). Bithorax alleles. *Dros. Inf. Serv.* **8,** 77.

Holohan, E. E., Kwong, C., Adryan, B., Bartkuhn, M., Herold, M., Renkawitz, R., Russell, S., and White, R. (2007). CTCF genomic binding sites in *Drosophila* and the organisation of the bithorax complex. *PLoS Genet.* **3,** e112.

Horard, B., Tatout, C., Poux, S., and Pirrotta, V. (2000). Structure of a Polycomb response element and *in vitro* binding of Polycomb group complexes containing GAGA factor. *Mol. Cell. Biol.* **20,** 3187–3197.

Ingham, P. W. (1988). The molecular genetics of embryonic pattern formation in *Drosophila*. *Nature* **335,** 25–34.

Irish, V. F., Martinez-Arias, A., and Akam, M. (1989). Spatial regulation of the *Antennapedia* and *Ultrabithorax* homeotic genes during *Drosophila* early development. *EMBO J.* **8,** 1527–1537.

Karch, F., Weiffenbach, B., Peifer, M., Bender, W., Duncan, I., Celniker, S., Crosby, M., and Lewis, E. B. (1985). The abdominal region of the bithorax complex. *Cell* **43,** 81–96.

Karch, F., Bender, W., and Weiffenbach, B. (1990). abdA expression in *Drosophila* embryos. *Genes Dev.* **4,** 1573–1587.

Karch, F., Galloni, M., Sipos, L., Gausz, J., Gyurkovics, H., and Schedl, P. (1994). *Mcp* and *Fab-7*: Molecular analysis of putative boundaries of *cis*-regulatory domains in the bithorax complex of *Drosophila melanogaster*. *Nucleic Acids Res.* **22,** 3138–3146 (ISSN: 0305-1048).

Kassis, J. A., VanSickle, E. P., and Sensabaugh, S. M. (1991). A fragment of engrailed regulatory DNA can mediate transvection of the white gene in *Drosophila*. *Genetics* **128,** 751–761 (ISSN: 0016-6731).

Kennison, J. A. (1993). Transcriptional activation of *Drosophila* homeotic genes from distant regulatory elements. *Trends Genet.* **9,** 75–78.

Kornberg, T. B., and Tabata, T. (1993). Segmentation of the *Drosophila* embryo. *Curr. Opin. Genet. Dev.* **3,** 585–594.

Kyrchanova, O., Toshchakov, S., Podstreshnaya, Y., Parshikov, A., and Georgiev, P. (2008). Functional interaction between the *Fab-7* and *Fab-8* boundaries and the upstream promoter region in the *Drosophila Abd-B* gene. *Mol. Cell. Biol.* **28,** 4188–4195.

Lewis, E. B. (1954). Pseudoallelism and the gene concept. *Proc. IX Int. Congr. Genet. Caryol. Suppl.* **100-105**.

Lewis, E. B. (1978). A gene complex controlling segmentation in *Drosophila*. *Nature* **276,** 565–570.

Lewis, E. B. (1985). Regulation of the genes of the bithorax complex in *Drosophila*. *Cold Spring Harb. Symp. Quant. Biol.* **50,** 155–164.

Lewis, E. B., Pfeiffer, B. D., Mathog, D. R., and Celniker, S. E. (2003). Evolution of the homeobox complex in the Diptera. *Curr. Biol.* **13,** R587–R588.

Lipshitz, H. D., Peattie, D. A., and Hogness, D. S. (1987). Novel transcripts from the *Ultrabithorax* domain of the bithorax complex. *Genes Dev.* **1,** 307–322.

Little, J. W., Byrd, C. A., and Brower, D. L. (1990). Effect of *abx*, *bx* and *pbx* mutations on expression of homeotic genes in *Drosophila* larvae. *Genetics* **124,** 899–908.

Macias, A., Casanova, J., and Morata, G. (1990). Expression and regulation of the *abd-A* gene of *Drosophila. Development* **110,** 1197–1207.

Maeda, R. K., and Karch, F. (2006). The ABC of the BX-C: The bithorax complex explained. *Development* **133,** 1413-1422.

Maeda, R. K., and Karch, F. (2007). Making connections: Boundaries and insulators *in Drosophila. Curr. Opin. Genet. Dev.* **17,** 394-399.

Martin, C. H., Mayeda, C. A., Davis, C. A., Ericsson, C. L., Knafels, J. D., Mathog, D. R., Celniker, S. E., Lewis, E. B., and Palazzolo, M. J. (1995). Complete sequence of the bithorax complex of *Drosophila. Proc. Natl. Acad. Sci. USA* **92,** 8398–8402 (ISSN: 0027-8424).

McGinnis, W., Levine, M. S., Hafen, E., Kuroiwa, A., and Gehring, W. J. (1984). A conserved DNA sequence in homoeotic genes of the *Drosophila Antennapedia* and bithorax complexes. *Nature* **308,** 428–433.

Mihaly, J., Hogga, I., Gausz, J., Gyurkovics, H., and Karch, F. (1997). In situ dissection of the *Fab-7* region of the bithorax complex into a chromatin domain boundary and a Polycomb-response element. *Development* **124,** 1809–1820.

Mihaly, J., Hogga, I., Barges, S., Galloni, M., Mishra, R. K., Hagstrom, K., Muller, M., Schedl, P., Sipos, L., Gausz, J., *et al.* (1998). Chromatin domain boundaries in the Bithorax complex. *Cell. Mol. Life Sci.* **54,** 60–70.

Mihaly, J., Barges, S., Sipos, L., Maeda, R., Cleard, F., Hogga, I., Bender, W., Gyurkovics, H., and Karch, F. (2006). Dissecting the regulatory landscape of the *Abd-B* gene of the bithorax complex. *Development* **133,** 2983–2993.

Modolell, J., Bender, W., and Meselson, M. (1983). *Drosophila melanogaster* mutations suppressible by the suppressor of Hairy-wing are insertions of a 7.3-kilobase mobile element. *Proc. Natl. Acad. Sci. USA* **80,** 1678–1682.

Mohan, M., Bartkuhn, M., Herold, M., Philippen, A., Heinl, N., Bardenhagen, I., Leers, J., White, R. A., Renkawitz-Pohl, R., Saumweber, H., *et al.* (2007). The *Drosophila* insulator proteins CTCF and CP190 link enhancer blocking to body patterning. *EMBO J.* **26,** 4203–4214.

Moon, H., Filippova, G., Loukinov, D., Pugacheva, E., Chen, Q., Smith, S. T., Munhall, A., Grewe, B., Bartkuhn, M., Arnold, R., *et al.* (2005). CTCF is conserved from *Drosophila* to humans and confers enhancer blocking of the *Fab-8* insulator. *EMBO Rep.* **6,** 165–170.

Muller, J., and Bienz, M. (1991). Long range repression conferring boundaries of Ultrabithorax expression in the *Drosophila* embryo. *EMBO J.* **10,** 3147–3155 (ISSN: 0261-4189).

Muller, J., and Bienz, M. (1992). Sharp anterior boundary of homeotic gene expression conferred by the fushi tarazu protein. *EMBO J.* **11,** 3653–3661 (ISSN: 0261-4189).

Muller, M., Hagstrom, K., Gyurkovics, H., Pirrotta, V., and Schedl, P. (1999). The *Mcp* element from the *Drosophila melanogaster* bithorax complex mediates long-distance regulatory interactions. *Genetics* **153,** 1333–1356.

Muravyova, E., Golovnin, A., Gracheva, E., Parshikov, A., Belenkaya, T., Pirrotta, V. V., and Georgiev, P. (2001). Loss of insulator activity by paired Su(Hw) chromatin insulators. *Science* **291,** 495–498.

Negre, B., and Ruiz, A. (2007). HOM-C evolution in *Drosophila*: Is there a need for *Hox* gene clustering? *Trends Genet.* **23,** 55–59.

Negre, B., Ranz, J. M., Casals, F., Caceres, M., and Ruiz, A. (2003). A new split of the *Hox* gene complex in *Drosophila*: Relocation and evolution of the gene labial. *Mol. Biol. Evol.* **20,** 2042–2054.

O'Kane, C. J., and Gehring, W. J. (1987). Detection in situ of genomic regulatory elements in *Drosophila. Proc. Natl. Acad. Sci. USA* **84,** 9123–9127.

Parkhurst, S. M., Harrison, D. A., Remington, M. P., Spana, C., Kelley, R. L., Coyne, R. S., and Corces, V. G. (1988). The *Drosophila su(Hw)* gene, which controls the phenotypic effect of the gypsy transposable element, encodes a putative DNA-binding protein. *Genes Dev.* **2,** 1205–1215.

Paro, R. (1990). Imprinting a determined state into the chromatin of *Drosophila. Trends Genet.* **6,** 416–421 (ISSN: 0168-9525).

Peifer, M., and Bender, W. (1988). Sequences of the gypsy transposon of *Drosophila* necessary for its effects on adjacent genes. *Proc. Natl. Acad. Sci. USA* **85,** 9650–9654.

Peifer, M., Karch, F., and Bender, W. (1987). The bithorax complex: Control of segmental identity. *Genes Dev.* **1,** 891–898.

Pirrotta, V. (1997). Chromatin-silencing mechanisms in *Drosophila* maintain patterns of gene expression. *Trends Genet.* **13,** 314–318.

Pirrotta, V., Chan, C. S., McCabe, D., and Qian, S. (1995). Distinct parasegmental and imaginal enhancers and the establishment of the expression pattern of the *Ubx* gene. *Genetics* **141,** 1439–1450.

Qian, S., Capovilla, M., and Pirrotta, V. (1991). The bx region enhancer, a distant *cis*-control element of the *Drosophila Ubx* gene and its regulation by hunchback and other segmentation genes. *EMBO J.* **10,** 1415–1425.

Regulski, M., Harding, K., Kostriken, R., Karch, F., Levine, M., and McGinnis, W. (1985). Homeobox genes of the *Antennapedia* and bithorax complexes of *Drosophila. Cell* **43,** 71–80.

Sanchez-Herrero, E. (1991). Control of the expression of the bithorax complex genes *abdominal-A* and *abdominal-B* by *cis*-regulatory regions in *Drosophila* embryos. *Development* **111,** 437–449.

Sanchez-Herrero, E., and Akam, M. (1989). Spatially ordered transcription of regulatory DNA in the bithorax complex of *Drosophila. Development* **107,** 321–329.

Sanchez-Herrero, E., Vernos, I., Marco, R., and Morata, G. (1985). Genetic organization of *Drosophila* bithorax complex. *Nature* **313,** 108–113.

Schmitt, S., Prestel, M., and Paro, R. (2005). Intergenic transcription through a Polycomb group response element counteracts silencing. *Genes Dev.* **19,** 697–708.

Schwartz, Y. B., and Pirrotta, V. (2007). Polycomb silencing mechanisms and the management of genomic programmes. *Nat. Rev. Genet.* **8,** 9–22.

Schwartz, Y. B., and Pirrotta, V. (2008). Polycomb complexes and epigenetic states. *Curr. Opin. Cell Biol.* **20,** 266–273.

Scott, M. P., and Weiner, A. J. (1984). Structural relationships among genes that control development: Sequence homology between the *Antennapedia, Ultrabithorax,* and *fushi tarazu* loci of *Drosophila. Proc. Natl. Acad. Sci. USA* **81,** 4115–4119.

Scott, M. P., Weiner, A. J., Hazelrigg, T. I., Polisky, B. A., Pirrotta, V., Scalenghe, F., and Kaufman, T. C. (1983). The molecular organization of the *Antennapedia* locus of *Drosophila. Cell* **35,** 763–776.

Shimell, M. J., Simon, J., Bender, W., and O'Connor, M. B. (1994). Enhancer point mutation results in a homeotic transformation in *Drosophila. Science* **264,** 968–971 (ISSN: 0036-8075).

Shimell, M. J., Peterson, A. J., Burr, J., Simon, J. A., and O'Connor, M. B. (2000). Functional analysis of repressor binding sites in the iab-2 regulatory region of the *abdominal-A* homeotic gene. *Dev. Biol.* **218,** 38–52.

Shippy, T. D., Ronshaugen, M., Cande, J., He, J., Beeman, R. W., Levine, M., Brown, S. J., and Denell, R. E. (2008). Analysis of the Tribolium homeotic complex: Insights into mechanisms constraining insect *Hox* clusters. *Dev. Genes Evol.* **218,** 127–139.

Simon, J. (1995). Locking in stable states of gene expression: Transcriptional control during *Drosophila* development. *Curr. Opin. Cell Biol.* **7,** 376–385 (ISSN: 0955-0674).

Simon, J., Peifer, M., Bender, W., and O'Connor, M. (1990). Regulatory elements of the bithorax complex that control expression along the anterior-posterior axis. *EMBO J.* **9,** 3945–3956 (ISSN: 0261-4189).

Simon, J., Chiang, A., Bender, W., Shimell, M. J., and O'Connor, M. (1993). Elements of the *Drosophila* bithorax complex that mediate repression by Polycomb group products. *Dev. Biol.* **158,** 131–144 (ISSN: 0012-1606).

Sipos, L., Kozma, G., Molnar, E., and Bender, W. (2007). In situ dissection of a Polycomb response element in *Drosophila melanogaster*. *Proc. Natl. Acad. Sci. USA* **104,** 12416–12421.

Struhl, G., and Akam, M. (1985). Altered distributions of *Ultrabithorax* transcripts in extra sex combs mutant embryos of *Drosophila*. *EMBO J.* **4,** 3259–3264.

Tiong, S., Bone, L. M., and Whittle, J. R. (1985). Recessive lethal mutations within the bithorax complex in *Drosophila*. *Mol. Gen. Genet.* **200,** 335–342.

Valenzuela, L., and Kamakaka, R. T. (2006). Chromatin insulators. *Annu. Rev. Genet.* **40,** 107–138.

Vazquez, J., Muller, M., Pirrotta, V., and Sedat, J. W. (2006). The *Mcp* element mediates stable long-range chromosome-chromosome interactions in *Drosophila*. *Mol. Biol. Cell* **17,** 2158–2165.

Von Allmen, G., Hogga, I., Spierer, A., Karch, F., Bender, W., Gyurkovics, H., and Lewis, E. (1996). Splits in fruitfly *Hox* gene complexes. *Nature* **380,** 116.

Wallace, J. A., and Felsenfeld, G. (2007). We gather together: Insulators and genome organization. *Curr. Opin. Genet. Dev.* **17,** 400–407.

White, R. A. H., and Wicox, M. (1985). Regulation of the distribution of *Ultrabithorax* proteins in *Drosophila*. *Nature* **318,** 563–567.

Yasukochi, Y., Ashakumary, L. A., Wu, C., Yoshido, A., Nohata, J., Mita, K., and Sahara, K. (2004). Organization of the *Hox* gene cluster of the silkworm, *Bombyx mori*: A split of the *Hox* cluster in a non-*Drosophila* insect. *Dev. Genes Evol.* **214,** 606–614.

Zhang, C. C., Muller, J., Hoch, M., Jackle, H., and Bienz, M. (1991). Target sequences for hunchback in a control region conferring *Ultrabithorax* expression boundaries. *Development* **113,** 1171–1179 (ISSN: 0950-1991).

Zhou, J., and Levine, M. (1999). A novel *cis*-regulatory element, the PTS, mediates an anti-insulator activity in the *Drosophila* embryo. *Cell* **99,** 567–575.

Zhou, J., Barolo, S., Szymanski, P., and Levine, M. (1996). The *Fab-7* element of the bithorax complex attenuates enhancer-promoter interactions in the *Drosophila* embryo. *Genes Dev.* **10,** 3195–3201.

Zhou, J., Ashe, H., Burks, C., and Levine, M. (1999). Characterization of the transvection mediating region of the *abdominal-B* locus in *Drosophila*. *Development* **126,** 3057–3065.

CHAPTER TWO

Evolution of the *Hox* Gene Complex from an Evolutionary Ground State

Walter J. Gehring,* Urs Kloter,* *and* Hiroshi Suga*,[1]

Contents

Abstract

In this chapter, we consider the question of how the ordered clusters of *Hox* genes arose during evolution. Since ordered Hox clusters are found in all major superphyla, we have to assume that the Hox clusters arose before the Cambrian "explosion" giving rise to all of these taxa. Based on his studies of the bithorax complex (BX-C) in *Drosophila* Lewis considered the ground state to be the mesothoracic segment (T2) since the deletion of all of the genes of the *BX-C* leads to a transformation of all segments from T3 to A8/9 (the last abdominal segment) into T2 segments. We define the developmental ground state genetically, by assuming that loss-of-function mutants lead to transformations toward the ground state, whereas gain-of-function mutants lead to homeotic transformations away from the ground state. By this definition, T2 also represents the developmental ground state, if one includes the anterior genes, that is, those of the Antennapedia complex. We have reconstructed the evolution of the Hox cluster on the basis of known genetic mechanisms which involve unequal crossover and lead from an *urhox* gene, first to an anterior and a posterior gene and subsequently to intermediate genes which are progressively inserted, between the anterior and posterior genes. These intermediate genes

* Department of Cell Biology, Biozentrum, University of Basel, Klingelbergstrasse Basel, Switzerland
[1] Present address: Barcelona Science Park, Universitat de Barcelona, Barcelona, Spain

Current Topics in Developmental Biology, Volume 88
ISSN 0070-2153, DOI: 10.1016/S0070-2153(09)88002-2

are recombinant due to unequal crossover, whereas the anterior and posterior genes are not affected and therefore had the longest time to diverge from the *urhox* gene. The molecular phylogenetic analysis strongly supports this model. We consider the ground state to be both developmental and evolutionary and to represent the prototypic body segment. It corresponds to T2 and is specified by Antennapedia or Hox6, respectively. Experiments in the mouse also suggest that the ground state is a thoracic segment. Evolution leads from the prototypic segment to segmental divergence in both the anterior and posterior direction. The most anterior head and tail segments are specified by homeobox genes localized outside of the cluster.

1. Introduction

Animals develop in a large variety of modes ranging from a very loosely defined cell lineage as in mice and humans where cellular interactions and cell migration play a predominant role, to an absolutely fixed cell lineage as for example in *Caenorhabditis* or *Ciona* in which cell fate is largely determined by cell lineage and differential cell divisions. Very few general principles underlying development have been found so far. The discovery of the homeobox genes (McGinnis *et al.*, 1984a,b; Scott and Weiner, 1984) has uncovered for the first time a universal principle of specification of the body plan. In Bilateria the anteroposterior axis is specified by the *Hox* genes. These genes are exceptional in that they are arranged in an ordered cluster. The *Hox* genes are arranged in the same order along the chromosome as they are expressed along the anteroposterior axis to specify the corresponding body regions. Ordered Hox clusters were first found in *Drosophila* (Kaufman *et al.*, 1980; Lewis, 1978) and subsequently in mammals (Boncinelli *et al.*, 1989; Duboule and Dolle, 1989) and also in Lophotrochozoa, like Lineus, a nemertean (Kmita-Cunisse *et al.*, 1998), and Euprymna, a cephalopod (Callaerts *et al.*, 2002), as well as in Amphioxus, a primitive chordate (Garcia-Fernandez and Holland, 1994). Since it is highly improbable that ordered Hox clusters could have evolved independently in all three bilaterian superphyla, Chordates, Lophotrochozoa, and Ecdysozoa, we have to assume that the *Urbilateria* already possessed an *ordered Hox cluster* prior to the Cambrian "explosion." In this chapter, we are trying to reconstruct the genetic events which lead to the evolution of the *Hox* gene clusters or complexes (HOX-C).

The genetic analysis of the bithorax complex (BX-C) by Lewis (1978, 1992) lead first to the discovery of the ordered homeotic genes which are arranged in the same sequence along the chromosome as they are expressed along the body axis, which he designated as *colinearity*.

In *Drosophila* the genes of the *BX-C* specify the posterior thoracic and the abdominal segments. In the genus *Drosophila*, the original Hox cluster has been split into two, the BX-C and the Antennapedia complex (ANT-C) (Kaufman *et al.*, 1980), by a chromosomal rearrangement. However, in *Drosophila melanogaster* and *D. virilis* the original cluster has been split at different positions in the cluster: In *D. melanogaster* between *Antennapedia* (*Antp*) and *Ultrabithorax* (*Ubx*), whereas in *D. virilis* the split has occurred between the *Ubx* and *abdominal-A* (*abd-A*) genes (Lewis *et al.*, 2003; Von Allmen *et al.*, 1996). Since in more primitive diptera-like Anopheles the HOX-C remained intact, this indicates that these chromosomal rearrangements have occurred within the genus *Drosophila* and, therefore, represent relatively recent evolutionary events. These findings raise the question of which selective forces keep the cluster in an ordered configuration over hundreds of millions of years, whereas other gene families have become dispersed throughout the genome. There have been several attempts to explain this phenomenon, and the coordinate regulation of the *Hox* genes certainly plays an important part (Kmita and Duboule, 2003), but the finding that the cluster can be split in *Drosophila* and fact that the Hox orthologs are largely dispersed in organisms like the nematode *C. elegans* (Aboobaker and Blaxter, 2003; Bürglin *et al.*, 1991), and in urochordates like *Ciona* (Ikuta and Saiga, 2005; Wada *et al.*, 2003) and *Oikopleura* (Edvardsen *et al.*, 2005; Seo *et al.*, 2004), indicate that these functional constraints can be overcome during evolution. In the following we are considering the phenomenon of how the HOX-C evolved in the first place.

2. The Lewis Model

The phenomenon of *colinearity* was first described by Lewis (1978) in *Drosophila* for the genes of the *BX-C* which specify the posterior thoracic and the abdominal segments. Subsequently, Kaufman *et al.* (1980) extended this notion to the anterior thoracic and head segments which are specified by the ANT-C. Molecular cloning of the *Antp* gene lead to the discovery of the homeobox, a highly conserved DNA segment of 180 bp encoding a conserved homeodomain, characteristic for all *Hox* genes contained in these complexes and for a number of dispersed homeobox genes as well (Gehring *et al.*, 1998). The *Hox* genes are not confined to insects and were found to be present in vertebrates including frogs, birds, mammals, and humans (Boncinelli *et al.*, 1989; Carrasco *et al.*, 1984; McGinnis *et al.*, 1984c) and subsequently in all bilateria analyzed so far (Duboule, 1994). The subsequent analysis of the murine and human genes revealed that the mammalian *Hox* genes are also arranged in a colinear order, but there are four complexes on different chromosomes, whereas the invertebrates studied so far have a single HOX-C. For the murine *Hox* genes it has been found (Kmita and Duboule,

2003) that the genes are also expressed in the same temporal order as they are arranged along the chromosomes which was designated as *temporal colinearity*.

To explain the ordered arrangement of the *bithorax* genes, Lewis has put forward a model which can be summarized as follows:

The BX-C consists of a battery of closely linked genes which in the course of evolution have arisen by tandem duplications. The genes of the *BX-C* specify the identity of the posterior thoracic and abdominal segments from T3 to A8 (+A9). One gene function is required for each segment. Each gene is primarily expressed in that particular segment which it specifies and in a subset of cells in all segments posterior to it, so that all genes of the *BX-C* are expressed in the last abdominal segment. The developmental ground state is the mesothoracic segment T2. Loss-of-function mutations lead toward the ground state (T2) and, for example, loss-of-function mutations in the *Ubx* gene convert T3 to T2 leading from a fly with two wings and two halteres to a four-winged fly (Lewis, 1995). In contrast, gain-of-function mutations lead to transformations away from the ground state, for example, from T2 to T3 in *Haltere-mimic* (*Hm*) mutants, resulting in a fly with four halteres (Lewis, 1995). The genes are arranged along the chromosome in the same order as they are expressed along the anteroposterior axis of the embryo which is designated as the *colinearity rule*. The function of the BX-C can only be understood on the basis of evolution, which has lead from homonomously segmented arthropods with a pair of legs on each segment to insects with only three pairs of legs (ventral appendages) and two pairs of wings (dorsal appendages) on the thoracic segments. Some fossil insects had three pairs of wings, with prothoracic (T1) winglets, albeit of smaller size (Carpenter, 1992). In dipteran flies, the metathoracic wings (T3) have been reduced secondarily to halteres. Each of these segmental transformations is primarily controlled by a homeotic gene or a combination of homeotic genes. Loss-of-function mutations in these genes turn the wheel of evolution backwards, so that additional legs can form on the abdominal segments or even on the genitalia in the last abdominal segments (Estrada and Sanchez-Herrero, 2001) and the ancestral evolutionary state is also restored in the four-winged fly.

In view of later findings, this model had to be revised. First, the genetic analysis by Morata and coworkers (Sanchez-Herrero *et al.*, 1985) and the molecular cloning (Bender *et al.*, 2003; Karch *et al.*, 2003) revealed that the BX-C contains three Hox protein coding genes only, so that Lewis modified his model from one gene per segment, to one enhancer per segment which he was able to map by chromosome rearrangements (Lewis, 1992). Also, the expression patterns of the *Hox* genes are primarily parasegmental rather than segmental (Simon *et al.*, 1990). More recently, micro-RNA (miRNA) genes have also been discovered in the BX-C (Bender, 2008; Ronshaugen *et al.*, 2005) which exert a homeotic control function. However, all of these data are still basically consistent with the Lewis model, of one regulatory genetic unit for each body segment.

3. The Developmental and Evolutionary Ground State

Since the ANT-C including the anterior thoracic and the posterior head genes also has to be taken into consideration, we have to re-examine the Lewis model. In particular the question of whether the second thoracic segment (T2) still has to be considered as the *developmental ground state* or whether the ground state is represented by one of the more anterior head segments. We have re-examined all loss-of-function mutants in the *Drosophila* HOX-C (Table 2.1) and found that all *Hox* mutant phenotypes lead toward T2 (arrows in Table 2.1). In contrast to the loss-of-function mutations in the BX-C which transform posterior segments into more anterior ones, the loss-of-function mutations of the ANT-C lead to posterior transformations. This is obvious in *proboscipedia* (*pb*) mutants which transform the labial palps into tarsal structures, which represents a posterior transformation. Loss-of-function mutants in *Deformed* (*Dfd*) show defects in the mandibulary and maxillary segments, where the maxillary sense organs and cirri are missing indicating a partial conversion of these head segments into thoracic segments. At the level of *Sex combs reduced* (*Scr*) the polarity of the transformations is inverted, T1 is converted toward T2 (posterior transformation as for *pb* and *Dfd*), whereas the labial segment is partially transformed into the more anterior maxillary segment.

In addition, we have constructed a new set of gain-of-function mutants by using the *nullo* promoter (Rose and Wieschaus, 1992). The *nullo* gal4 promoter drives gene expression exclusively at preblastoderm and blastoderm stages, when the body plan of *Drosophila* is laid down (Chan and Gehring, 1971). The data are shown in Fig. 2.1 and Table 2.2. Gain of function of the *Hox* genes of the *ANT-C* leads to transformations in the anterior direction (arrows in Table 2.2), that is, away from the ground state T2. Gain-of-function mutants of *pb* lead to the loss of Keilin's organs. This reflects reduction of legs, since the leg imaginal disks originate in intimate association with the Keilin's organs (Bate and Martinez-Arias, 1991). In the adult fly gain-of-function mutants of *pb* lead to homeotic transformations of legs to mouthparts (Aplin and Kaufman, 1997). *Dfd* expression allover the blastoderm embryo leads to the formation of additional maxillary cirri on T1 and T2 (Fig. 2.1C), and *Scr* overexpression converts T2 and T3 toward T1 (Fig. 2.1D). All of these transformations lead away from the ground state and strongly support the notion that T2 (or parasegment 4) represents the developmental ground state.

This interpretation is strengthened by the phenotypes resulting from mutations in the *Polycomb* (*Pc*) group genes, which are involved as repressors in the maintenance of homeotic gene expression (Duncan and Lewis,

Table 2.1 Loss-of-function phenotypes

Hox gene	Ortholog	Phenotype	Transformation
Labial (*lab*)	(Hox1)	Defects in all gnathocephalic segments. Expression in the intercalary or most anterior gnathal segments	Head defects
Proboscipedia (*pb*)	(Hox3)	Labial palps transformed to tarsal structures (T1)	Posterior ↓
Deformed (*Dfd*)	(Hox4)	Defects in mandibular and maxillary segments. Maxillary sense organ mouth hooks, maxillary cirri (missing). Head → thorax	Head defects posterior ↓
Sex combs reduced (*Scr*)	(Hox5)	T1 → T2	Posterior ↓
		Mx ← Lb	Anterior ↑
Antennapedia (*Antp*)	(Hox6)	T2 → T1 (PS4 and 5)	Anterior ↑
Ultrabithorax (*Ubx*)	(Hox7)	A1, T3 → T2	Anterior ↑
abdominal-A (*abd-A*)	(Hox8)	A2-A8 → A1	Anterior ↑
Abdominal-B (*Abd-B*)	(Hox9)	A5, A6, A7 → anterior transformation	Anterior ↑

1982), and the *trithorax* (*trx*) group genes which maintain homeotic genes in an active configuration. Mutations in *Pc* group genes show gain-of-function phenotypes of *Hox* genes, due to derepression of the respective *Hox* genes. By contrast *trithorax* mutations result in the transformation of all imaginal disks, including those of the anterior thoracic and the head region toward T2 (Ingham, 1998), which is in line with the loss-of-function mutations in the *Hox* genes. Therefore, the *Pc* and *trx* group genes strongly support the notion that T2 is the developmental ground state.

For the BX-C Lewis proposed an evolutionary interpretation and considered the developmental ground state also as an *evolutionary ground state*; evolution leading from the ground state, a prototypic segment with a pair of legs and a pair of wings, to progressively more segmental diversification by adding one gene after the other by tandem duplications. The addition of *Ultrabithorax* leads to the transformation of the hindwings into halteres, whereas *abd*-A removes the legs from the abdominal segments and *Abdominal*-B (*Abd-B*) which has two different splice forms, also removes the legs and specifies the genitalia and the last abdominal segments. Based on our nullo gal4 constructs (Fig. 2.1) we can now extend this model to the

Table 2.2 Gain-of-function phenotypes

Hox gene	Ortholog	Phenotype	Transformation
Labial (*lab*)	(Hox1)	Head defects, missing abdominal segments	Head defects
Proboscipedia (*pb*)	(Hox3)	Additional antennomaxillary sense organs on thoracic and abdominal segments	Anterior ↑
Deformed (*Dfd*)	(Hox4)	Additional maxillary cirri on T1 and T2	Anterior ↑
Sex combs reduced (*Scr*)	(Hox5)	T2, T3 → T1	Anterior ↑
Antennapedia (*Antp*)	(Hox6)	T1 → T2	Posterior ↓
Ultrabithorax (*Ubx*)	(Hox7)	H, T1, T2, T3 → A1	Posterior ↓
abdominal-A (*abd-A*)	(Hox8)	T1, T2, T3 → A	Posterior ↓
Abdominal-B (*Abd-Br*)	(Hox9)	T1-A7 → A8/9	Posterior ↓
		T1-A7 → A8/9	
(*Abd-Bm*)		T1-A8 → A9 (Filzkörper)	Posterior ↓

anterior thoracic segments and the gnathal and anterior head segments: *Scr* converts the middle (T2) legs into fore (T1) legs (in combination with *Antp*); *Dfd*, *pb*, and *labial* (*lab*) modify the legs into gnathal mouth parts, whereas the most anterior head structures are specified by *twin of eyeless* (*toy*), *eyeless* (*ey*), *orthodenticle* (*otd*), and *empty spiracles* (*ems*) which are located outside of the Hox cluster. Similarly the most posterior segment forming the analia is specified by the *caudal* (*cad*) gene which is also located outside the HOX-C (Moreno and Morata, 1999).

In *Drosophila* the developmental ground state is specified by *Antp* and we have analyzed *Antp* gain-of-function mutants for their capacity to induce a complete T2 segment. To install the T2 program in the head region, the resident homeotic gene first has to be repressed. This can be accomplished in *toy* mutants which lead in the extreme case to a "headless" phenotype. Homozygous $toy^{-/-}$ mutants are pupal lethals lacking all the structures derived from the eye-antennal disks including the head capsule, the antennae, and the maxillary palps. The lethal pharate adults form a proboscis only consisting of the clypeolabrum (Fig. 2.2) and the labial palps, attached to thorax of the fly. By combining a hypomorphic *toy* mutation ($toy^{D3.3}$) which forms a greatly reduced head, with an *Antp* gain-of-function mutation ($Antp^{73b}$) we have obtained a transformation of antennae into middle legs and of the dorsal head capsule into wing structures (Fig. 2.2) lending further support for the hypothesis that T2 represents the developmental ground state which is specified by *Antp*.

The genetic analysis of these loss- and gain-of-function mutants in mammals is complicated by the fact that mammals have four Hox clusters and there is extensive functional redundancy. However, Wellig and Capecchi (2003) present evidence which points in the same direction for

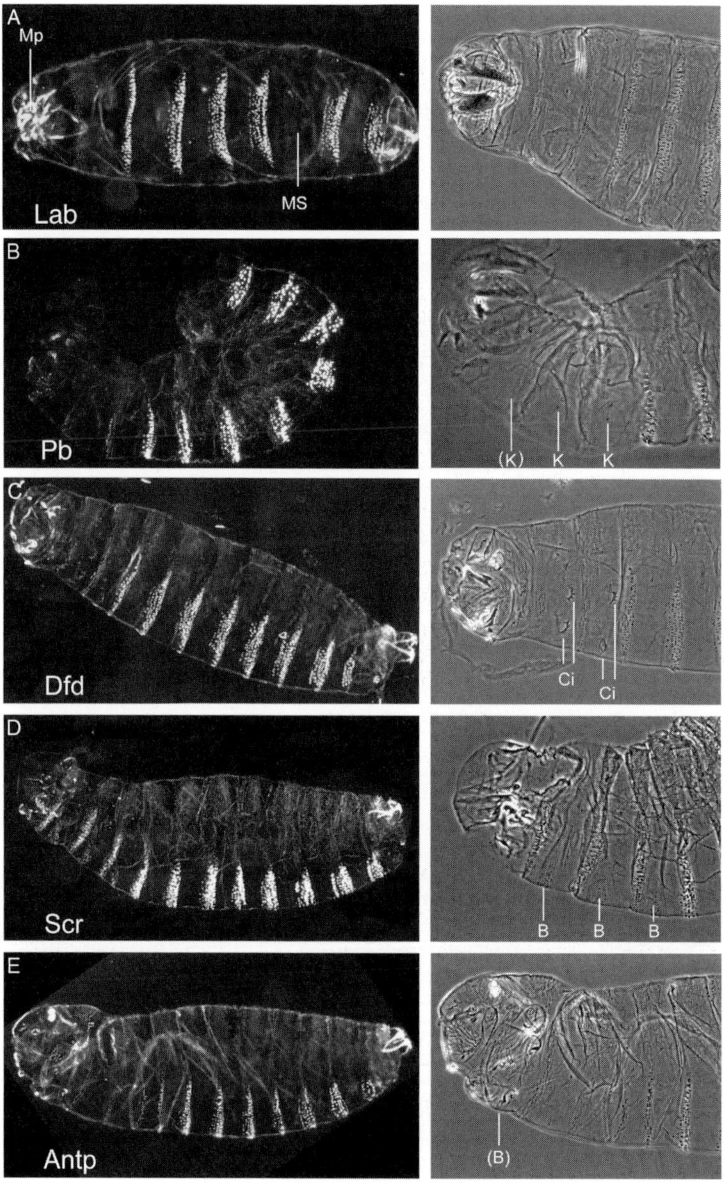

Figure 2.1 (Continued)

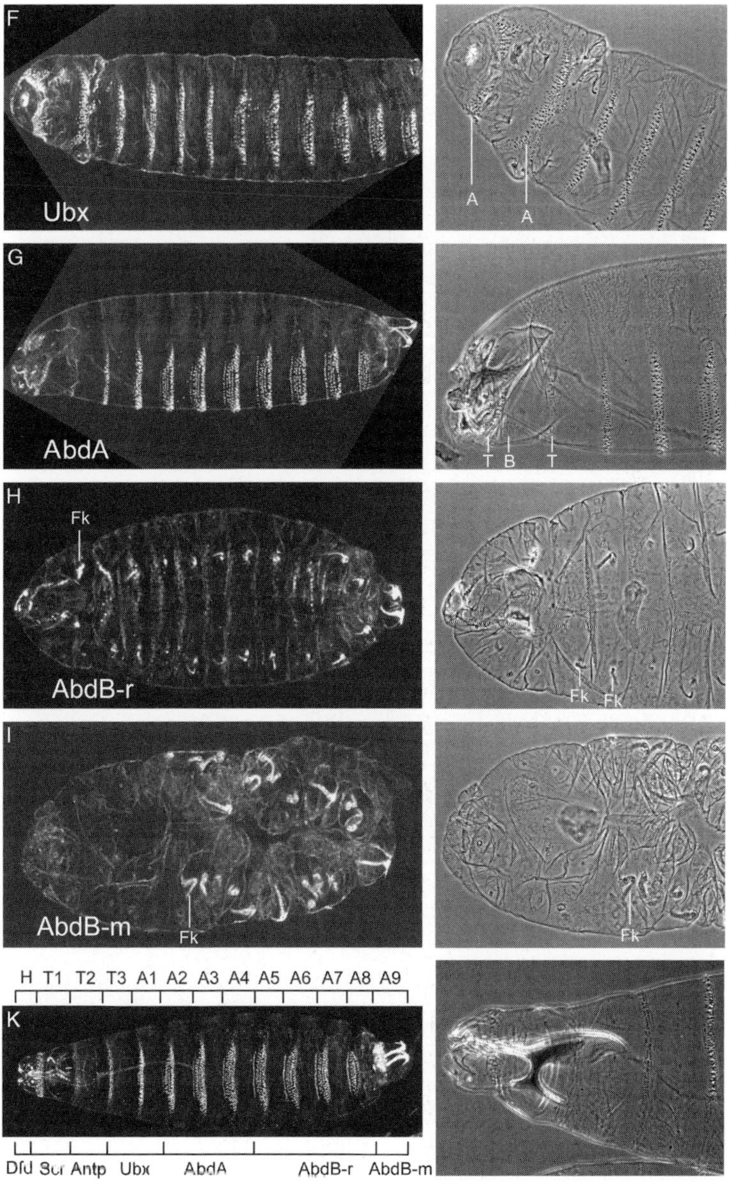

Figure 2.1 Late embryonic phenotypes of Hox gain-of-function constructs in *Drosophila*. The nullo-gal4 promoter was used to express the respective UAS-*Hox* genes all over the preblastoderm and blastoderm embryos, when the body plan is laid down. The head of all embryos is severely affected since the anterior head genes which reside outside of the Hox complex are competitively inhibited by the ectopic expression of the respective Hox proteins of the complex: (A–E) ANTP-C, (F–I) BX-C, and (K) Control embryo. Left: overview in dark field. Right: enlarged anterior segments in

axial patterning in the mouse. By disrupting all paralogs of Hox10 and Hox11 they show that these genes are required for the global patterning of the mammalian skeleton. The normal axial formula is 7 cervical, 13 thoracic, 6 lumbar, 4 sacral, and a slightly variable number of some 20 caudal vertebrae (Burke *et al.*, 1995). Rib formation is confined to the thoracic vertebrae. *Hox10* triple mutants lack lumbar vertebrae and form ribs extending beyond the thoracic (T13) region through the lumbar and sacral regions which represents anterior transformations. *Hox11* triple mutant skeletons exhibit normal thoracic and lumbar vertebrae, but they lack a sacral region, which is replaced by lumbar vertebrae, which also represents an anterior transformation. Since many primitive tetrapods have ribs projecting from all vertebrae from cervix to tail, it has been proposed that the ground state is a body segment with vertebrae including rib projections (Cohn and Tickle, 1999; Hildebrand, 1995). The data on *Hox10* and *Hox11* loss-of-function mutants support this idea, if we assume that *Hox* genes have been used in the course of evolution to suppress and modify rib formation in the lumbosacral region.

Gain-of-function mutations of *Hox10* result in homeotic transformations in the opposite direction (Carapuco *et al.*, 2005). By expression of Hoxa10 under the control of the *distalless* (*Dll*) 1 promoter which has been shown to be active in the presomitic mesoderm at the presomitic stage and in early somites, but not at later stages of somite development, the total absence of ribs was induced which represents a homeotic transformation of the thoracic into lumbar vertebrae. Thus, gain- and loss-of-function mutations give rise to homeotic transformations in opposite directions. As in *Drosophila* loss of function of Hox10 leads to transformation toward the putative thoracic ground state, whereas the corresponding gain-of-function mutation leads away from the ground state.

With respect to the putative thoracic ground state in vertebrates, interesting experiments have been performed on Hoxb6, the putative *Antp* homolog of mice. By expression of Hoxb6 under the *Dll* promoter, transgenic mice were obtained which had ribs on all of their first 26 vertebrae; the first seven on the cervical region, the next 13 in the thoracic region (as in

phase contrast. Anterior is to the left. (A) *Labial*: Mp, reduced mouth parts; MS, missing abdominal segment. (B) *Proboscipedia*: dorsal closure incomplete: K, Keilin's organ; (K), missing Keilin's organ. (C) *Deformed*: Ci, extra cirri in T2 and T3. (D) *Sex combs reduced*: B, additional beards in T2 and T3. (E) *Antennapedia*: (B), missing beard in T1. (F) *Ultrabithorax*: A, two additional abdominal (A1) denticle belts in the head region. (G) *Abdominal-A*: T, thoracic denticle belt; B, beard, A4-8 transformed toward A4. (H) *Abdominal-Br*: T1-A8 are transformed toward A8 with Filzkörper (Fk). (I) *Abdominal-Bm*: the segmental boundaries are not formed. All thoracic and abdominal segments are transformed toward A9 with 11 pairs of Filzkörper (Fz). (K) Wild-type control embryo with segment designations and the major expression domains of the various *Hox* genes.

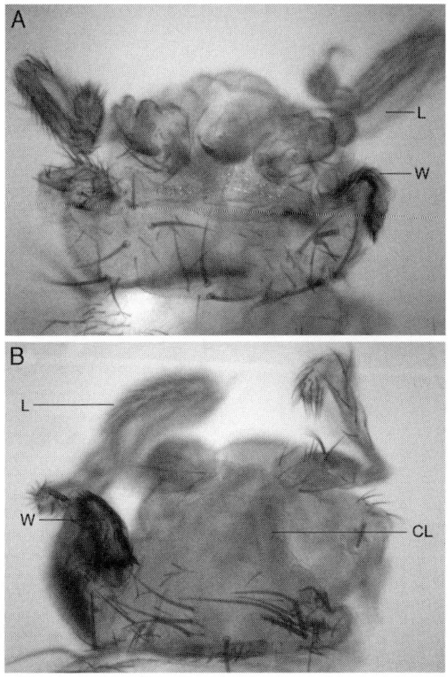

Figure 2.2 Homeotic transformation of the head into a second thoracic segment with a pair of middle legs and pair of wings by an *Antennapedia* gain-of-function mutation in a *twin of eyeless* hypomorphic mutant background. The *twin of eyeless (toy)* gene specifies the adult head including the antennae. $toy^{-/-}$ flies are essentially headless; they possess a normal thorax to which the proboscis including the clypeolabrum (CL) is attached. To "install" the T2 program, the *Antp* gene first has to repress the resident head program. Most *Antp* gain-of-function mutations do not repress the head program sufficiently to induce a complete transformation. Therefore, we have combined the dominant $Antp^{73b}$ mutation with a recessive hypomorphic mutation in *toy*, that is, $toy^{D3.3}$ and obtained flies with a pair of middle legs (L) and wings (W) in the head region. This indicates that *Antp* specifies the entire T2 segment with a pair of legs and pair of wings.

normal mice), and the last six in the lumbar region (Mallo *et al.*, in preparation). This corresponds to the transformation of both the cervical and the lumbar vertebrae into thoracic ones. This is in line with the notion that the ground state is a thoracic vertebra with a pair of ribs. Hox6 is in fact the homolog of *Antp* (Malicki *et al.*, 1990; L. Michaut, unpublished data) which specifies the ground state T2 in *Drosophila*.

It should be emphasized that our notion of the ground state being specified by the *urhox gene* differs from that used by other authors (Duboule and Morata, 1994) who assume that the ground state is defined by the total absence of any *Hox* gene expression. If all the genes from the ANT-C and BX-C are deleted, the homeobox-containing genes outside of these clusters (*toy*, *ey*, *otd*, *ems*, and *cad*) are still present and become epistatic, which is reflected by the fact that for example, *ems* which specifies mouth hooks, becomes derepressed in all segments which leads to the formation of rudimentary mouth hooks in every segment.

There is a significant difference between the legs of insects and the limbs of vertebrates; whereas *Antp* specifies the ground state segment T2 with a pair of legs and a pair of wings, that is, the legs are formed from one segment, whereas the vertebrate limbs derive from several body segments (six myotomes), and therefore, are not directly comparable to insect legs. However, several *Hox* genes are also expressed in a colinear order along the proximo-distal axis of the limbs (Kmita *et al.*, 2002). Therefore, the basic control features of the body plan, including the ground state are apparently conserved between insects and mammals, but certain special features are found in vertebrates only.

4. Mechanisms of Epistatic Hox-Hox Interactions

Genetic interactions between *Hox* genes are essential for the establishment of the body plan. Early studies showed that the posterior genes of the *BX-C* repress the more anteriorly expressed *Antp* gene at the transcriptional level. Removal of *Ubx* which is normally expressed in T3 and A1, leads to the derepression of *Antp* transcription in T3 and A1 and to the transformation of these two segments into additional T2 segments. A deletion which removes the entire BX-C, Df (3R) P9 (Lewis, 1978), leads to the derepression of *Antp* transcription in all posterior segments, from T3 to A8 and to their transformation into T2 segments (Hafen *et al.*, 1984). Therefore, the posterior *Hox* genes repress the more anterior ones in the wild-type. The same general rule was found in mice and variously termed posterior prevalence or phenotypic suppression. However, the conventional terms for interactions between nonallelic genes are epistasis and hypostasis.

Subsequent studies suggested that the cross-regulatory interactions between homeotic genes at the transcriptional level did not fully explain the establishment of segmental identity (Gonzalez-Reyes *et al.*, 1990). The coexpression of both ANTP and UBX proteins under a heat shock promoter leads to the repression of ANTP function in T1 and head segments and suggests a posttranslational mechanism based on competitive protein-protein interactions. Such a mechanism was first demonstrated at the molecular level for the interaction of *Antp* with *ey* in eye development (Plaza *et al.*, 2001, 2008). Upon coexpression of EY and ANTP the two proteins interact competitively by binding of the homeodomain (HD) of *Antp* to the paired domain (PD) and/or the HD of *ey*. By mutational analysis of the homeobox of *Antp* we have been able to dissociate DNA-binding from protein-binding amino acid residues. By using glutathion-*S*-tranferase fusion (GST) proteins, the direct protein-protein interaction was demonstrated *in vitro* (Plaza *et al.*, 2001) and more recently, these Hox-Pax and Hox-Hox interactions have also been shown to occur in imaginal disks *in vivo* by bimolecular fluorescence complementation (BMFC) (Plaza *et al.*, 2008). Using the nullo gal4 driver to overexpress various *Hox* genes allover the preblastoderm and blastoderm embryo, for example, *AbdB*, we find that transcription and translation of *Antp* are not affected, even though all three thoracic and eight abdominal segments are converted into A8 segments (Fig. 2.1h) (Y. Adachi, U. Kloter, and W. Gehring, in preparation). Using BMFC the direct interaction of the two homeodomain proteins can be demonstrated *in vivo* in imaginal disks (Y. Adachi, unpublished data). Therefore, the identity of the various body segments not only depends on transcriptional regulation, but also on competitive protein-protein interaction. The homeodomain is not only a DNA-binding domain, but, for example, in the case of *bicoid* (*bcd*) also capable of RNA binding (Rivera-Pomar *et al.*, 1996) and as demonstrated by Plaza *et al.* (2001, 2008) involved in competitive protein-protein interactions. The analysis of the gain-of-function mutants generated by ectopic overexpression using the nullo gal4 driver (Fig. 2.1) clearly indicates that the anterior *Hox* genes of the *ANT-C* are epistatic over the more posterior genes, for example, *Scr* induces additional beards in T2 and T3, *Dfd* induces additional maxillary cirri in T1 and T2, and *pb* gain-of-function constructs transform legs into maxillary and/or labial palps (Aplin and Kaufman, 1997). In all of these cases there is anterior, rather than posterior, epistasis (phenotypic suppression). The same conclusion was reached by M. Müller (unpublished data) by using the scabrous gal4 driver rather than the nullo gal4, which confirms our conclusion.

These Hox-Hox interactions are combinatorial as proposed by the Lewis model. This can be clearly demonstrated for the three genes specifying the thoracic segments: *Scr* for TI, *Antp* for T2, and *Ubx* for T3/A1. By removing both *Scr* and *Ubx* genes in genetic mosaics ($Scr^{-/-}Ubx^{-/-}$), flies are obtained with six middle legs and four wings (Fig. 2.3). This indicates that the combination of Scr^{+} and $Antp^{+}$ is required to convert a middle leg (T2) into a foreleg (T1) and $Antp^{+}$ plus Ubx^{+} are required to form a hind leg and a haltere. The fact that Antp is already strongly expressed in the dorsal prothoracic disk in the wild-type is a likely explanation that we do not obtain a six-winged fly in this experiment.

5. The Evolutionary Origin of the Hox Cluster

The Urbilateria presumably possessed at least eight homeobox genes in addition to the core *Hox* genes (Butts *et al.*, 2008) which could at least in part absorb the selective pressure to develop normally when the *Hox* genes

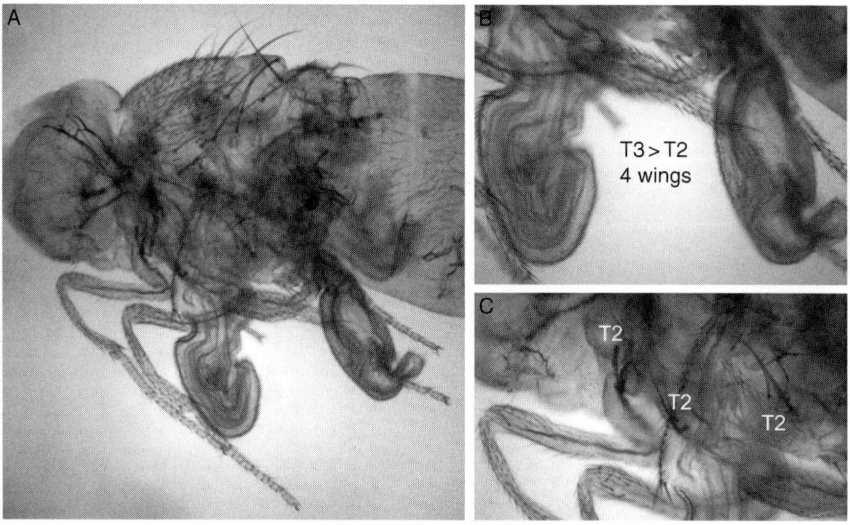

Figure 2.3 Phenotype of an adult fly lacking two *Hox* genes *Scr*$^-$ *Ubx*$^-$ with six middle legs and four wings. (A) Overview. (B) Transformation of the halteres (T3) into wings (T2). (C) Transformation of all three pairs of legs into middle legs (T2) with stemopleurae. Genotype: *Antp* P1 gal4; UAS-Flp; FRT, M(3)w/FRT, *Scr*, *Ubx*.

began to diverge. There are at least two homeobox genes located outside of the HOX-C which specify the anterior head segments *orthodenticle* (*otd*) (Finkelstein *et al.*, 1990) and *empty spiracles* (*ems*) (Dalton *et al.*, 1989) that were subsequently also found in vertebrates (Simeone *et al.*, 1992), and *caudal* (*cad*) (Mlodzik *et al.*, 1985) which specifies the most posterior abdominal segments (Moreno and Morata, 1999). The caudal orthologs, the *cdx* genes also belong to the core *Hox* genes. These genes must have absorbed the selective pressure to maintain the most anterior and posterior body segments, while the HOX-C evolved by tandem duplications. It is interesting to note that these genes share certain characteristics with the clustered *Hox* genes, as, for example, the YPWM motif which is found in both *ems* and *cad*. This might suggest that these genes originally were part of the cluster before they were translocated. However, they are not an integral part of the cluster, neither in insects nor in mammals.

The ordered colinear arrangement of the *Hox* genes can be explained by the mechanism of unequal crossing over which leads to progressive gene duplications starting from an *urhox* gene (Gehring, 1998; Zhang and Nei, 1995). The first duplication was presumably based on a repetitive DNA sequence, like a transposon located on either side of the *urhox* gene which allowed displaced pairing and recombination to occur (Goldberg *et al.*, 1983), giving rise to the first pair of *Hox* genes which began to diverge from one another (Fig. 2.4) into an anterior and a posterior *Hox* gene. These

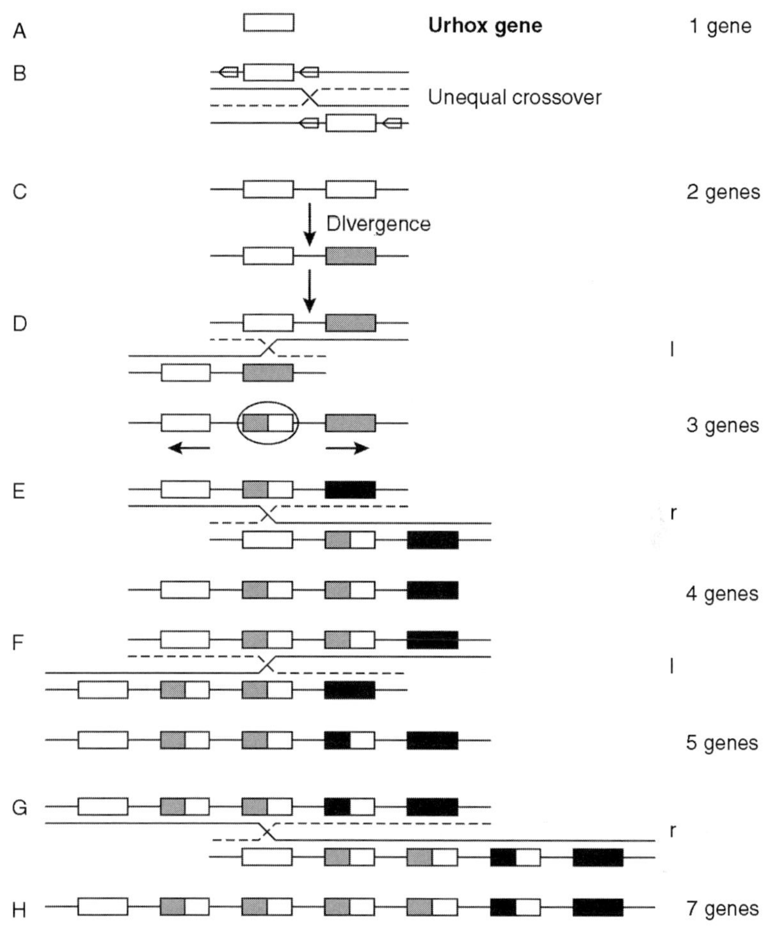

Figure 2.4 Generation of the *Hox* cluster by unequal crossover. (A) *Urhox* gene. (B) A transposon (arrow) flanking the *Urhox* gene on either side and in the same orientation allows for displaced chromosome pairing and unequal crossover generating a first gene duplication. (C) The two first *Hox* genes diverge into an anterior (white) and a posterior (gray) gene. (D) Displaced pairing between the duplicated genes generates a third gene which is a hybrid between the anterior and posterior genes (encircled) which resembles the original *Urhox* gene most closely. The outer genes are not affected by the unequal crossover and continue to diverge in the anterior and posterior direction (arrows), respectively, leading to an anterior (white) gene, an intermediate hybrid gene, and a posterior (black) gene. (E) The next unequal crossover leads to four genes. The pairing can be displaced either to the left (l) as in the (D) or to the right as in (E). The probability for displacement to the left and to the right is the same. (F) The new genes are always added in the middle of the cluster, and the flanking anterior (white) and posterior (black) genes, which arose first during evolution, are not affected. They have the longest time to diverge. The intermediate genes in the middle of the cluster are homogenized by unequal crossover. Therefore, their sequences most closely resemble those of the *Urhox* gene, even though they arose later in evolution. (G) The chromosome pairing can also be displaced by two genes leading from five genes to seven genes in (H). The clusters of protostomes generally contain nine genes, whereas the chordates have 13 genes per cluster, or 14 in the case of *Amphioxus*.

two genes are the most ancestral *Hox* genes which had the longest time to diverge during evolution. The second unequal crossing over is postulated to have occurred by displaced pairing and recombination between the first two *Hox* genes, resulting in three genes which evolved into an anterior, and intermediate and a posterior gene, the intermediate gene being a hybrid between the anterior and posterior gene most closely resembling the *urhox* gene (see Fig. 2.4). We propose that the intermediate gene absorbs the selection pressure and allows the anterior and posterior genes to diverge in opposite directions. In all subsequent rounds of unequal crossing over, the anterior- and posterior-most genes are not affected (Fig. 2.4) and therefore, had the largest time to diverge from each other. Since the displaced pairing has an equal probability to occur to the left or to the right, the newly formed genes arise in the middle of the cluster where the sequences are recombinant due to unequal crossing over. This model is supported by measuring the horizontal distances, that is, the number of amino acid substitutions between homeodomains of the cluster, both in *Drosophila* and in mammals which also takes the ordered arrangement of the genes along the chromosome into account. The horizontal distances gradually increase from Hox6, which specifies the ground state, in both directions toward Hox1 and Hox9 (or Hox13), respectively (Fig. 2.5A and B). Also the vertical distances between the *Drosophila* and the mammalian genes increase gradually in both directions (Fig. 2.5C and D). Therefore, the selective pressure to keep the amino acid sequences constant, are highest in the center of the Hox cluster (around Hox6), that is, on those genes which resemble in their DNA sequence the original *urhox* gene most closely. This hypothesis was tested in two ways, by constructing phylogenetic trees and phylogenetic networks of the *Hox* cluster genes. Two unrooted maximum likelihood (ML) trees were inferred from comparisons of the homeodomain amino acid sequences (Fig. 2.6). We chose the ordered Hox clusters of Homo (vertebrate), Amphioxus (cephalochordate), Euprymna (cephalopod, lophotrochozoa), Lineus (Nemertean, lophotrochozoa), and *Drosophila* (insect, ecdysozoa), and a broad range of dispersed non-*Hox* homeobox-containing genes including those from the NK, PRD, Lim, and ZF classes (Bürglin, 1994, 2005) as an outgroup. In the Hox tree the assignment of the Hox1, Hox2, Hox3, Hox4, and Hox5 subfamilies was possible, since the orthologs of the genes belonging to these clusters had been well established by previous publications (Callaerts *et al.*, 2002; Holland and Garcia-Fernandez, 1996; Kmita-Cunisse *et al.*, 1998; Wada *et al.*, 1999). Similarly, each of the human amphioxus pairs was assigned to the Hox1, Hox2, Hox3, Hox4, Hox5, Hox6, Hox7, and Hox8 subfamilies (Garcia-Fernández and Holland, 1994; Holland and Garcia-Fernandez, 1996). We assessed the root position of the Hox tree by using the non-Hox tree as an outgroup. The branch to which the outgroup tree is most likely to attach is shown by a bold line and corresponds to Hox1. By using the *p* values of the AU (approximately unbiased) test (Shimodaira, 2002) of the CONSEL program

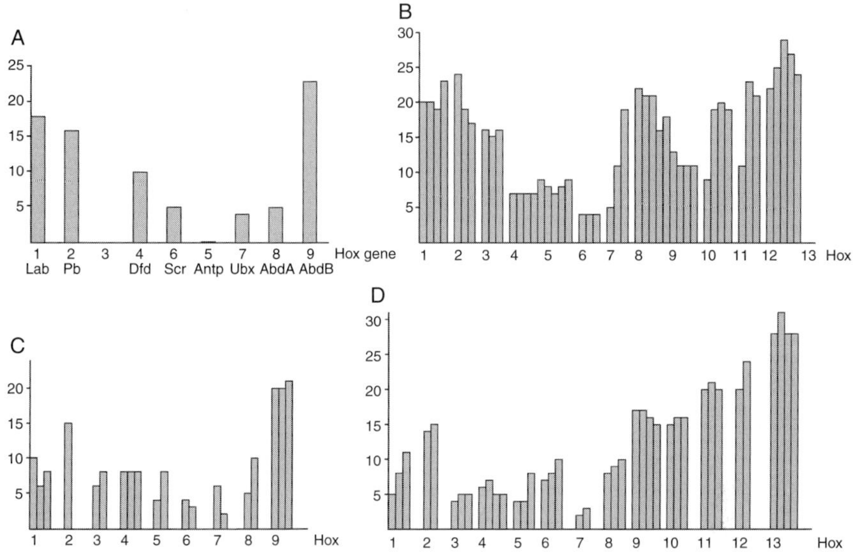

Figure 2.5 Horizontal and vertical evolutionary distances between the *Hox* gene complexes of *Drosophila, Lineus, Euprymna, Amphioxus*, and the four human complexes. (A) Cumulative horizontal distances between the amino acid sequences of the eight homeodomains of *Drosophila* relative to *Antennapedia: lab, Labial (Hox1); pb, Proboscipedia (Hox2); Dfd, Deformed (Hox4); Scr, Sex combs reduced (Hox5); Antp, Antennapedia (Hox6); Ubx, Ultrabithorax (Hox7); abdA, Abdominal-A (Hox8)*; and *Abd-B, Abdominal-B (Hox9)*. There is no *Hox3* homolog with a homeotic gene function in *Drosophila*, since *zerknüllt* (*zen* and *zen2*) and *bicoid* (*bcd*) have diverged and serve a different function. The horizontal distances increase progressively from *Antp* to both the anterior and posterior end, that is, Hox1 and Hox9, respectively. (B) Horizontal distances between the homeodomains of neighboring *Hox* genes of *Amphioxus* and the four human Hox complexes. The human clusters have undergone some gene losses, so that there are between three and five values per *Hox* gene. The low point is at Hox6 and the distances increase progressively toward both ends of Hox complex. (C) Vertical distances between the orthologous homeodomains of *Drosophila, Lineus*, and *Euprymna* relative to the *Amphioxus* sequences. The low point is at Hox6 (*Antp*) and the number of amino acid substitutions increases toward both ends of the Hox complex. (D) Vertical distances between the homeodomains of the chordate *Hox* genes relative to those of *Amphioxus*. The low point is located at Hox7 and the number of amino acid substitutions increases toward both ends Hox1/2 and Hox13. Abscissa: *Hox* gene number. Ordinate: number of amino acid substitutions.

(Shimodaira and Hasegawa, 2001); however, we cannot exclude the possibility that the outgroup tree attaches to other branches, except for those six branches with a p value of <0.05, which are shown by dotted lines in Fig. 2.6. The most likely position of the root suggests that the Hox1 subfamily is the most ancient, which is in agreement with our above hypothesis which assumes that the first gene duplication generated an anterior and a posterior gene, and that the intermediate genes were filled in later. Since the anterior *Hox1* gene

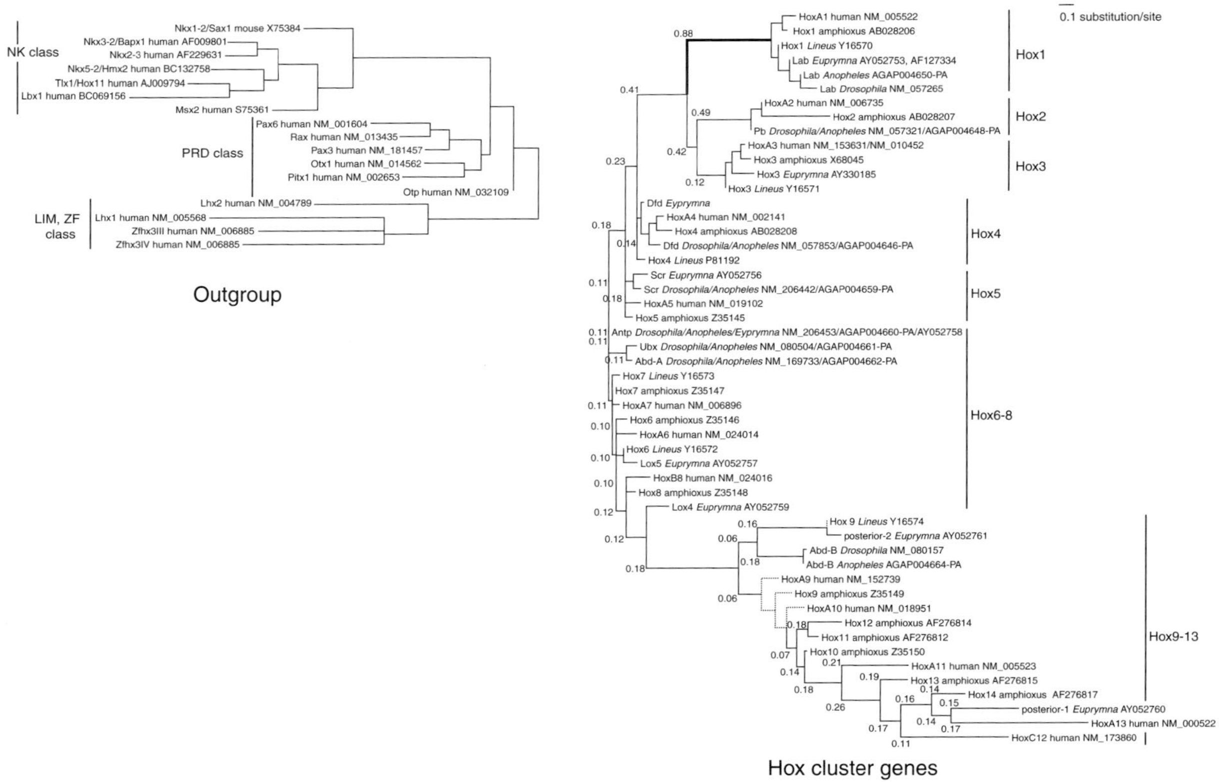

Figure 2.6 Phylogenetic tree of the *Hox* cluster genes and of dispersed homeobox genes based on maximum likelihood (ML) methods. Two unrooted ML trees were inferred from the comparisons of the homeodomain amino acid sequences (60 amino acids). (A) As an outgroup, a range of dispersed homeobox genes including those of the NK class, PRD class, Lim class, and Zf class was used. (B) The homeodomains of the *Hox* cluster genes of *Drosophila, Lineus, Euprymna, Amphioxus*, and the four human clusters were analyzed. The orthologues were

presumably is the most ancient daughter gene of the *urhox* gene, it had the longest time to accumulate mutations and to deviate most from original urhox sequence. The same applies to the posterior daughter *Hox* gene, which has exactly the same age as Hox1, since they arose by duplication. However, the posterior *Hox* genes of chordates have undergone a later expansion after vertebrates and invertebrates have separated (Zhang and Nei, 1995); amphioxus has six posterior *Hox* genes (Hox9–14) and vertebrates have five (Hox9–13), whereas invertebrates, with the exception of urochordates generally have only one (Hox9). Our hypothesis assumes that the intermediate genes are hybrid genes generated by the unequal crossover events and that they absorb the selective pressure. Thus, they deviate the least from the original amino acid sequence encoded by the *urhox* gene. Therefore, we propose that the *urhox* gene originally specified the developmental ground state (a thoracic segment) and that this function has been continuously maintained by the intermediate genes, while the anterior and posterior genes have progressively diverged in opposite directions.

This interpretation is also supported by our reconstruction of the phylogenetic network of the *Hox* cluster genes using the split Tree4 program (Huson and Bryant, 2006) shown in Fig. 2.7. The Hox1, Hox2, Hox3, Hox4, and Hox5 can clearly be resolved, whereas the intermediate *Hox* genes *6–8* are very similar to one another and cannot be resolved, which is in line with our hypothesis that they have been scrambled by the unequal crossover events. The posterior genes *Hox9–13* are clearly separated from the anterior and intermediate subfamilies.

Studies of *Hox* genes in the most primitive metazoan phyla are also in agreement with our hypothesis, although the data are still incomplete. A recent survey of the *Hox* genes found in Cnidaria, in particular *Nematostella vectensis* in which the entire genome has been sequenced, reveals the presence of anterior (paralog group (PG) 1 and 2) and posterior *Hox* genes (*PG 9*) only (Finnerty and Martindale, 1999; Quiquand *et al.*, 2009). In the most basal bilaterian clade, the acoel flatworms in addition to a *PG 1* and a *PG 9* group gene, an intermediate class gene similar to Hox4 and Hox5 have been identified, suggesting that the acoels may have diverged from other bilateria before an *8–10 Hox* gene cluster had evolved (Cook *et al.*, 2004). These data

determined on the basis of previous publications including sequences flanking the homeodomain (Callaerts *et al.*, 2002; Duboule, 1994; Garcia-Fernandez and Holland, 1994, 1996; Holland and Garcia-Fernandez, 1996; Kmita-Cunisse *et al.*, 1998). The root position of the tree in (B) was assessed by using the non-Hox tree in (A) as an outgroup. The branch to which the outgroup tree is most likely to be attached with 80% probability is shown by a bold line. However, using the p values of the AU test, the possibility that the outgroup tree attaches to other braches of the Hox tree cannot be excluded, except for five branches with p values < 0.05 which are indicated by dotted lines. The GenBank or RefSeq accession numbers are shown together with the gene and species names.

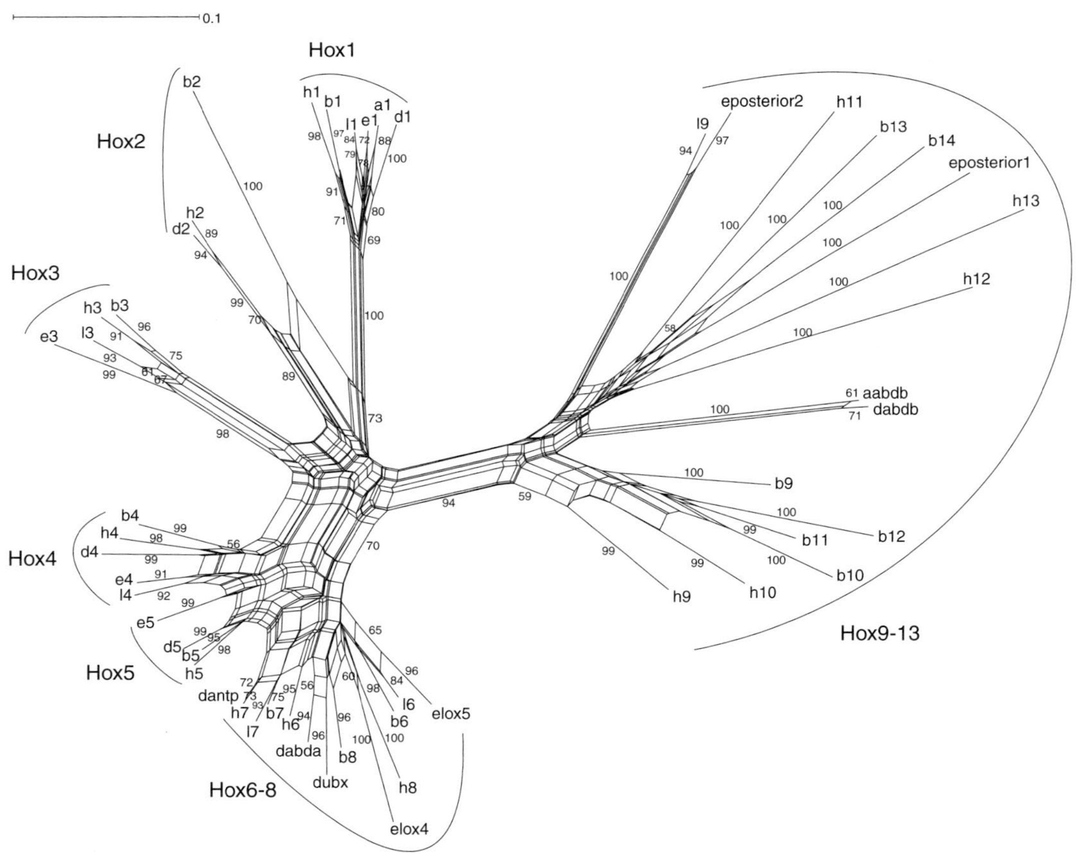

Figure 2.7 Phylogenetic network of the *Hox* cluster genes using the split-tree program (Huson and Bryant, 2006). The anterior (*Hox1-8*) and posterior *Hox* genes (*Hox9-13*) are clearly separated. Whereas the *Hox1-5* genes are clearly separated, the intermediate genes *Hox6-8* are not resolved. This is in line with the assumption that intermediate *Hox* genes have arisen more recently in evolution and, therefore, have diverged the least.

are consistent with our hypothesis that the anterior and posterior genes arose first during evolution followed by the intermediate genes.

6. Duplication and Divergence as a General Evolutionary Principle

Selection is generally conservative and leads to stabilization of DNA sequences. The stronger the selective pressure, the higher the degree of conservation. The selective pressure is high on the coding region which encodes an RNA or protein which has been functionally optimized over numerous generations and in most cases carries out vital functions. Selective pressure is also high on *cis*-regulatory elements which define the spatial and temporal expression pattern of the respective gene. This raises the question of how new functions can evolve, if the original gene function is of vital importance. Over the last few years, it has become increasingly clear evolution proceeds more effectively at the level of the *cis*-regulatory elements which define the pattern of gene expression, rather than at the level of the coding regions. This has been clearly illustrated by the analysis of the stripe 2 enhancer of the *even-skipped* (*eve* gene) in 12 different *Drosophila* species (see Carroll *et al.*, 2001). This is in part due to the fact that the target-binding sites for the various transcription factors regulating *eve* are redundant. This allows the maintenance of the pattern of gene expression in various combinations of these *cis*-regulatory elements, while at the same time they can also acquire new functions.

In his book on "Evolution by Gene Duplication," Ohno (1970) has pointed out that gene duplication is of fundamental importance for the evolution of new functions, and the *Hox* genes are a perfect example. Duplicated genes increase the diversity of gene expression (Gu *et al.*, 2005) due to the fact that one copy of gene can retain the original function and the other has a certain degree of freedom to diverge from the original pattern of gene expression. This leads to neofunctionalization (Ohno, 1970). Furthermore, gene duplication allows subfunctionalization (Bürglin, 1995; Force *et al.*, 1999), since some of the *cis*-regulatory motifs may get lost in one or the other copy of the duplicated genes. Both of these phenomena are found in the Hox clusters. Neofunctionalization occurs when a *Hox* gene specifying a walking leg is duplicated and diverges to specify an antenna. Subfunctionalization is likely to have occurred in the abdominal *Hox* genes, which remove the legs from the posterior (abdominal) segments. Gene duplication allows the differentiation of cells, tissues, and organs by increasing expression diversity, and the differential expression pattern between duplicated genes increases with evolutionary time (Li *et al.*, 2005).

The evolution to increasing complexity by duplication (or replication) and subsequent divergence appears to be a general principle of evolution that applies to all levels of biological organization. It applies to DNA molecules which replicate and diverge by mutation. It forms the basis for the widespread occurrence of diploidy in higher organisms. One copy of the gene retains the normal wild-type function, whereas its homolog can mutate to provide the necessary variation upon which selection can act. A clear example of gene duplication is provided by the *ey* and *toy* genes of *Drosophila*. Whereas most animals have a single *Pax6* gene, the *Pax6* gene has been duplicated during insect evolution (Czerny *et al.*, 1999; Punzo *et al.*, 2004). They partly substitute for each other; however, they have diverged in that *toy* specifies the entire head capsule with compound eyes, ocelli, and antennae, whereas the function of *ey* is more restricted to the compound eyes.

The duplication and divergence principle also applies to cells. The larva of the jellyfish *Tripedalia* contains single-celled photoreceptors which serve both functions, the photoreceptor function as well as the function of a shielding pigment cell (Nordstrom *et al.*, 2003). In the adult jellyfish the differentiation into neuronal photoreceptor and pigment cells has occurred. It is conceivable that photoreceptor and pigment cells might have evolved independently by differentiation from another cell type. However, the fact the optic cup in vertebrates gives rise to both retinal photoreceptor cells and retinal pigment cells supports the notion of a common evolutionary origin. Whereas most melanocytes in vertebrates originate from the neural crest, the retinal pigment cells are of neural origin and their differentiation specifically depends on the Mitf transcription factor (Steingrímsson *et al.*, 2004). It is important to note that the ortholog of Mitf is already present in *Tripedalia* (Kozmik *et al.*, 2008) and that it is found in all major branches of the evolutionary tree. The principle of duplication and divergence even applies to multicellular entities like body segments, and also to complete organisms like the different sexes in sexually dimorphic organisms, or the casts in social insects and even to different populations of organisms. The diversification between two populations, for example, a mainland and an island population may lead to speciation. The selective advantage of duplication or replication prior to divergence, over direct conversion, lies in the fact that the original function or parts thereof may be retained by one of the duplicated entities, which meets the functional constraints, that is, the selective pressure, whereas the duplicate is less constrained to diverge.

7. Conclusion

In this chapter, we have tried to reconstruct the evolution of the *Hox* gene cluster from an *urhox* gene. Homeobox genes specifying the most anterior head segments (*otd* and *ems*) and the most posterior abdominal segments (*cad*) have arisen prior to the formation of the Hox cluster. The Hox cluster is thought to have originated from an *urhox* gene by a series of tandem duplications due to *unequal crossing over*, which first generated an anterior and a posterior gene. These two genes diverged from each other and generated an intermediate hybrid gene by a second unequal crossover. Subsequent rounds of unequal crossover do not affect the flanking anterior and posterior genes, since the additional genes are added in the middle of the cluster (Fig. 2.4). Repeated unequal crossovers lead to recombination between the intermediate genes and therefore, the flanking genes have had the longest time to diverge. This model is strongly supported by the molecular phylogenetic analysis. The horizontal and vertical distances between the amino acid sequences of the homeodomains increase in both directions from the center to the edges of the cluster. In the center, the selective pressure to maintain the amino acid sequences constant is highest. We therefore propose that the amino acid sequence of the central homeodomain (Hox6) resemble that of the *urhox* gene most closely. Furthermore, we propose that the *urhox* gene specifies the developmental and evolutionary ground state which corresponds to the mesothoracic segment in *Drosophila* and a thoracic segment in mice. The *developmental ground state* in *Drosophila* is clearly defined by both loss- and gain-of-function mutations leading toward and away from the ground state, respectively. From the ground state which may be considered as the *prototypic body segment*, evolutionary changes have given rise to progressive diversification both in the anterior direction (e.g., conversion of walking legs into mouth parts) and in the posterior direction (removal of the legs from the abdominal segments). Both of these changes can be reverted by loss-of-function mutations which lead back to the prototypic segment.

The general evolutionary principle of duplication (or replication) prior to diversification applies not only to genes, but to all levels of biological organization from DNA molecules to genes, cells, segments, individuals, and even to populations, in which it provides a basis for speciation.

ACKNOWLEDGMENTS

This work was supported by a grant from the Swiss National Science Foundation (SNF) and by the Kantons of Basel Stadt and Basel Landschaft.

REFERENCES

Aboobaker, A. A., and Blaxter, M. L. (2003). *Hox* gene loss during dynamic evolution of the nematode cluster. *Curr. Biol.* **13,** 37–40.

Aplin, A. C., and Kaufman, T. C. (1997). Homeotic transformation of legs to mouthparts by proboscipedia expression in *Drosophila* imaginal discs. *Mech. Dev.* **62,** 51–60.

Bate, M., and Martinez Arias, A. (1991). The embryonic origin of imaginal discs in *Drosophila. Development* **112,** 755–761.

Bender, W. (2008). MicroRNAs in the *Drosophila bithorax* complex. *Genes Dev.* **22,** 14–19.

Bender, W., Akam, M., Karch, F., Beachy, P. A., Peifer, M., Spierer, P., Lewis, E. B., and Hogness, D. S. (2003). Molecular genetics of the bithorax complex in *Drosophila melanogaster. In* "Genes, Development and Cancer. The Life and Work of Edward B. Lewis." Kluwer Academic Publishers, Norwell, MA.

Boncinelli, E., Acampora, D., Pannese, M., D'Esposito, M., Somma, R., Gaudino, G., Stornaiuolo, A., Cafiero, M., Faiella, A., and Simeone, A. (1989). Organization of human class I homeobox genes. *Genome* **31,** 745–756.

Bürglin, T. R. (1994). A comprehensive classification of homeobox genes. *In* "Guidebook to the Homeobox Genes" (D. Duboule, Ed.), pp. 25–71. Oxford University Press, Oxford.

Bürglin, T. R. (1995). The evolution of homeobox genes. *In* "Biodiversity and Evolution" (R. Arai, M. Kato, and Y. Doi, Eds.) pp. 291–336. The National Science Museum Foundation, Tokyo.

Bürglin, T. R. (2005). Homeodomain proteins. *In* "Encyclopedia of Molecular Cell Biology and Molecular Medicine" (R.A. Meyers, Ed.), vol. 6, pp. 179–222. Wiley-VCH Verlag GmbH & Co., Weinheim.

Bürglin, T. R., Coulson, A., Hawkins, N. C., McGee, J. D., Schaller, D., Wittmann, C., Müller, F., and Waterston, R. H. (1991). Nematode homeobox cluster. *Nature* **351,** 703.

Burke, A. C., Nelson, C. E., Morgan, B. A., and Tabin, C. (1995). *Hox* genes and the evolution of vertebrate axial morphology. *Development* **121,** 333–346.

Butts, T., Holland, P. W., and Ferrier, D. E. (2008). The urbilaterian Super-Hox cluster. *Trends Genet.* **24,** 259–262.

Callaerts, P., Lee, P. N., Hartmann, B., Farfan, C. B., Choy, D. W., Fischbach, K.-F., Gehring, W. J., and de Couet, H. G. (2002). *HOX* genes in the sepiolid squid *Euprymna scolopes*: Implications for the evolution of complex body plans. *Proc. Natl. Acad. Sci. USA* **99,** 2088–2093.

Carapuco, M., Novoa, A., Bobola, N., and Mallo, M. (2005). *Hox* genes specify vertebral types in the presomitic mesoderm. *Genes Dev.* **19,** 2116–2121.

Carpenter, F. M. (1992). Order Protorthoptera. *In* "Treatise on Invertebrate Paleontology" (F.M. Carpenter, Ed.), vol. 3, pp. 97–134. The University of Kansas and the Geological Society of America, KansasArthropoda 4, Superclass Hexapoda.

Carrasco, A. E., McGinnis, W., Gehring, W. J., and De Robertis, E. M. (1984). Cloning of an *X. laevis* gene expressed during early embryogenesis that codes for a peptide region homologous to *Drosophila* homeotic genes. *Cell* **37,** 409–414.

Carroll, S. B., Grenier, J. K., and Weatherbee, S. D. (2001). From DNA to diversity: The primacy of regulatory evolution. *In* "From DNA to Diversity: Molecular Genetics and the Evolution of Animal Design." Blackwell Science, Inc., Malden, MA.

Chan, L.-N., and Gehring, W. J. (1971). Determination of blastoderm cells in *Drosophila melanogaster. Proc. Natl. Acad. Sci. USA* **68,** 2217–2221.

Cohn, M. J., and Tickle, C. (1999). Developmental basis of limblessness and axial patterning in snakes. *Nature* **399,** 474–479.

Cook, C. E., Jimenez, E., Akam, M., and Salo, E. (2004). The *Hox* gene complement of acoel flatworms, a basal bilaterian clade. *Evol. Dev.* **6,** 154–163.

Czerny, T., Halder, G., Callaerts, P., Kloter, U., Souabni, A., Gehring, W. J., and Busslinger, M. (1999). Twin of eyeless, a second *Pax-6* gene of *Drosophila*, acts upstream of eyeless in the control of eye development. *Mol. Cell* **3,** 297–307.

Dalton, D., Chadwick, R., and McGinnis, W. (1989). Expression and embryonic function of empty spiracles: A *Drosophila* homeo box gene with two patterning functions on the anterior-posterior axis of the embryo. *Genes Dev.* **3,** 1940–1956.

Duboule, D. (Ed.), (1994). "Guidebook to the Homeobox Genes."Oxford University Press, Oxford.

Duboule, D., and Dolle, P. (1989). The structural and functional organization of the murine *HOX* gene family resembles that of *Drosophila* homeotic genes. *EMBO J.* **8,** 1497–1505.

Duboule, D., and Morata, G. (1994). Colinearity and functional hierarchy among genes of the homeotic complexes. *Trends Genet.* **10,** 358–364.

Duncan, I., and Lewis, E. B. (1982). Genetic control of body segment differentiation in *Drosophila*. *In* "Developmental Order: Its Origin and Regulation" (S. Subtelny and P.B. Green, Eds.), pp. 533–554. Alan R. Liss, Inc., New York.

Edvardsen, R. B., Seo, H. C., Jensen, M. F., Mialon, A., Mikhaleva, J., Bjordal, M., Cartry, J., Reinhardt, R., Weissenbach, J., Wincker, P., and Chourrout, D. (2005). Remodelling of the homeobox gene complement in the tunicate *Oikopleura dioica*. *Curr. Biol.* **15,** R12–R13.

Estrada, B., and Sanchez-Herrero, E. (2001). The *Hox* gene Abdominal-B antagonizes appendage development in the genital disc of *Drosophila*. *Development* **128,** 331–339.

Finkelstein, R., Smouse, D., Capaci, T. M., Spradling, A. C., and Perrimon, N. (1990). The *orthodenticle* gene encodes a novel homeo domain protein involved in the development of the *Drosophila* nervous system and ocellar visual structures. *Genes Dev.* **4,** 1516–1527.

Finnerty, J. R., and Martindale, M. Q. (1999). Ancient origin of axial patterning genes. *Evol. Dev.* **1,** 16–23.

Force, A., Lynch, M., Pickett, F. B., Amores, A., Yan, Y. L., and Postlethwait, J. (1999). Preservation of duplicate genes by complementary, degenerative mutations. *Genetics* **151,** 1531–1545.

Garcia-Fernandez, J., and Holland, P. W. (1994). Archetypal organization of the amphioxus *Hox* gene cluster. *Nature* **370,** 563–566.

Gehring, W. J. (1998). "Master Control Genes in Development and Evolution: The Homeobox Story." Yale University Press, New Haven, CT.

Goldberg, M. L., Sheen, J.-Y., Gehring, W. J., and Green, M. M. (1983). Unequal crossing-over associated with asymmetrical synapsis between nomadic elements in the *Drosophila melanogaster* genome. *Proc. Natl. Acad. Sci. USA* **80,** 5017–5021.

Gonzales-Reyes, A., Urquia, N., Gehring, W. J., Struhl, G., and Morata, G. (1990). Are cross-regulatory interactions between homoeotic genes functionally significant? *Nature* **344,** 78–80.

Gu, X., Zhang, Z., and Huang, W. (2005). Rapid evolution of expression and regulatory divergences after yeast gene duplication. *Proc. Natl. Acad. Sci. USA* **102,** 707–712.

Hafen, E., Levine, M., and Gehring, W. J. (1984). Regulation of Antennapedia transcript distribution by the bithorax complex in *Drosophila*. *Nature* **307,** 287–289.

Hildebrand, M. (1995). "Analysis of Vertebrate Structure" 4th Edn. Wiley, New York.

Holland, P. W., and Garcia-Fernandez, J. (1996). *Hox* genes and chordate evolution. *Dev. Biol.* **173,** 382–395.

Huson, D. H., and Bryant, D. (2006). Application of phylogenetic networks in evolutionary studies. *Mol. Biol. Evol.* **23,** 254–267.

Ikuta, T., and Saiga, H. (2005). Organization of *Hox* genes in ascidians: Present, past, and future. *Dev. Dyn.* **233,** 382–389.

Ingham, P. W. (1998). *Trithorax* and the regulation of homeotic gene expression in *Drosophila*: A historical perspective. *Int. J. Dev. Biol.* **42,** 423–429.

Karch, F., Weiffenbach, B., Peifer, M., Bender, W., Duncan, I., Celniker, S., Crosby, M., and Lewis, E. B. (2003). "Genes, Development and Cancer. The Life and Work of Edward B. Lewis." Kluwer Academic Publishers, Norwell, MA.

Kaufman, T. C., Lewis, R., and Wakimoto, B. (1980). Cytogenetic analysis of chromosome 3 in *Drosophila melanogaster*: The homoeotic gene complex in polytene chromosome interval 84A-B. *Genetics* **94,** 115–133.

Kmita, M., and Duboule, D. (2003). Organizing axes in time and space; 25 years of colinear tinkering. *Science* **301,** 331–333.

Kmita, M., Fraudeau, N., Herault, Y., and Duboule, D. (2002). Serial deletions and duplications suggest a mechanism for the collinearity of *Hoxd* genes in limbs. *Nature* **420,** 145–150.

Kmita-Cunisse, M., Loosli, F., Bierne, J., and Gehring, W. J. (1998). Homeobox genes in the ribbonworm *Lineus sanguineus*: Evolutionary implications. *Proc. Natl. Acad. Sci. USA* **95,** 3030–3035.

Kozmik, Z., Ruzickova, J., Jonasova, K., Matsumoto, Y., Vopalensky, P., Kozmikova, I., Strnad, H., Kawamura, S., Piatigorsky, J., Paces, V., and Vlcek, C. (2008). Assembly of the cnidarian camera-type eye from vertebrate-like components. *Proc. Natl. Acad. Sci. USA* **105,** 8989–8993.

Lewis, E. B. (1978). A gene complex controlling segmentation in *Drosophila*. *Nature* **276,** 565–570.

Lewis, E. B. (1992). The 1991 Albert Lasker medical awards. Clusters of master control genes regulate the development of higher organisms. *JAMA* **267,** 1524–1531.

Lewis, E. B. (1995). Nobel lecture, December 8, 1995. The bithorax complex: The first fifty years. *In* "Genes, Development and Cancer. The Life and Work of Edward B. Lewis" (H. Lifshytz, Ed.). Kluwer Academic Publishers, Norwell, MA.

Lewis, E. B., Pfeiffer, B. D., Mathog, D. R., and Celniker, S. E. (2003). Evolution of the homeobox complex in the Diptera. *Curr. Biol.* **13,** R587–R588.

Li, W.-H., Yang, J., and Gu, X. (2005). Expression divergence between duplicate genes. *Trends Genet.* **21,** 602–607.

Malicki, J., Schughart, K., and McGinnis, W. (1990). Mouse *Hox-2.2* specifies thoracic segmental identity in *Drosophila* embryos and larvae. *Cell* **63,** 961–967.

McGinnis, W., Levine, M. S., Hafen, E., Kuroiwa, A., and Gehring, W. J. (1984a). A conserved DNA sequence in homeotic genes of the *Drosophila* Antennapedia and *bithorax* complex. *Nature* **308,** 428–433.

McGinnis, W., Garber, R. L., Wirz, J., Kuroiwa, A., and Gehring, W. J. (1984b). A homologous protein-coding sequence in *Drosophila* homeotic genes and its conservation in other metazoans. *Cell* **37,** 403–408.

McGinnis, W., Hart, C. P., Gehring, W. J., and Ruddle, F. H. (1984c). Molecular cloning and chromosome mapping of a mouse DNA sequence homologous to homeotic genes of *Drosophila*. *Cell* **38,** 675–680.

Mlodzik, M., Fjose, A., and Gehring, W. J. (1985). Isolation of *caudal*, a *Drosophila* homeo box-containing gene with maternal expression, whose transcripts form a concentration gradient at the pre-blastoderm stage. *EMBO J.* **4,** 2961–2969.

Moreno, E., and Morata, G. (1999). *Caudal* is the *Hox* gene that specifies the most posterior *Drosophile* segment. *Nature* **400,** 873–877.

Nordstrom, K., Wallen, R., Seymour, J., and Nilsson, D. (2003). A simple visual system without neurons in jellyfish larvae. *Proc. Biol. Sci.* **270,** 2349–2354.

Ohno, S. (1970). "Evolution by Gene Duplication." Springer, New York.

Plaza, S., Prince, F., Jaeger, J., Kloter, U., Flister, S., Benassayag, C., Cribbs, D., and Gehring, W. J. (2001). Molecular basis for the inhibition of *Drosophila* eye development by *Antennapedia*. *EMBO J.* **20,** 802–811.

Plaza, S., Prince, F., Adachi, Y., Punzo, C., Cribbs, D., and Gehring, W. J. (2008). Cross-regulatory protein-protein interactions between Hox and Pax transcription factors. *Proc. Natl. Acad. Sci. USA* **105,** 13439–13444.

Punzo, C., Plaza, S., Seimiya, M., Schnupf, P., Kurata, S., Jaeger, J., and Gehring, W. J. (2004). Functional divergence between eyeless and twin of eyeless in *Drosophila melanogaster*. *Development* **131,** 3943–3953.

Quiquand, M., Yanze, N., Schmich, J., Schmid, V., Galliot, B., and Piraino, S. (2009). More constraint on *ParaHox* than *Hox* gene families in early metazoan evolution. *Dev. Biol.* (in press).

Rivera-Pomar, R., Niessing, D., Schmidt-Ott, U., Gehring, W. J., and Jackle, H. (1996). RNA binding and translational suppression by bicoid. *Nature* **379,** 747–749.

Ronshaugen, M., Biemar, F., Piel, J., Levine, M., and Lai, E. C. (2005). The *Drosophila* microRNA iab-4 causes a dominant homeotic transformation of halteres to wings. *Genes Dev.* **19,** 2947–2952.

Rose, L. S., and Wieschaus, E. (1992). The *Drosophila* cellularization gene *nullo* produces a blastoderm-specific transcript whose levels respond to the nucleocytoplasmic ratio. *Genes Dev.* **6,** 1255–1268.

Sanchez-Herrero, E., Vernos, I., Marco, R., and Morata, G. (1985). Genetic organization of *Drosophila bithorax* complex. *Nature* **313,** 108–113.

Scott, M. P., and Weiner, A. J. (1984). Structural relationships among genes that control development: Sequence homology between the Antennapedia, Ultrabithorax, and fushi tarazu loci of *Drosophila. Proc. Natl. Acad. Sci. USA* **84,** 4115–4119.

Seo, H.-C., Edvardsen, R. B., Maeland, A. D., Bjordal, M., Jensen, M. F., Hansen, A., Flaat, M., Weissenbach, J., Lehrach, H., Wincker, P., Reinhardt, R., and Chourrout, D. (2004). *Hox* cluster disintegration with persistent anteroposterior order of expression in *Oikopleura dioica. Nature* **431,** 67–71.

Shimodaira, H., and Hasegawa, M. (2001). CONSEL: For assessing the confidence of phylogenetic tree selection. *Bioinformatics* **17,** 1246–1247.

Simeone, A., Gulisano, M., Acampora, D., Stornaiuolo, A., Rambaldi, M., and Boncinelli, E. (1992). Two vertebrate homeobox genes related to the *Drosophila* empty spiracles gene are expressed in the embryonic cerebral cortex. *EMBO J.* **11,** 2541–2550.

Simodaira, H. (2002). An approximately unbiased test of phylogenetic tree selection. *Syst. Biol.* **51,** 492–508.

Simon, J., Peifer, M., Bender, W., and O'Connor, M. (1990). Regulatory elements of the bithorax complex that control expression along the anterior-posterior axis. *EMBO J.* **9,** 3945–3956.

Steingrimsson, E., Copeland, N. G., and Jenkins, N. A. (2004). Melanocytes and the microphthalmia transcription factor network. *Annu. Rev. Genet.* **38,** 365–411.

Von Allmen, G., Hogga, I., Spierer, A., Karch, F., Bender, W., Gyurkovics, H., and Lewis, E. (1996). Splits in fruitfly *Hox* gene complexes. *Nature* **380,** 116.

Wada, H., Garcia-Fernandez, J., and Holland, P. W. (1999). Colinear and segmental expression of amphioxus *Hox* genes. *Dev. Biol.* **213,** 131–141.

Wada, S., Tokuoka, M., Shoguchi, E., Kobayashi, K., Di Gregorio, A., Spagnuolo, A., Branno, M., Kohara, Y., Rokhsar, D., Levine, M., Saiga, H., Satoh, N., *et al.* (2003). A genomewide survey of developmentally relevant genes in *Ciona intestinalis*. II. Genes for homeobox transcription factors. *Dev. Genes Evol.* **213,** 222–234.

Wellig, D. M., and Capecchi, M. R. (2003). *Hox10* and *Hox11* genes are required to globally pattern the mammalian skeleton. *Science* **301,** 363–367.

Zhang, J., and Nei, M. (1995). Evolution of *Antennapedia*-class homeobox genes. *Genetics* **142,** 295–303.

CHAPTER THREE

Hox Specificity: Unique Roles for Cofactors and Collaborators

Richard S. Mann,* Katherine M. Lelli,† *and* Rohit Joshi*

Contents

Abstract

Hox proteins are well known for executing highly specific functions *in vivo*, but our understanding of the molecular mechanisms underlying gene regulation by these fascinating proteins has lagged behind. The premise of this review is that an understanding of gene regulation—by any transcription factor—requires the dissection of the *cis*-regulatory elements that they act upon. With this goal in mind, we review the concepts and ideas regarding gene regulation by Hox proteins and apply them to a curated list of directly regulated Hox *cis*-regulatory elements that have been validated in the literature. Our analysis of the Hox-binding sites within these elements suggests several emerging generalizations. We distinguish between Hox cofactors, proteins that bind DNA cooperatively with Hox proteins and thereby help with DNA-binding site selection, and Hox collaborators, proteins that bind in parallel to Hox-targeted *cis*-regulatory elements and dictate the sign and strength of gene regulation. Finally, we summarize insights that come from examining five X-ray crystal structures of

* Department of Biochemistry and Molecular Biophysics, Columbia University, New York, NY, USA
† Department of Genetics and Development, Columbia University, New York, NY, USA

Current Topics in Developmental Biology, Volume 88
ISSN 0070-2153, DOI: 10.1016/S0070-2153(09)88003-4

Hox-cofactor-DNA complexes. Together, these analyses reveal an enormous amount of flexibility into how Hox proteins function to regulate gene expression, perhaps providing an explanation for why these factors have been central players in the evolution of morphological diversity in the animal kingdom.

1. An Introduction to the Problem

Hox proteins are homeodomain-containing transcription factors that have the capacity to carry out exquisitely precise functions *in vivo* that are critical for many aspects of animal morphogenesis. Most typically, each *Hox* gene is expressed in a subset of the anterior-posterior (AP) body axis, where it specifies cellular and tissue identities. Famous examples of the power that *Hox* genes have to sculpt animal morphogenesis include the antenna-to-leg transformation caused by the *Antennapedia* (*Antp*) mutation in *Drosophila* and several polydactyly syndromes in humans (Goodman, 2002; Lewis, 1978; Randazzo *et al.*, 1991). These types of phenotypes have been a long-standing source of fascination for both biologists and lovers of science fiction.

An important and long-debated question for Hox biologists has been how these proteins achieve this apparently high degree of *in vivo* specificity. In this review, we summarize ideas and recent data bearing on the question of Hox specificity, with a special emphasis on what can be learned by analyzing native *cis*-regulatory elements that are directly bound and regulated by Hox proteins. Excellent reviews discussing the range of Hox target genes that have been identified using genome-wide and traditional approaches complement the emphasis of this chapter (Hueber and Lohmann, 2008; Pearson *et al.*, 2005).

When *Hox* genes were first cloned and shown to encode homeodomain-containing proteins (Akam, 1989; Regulski *et al.*, 1985), researchers initially speculated that Hox proteins would bind and regulate the correct subset of target genes according to the DNA recognition properties of their homeodomains. However, early work from a number of labs quickly established that homeodomains were not likely to be up to the task of dictating Hox-DNA-binding specificities on their own (Affolter *et al.*, 1990; Desplan *et al.*, 1988; Ekker *et al.*, 1991, 1994; Hoey and Levine, 1988). Indeed, homeodomains, particularly the subset present in the Hox protein family, all bind to a very similar set of "AT"-rich DNA-binding sites, raising the fundamental question of how specificity is achieved. In addition to this classical problem of degenerate binding site recognition, experiments using chimeric Hox proteins—where bits of one Hox protein were replaced with homologous bits of another—highlight an additional complication. As expected, specific Hox functions required the homeodomain. In some cases, however, specificity also required nonhomeodomain residues, in particular, those that lie immediately N- or C-terminal to the homeodomain (Chan *et al.*, 1994; Dessain *et al.*, 1992; Furukubo-Tokunaga *et al.*, 1993; Gibson *et al.*, 1990; Kuziora and McGinnis, 1989, 1990; Lin and McGinnis, 1992; Mann and Hogness, 1990; Zhao and Potter, 2001, 2002). How these nonhomeodomain residues may impact Hox specificity is just now coming into focus.

Because they are DNA-binding transcription factors, it is probably a safe bet that Hox proteins carry out the majority of their functions by binding to *cis*-regulatory elements (although alternative mechanisms have been proposed; Plaza *et al.*, 2008). Because eukaryotic transcription is governed by *cis*-regulatory elements that typically integrate multiple inputs, each Hox-targeted element is likely to have binding sites for many transcription factors. Therefore, to understand how Hox proteins ultimately function to control target gene expression, it is necessary to consider two broad questions. First, how do Hox proteins recognize their DNA-binding sites and second, how do they interact with other transcriptional inputs that feed into the same *cis*-regulatory element? We suggest that it is helpful to break the problem of Hox specificity down into two conceptually separable steps (Fig. 3.1). In the first step, the question can be rephrased to ask: How do Hox proteins find the right DNA-binding sites *in vivo*? Many examples exist in the literature suggesting that Hox proteins solve this initial "DNA-binding specificity" step in multiple ways. As will be explored more fully below, one solution is by the use of cooperatively binding cofactors such as Extradenticle (Exd), Pbx, Homothorax (Hth), and Meis that increase Hox-DNA-binding specificities (previously reviewed by Mann and Affolter, 1998; Mann and Chan, 1996; Moens and Selleri, 2006). However, it is also increasingly clear that Hox proteins regulate many genes without the help of these cofactors. In the second step, the question is: Once bound, how do Hox proteins orchestrate a transcriptional response? As the same Hox protein can activate some target genes, and repress others, it is clear that this "activity regulation" step is also critical for how Hox proteins execute their *in vivo* functions. In fact, as will be described more fully below, there is now good evidence for both of the steps outlined in Fig. 3.1 playing critical roles in Hox specificity.

2. Too Many Binding Sites, Not Enough Specificity

Because all Hox proteins have a homeodomain, understanding how Hox proteins recognize their DNA-binding sites *in vivo* certainly depends, at least in part, on how this 60 amino acid domain recognizes DNA sequences. The basic DNA recognition principles for homeodomains

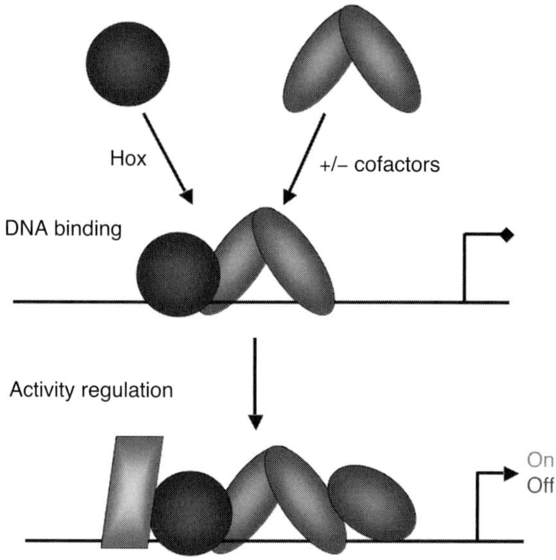

Figure 3.1 Two contributing steps in Hox specificity. In principle, Hox specificity can be broken down into two separate steps. The first step is DNA binding by Hox proteins, which can occur either with or without cooperatively binding cofactors. The second step involves the recruitment of additional factors, Hox collaborators, to the *cis*-regulatory element. The recruitment of these factors may depend on contacts between them and the DNA and/or protein-protein contacts between them and the Hox-cofactor complex. It is the recruitment of these collaborators, which we suggest depends on the architecture of the entire *cis*-regulatory element (including the details of the Hox-binding site) that ultimately determines the sign of the transcriptional regulation. (See Color Insert.)

were established from biochemical and structural studies (reviewed previously by Gehring *et al.*, 1994). These studies show that all homeodomains fold into a bundle of three alpha-helices and an unstructured "N-terminal" arm. DNA contacts are formed primarily by residues 47, 50, 51, and 54 in the third alpha-helix (the so-called recognition helix) and by an arginine in position 5 of the N-terminal arm. While these studies provided a high resolution view of how homeodomains generally bind to DNA, they did not provide much insight into the problem of Hox specificity for three reasons. First, nearly all Hox homeodomains, even those with very disparate *in vivo* functions, have the same residues in the DNA-contacting residues visualized in these structures (Mann, 1995). Second, although nonhomeodomain residues were known to play a role in Hox specificity from studies of chimeric Hox proteins (see above), these domains were not present in any of the initial structural studies. Third, the DNA sequences used in these early structural studies were not *in vivo* binding sites. Instead, these structures used high-affinity consensus sites that would not be expected to reveal insights into homeodomain specificity.

Another limitation in the early studies on homeodomain-DNA recognition was that only a small subset of homeodomain proteins were studied. Thanks to the advent of new, powerful methodologies, two large-scale studies have recently defined the individual DNA-binding specificities for nearly all homeodomains, including the subset present in the Hox proteins. One group used a bacterial one-hybrid approach (B1H) to analyze the DNA-binding preferences for all of the homeodomains encoded in the *Drosophila melanogaster* genome (Noyes *et al.*, 2008) while the second group used an entirely *in vitro* platform, protein-binding microarrays (PBMs), to answer the same question for all mouse homeodomains (Berger *et al.*, 2008). Although there are pros and cons for each approach (Affolter *et al.*, 2008), these studies confirmed that Hox homeodomains like to bind "AT"-rich DNA sequences (Fig. 3.2). In particular, the so-called "Antennapedia (Antp) Group" of homeodomains, which includes all Hox homeodomains except for those of the Abdominal-B (Abd-B) class, like to bind the sequence TAAT[t/g][a/g]. There are 87,307 copies of the sequence TAATTA and 86,201 copies of the sequence TAATGA in the *D. melanogaster* genome, each more than five times the number of annotated protein-coding genes. Clearly, the presence of TAAT[t/g][a/g] cannot be sufficient information for Hox regulation. Moreover, as elegantly illustrated by the homeodomain-binding site survey studies (Berger *et al.*, 2008; Noyes *et al.*, 2008), TAAT[t/g][a/g] is readily bound by most Hox homeodomains, as well as many non-Hox homeodomains. Therefore, this and related binding sites cannot be sufficient to distinguish between Hox family members that carry out distinct functions *in vivo*.

3. How Specific Do Hox Proteins Need to be?

Hox biologists can readily point to highly specific functions that are uniquely specified by individual Hox proteins. For example, in *Drosophila*, only the *Hox* gene *Sex combs reduced* (*Scr*) can orchestrate the development of a salivary gland, presumably by regulating a network of salivary gland-promoting genes (Bradley *et al.*, 2001). The flip view that multiple Hox proteins probably share many targets is typically given less attention. We believe this discussion is highly relevant to how one thinks about Hox specificity, because it may be that only a subset of Hox targets for any particular Hox protein need be highly specific.

One example of a morphological process that is controlled by multiple *Hox* genes is appendage development in *Drosophila*. Leg development is limited to the thoracic segments due to repression of *Distalless* (*Dll*) by the abdominal *Hox* genes, *Ultrabithorax* (*Ubx*), *abdominal-A* (*abd-A*), and *Abdominal-B* (*Abd-B*) (Estrada and Sánchez-Herrero, 2001; Vachon *et al.*, 1992).

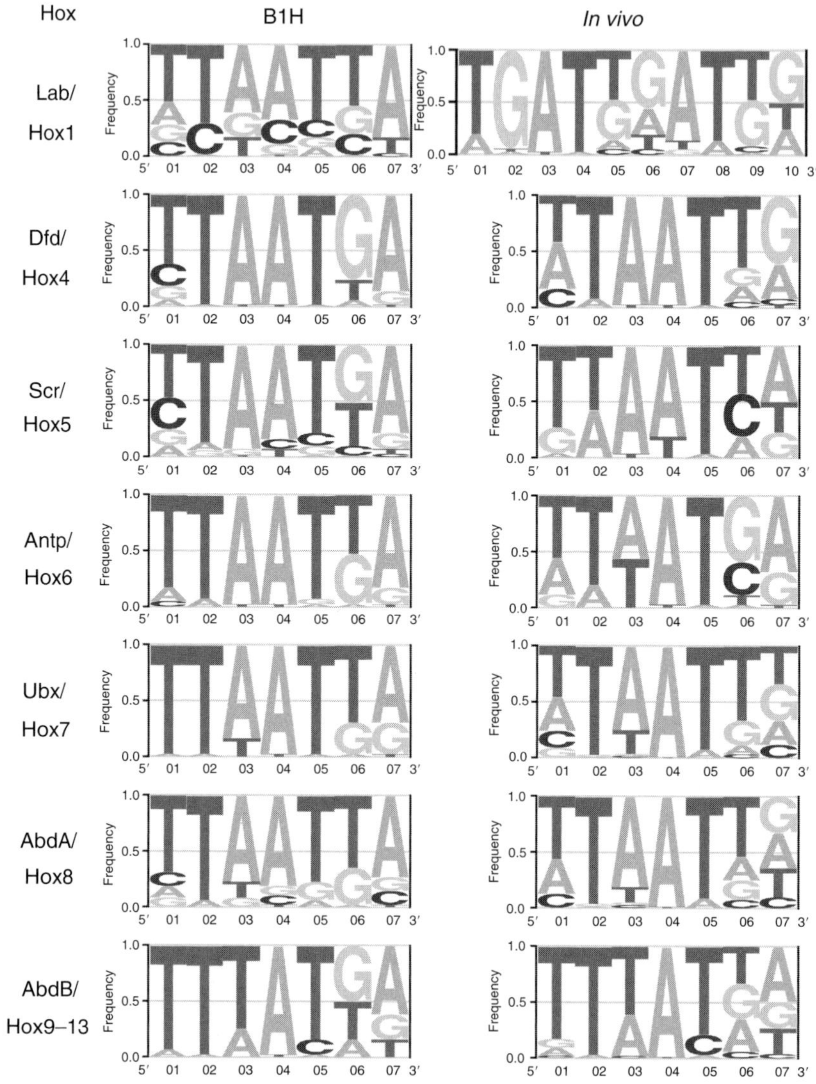

Figure 3.2 Comparison of *in vitro* and *in vivo* Hox-binding site preferences. Shown are LOGO diagrams summarizing Hox-binding site preferences for most paralogs. The column on the left lists the LOGOs generated using the binding sites identified by the bacterial 1-hybrid (B1H) method (Noyes *et al.*, 2008). The column on the right lists the LOGOs generated using the *in vivo* binding sites in Table 3.1. To generate these LOGOs, we used CONSENSUS (as part of Target Explorer; http://luna.bioc.columbia.edu/Target_Explorer/) to generate position weight matrices (PWMs). PWMs that maximized alignment of an "AT" sequence were converted to Transfac format using the phiSITE conversion server (http://www.phisite.org/main/index.php?nav=home). enoLOGOS (http://chianti.ucsd.edu/cgi-bin/enologos/enologos.cgi/) was then used to generate the LOGOs using nucleotide frequency for the *Y*-axis.

Moreover, *Ubx* and *abd-A* work, at least in part, through common binding sites in a *Dll cis*-regulatory element (Gebelein *et al.*, 2002, 2004). Thus, *Dll* is a shared target for the abdominal *Hox* genes. Similarly, the antenna-specifying gene *homothorax* (*hth*) can be repressed by all of the trunk Hox proteins, suggesting that *hth* is also a shared target (Casares and Mann, 1998, 2001; Yao *et al.*, 1999). A third example is the *Drosophila* head-promoting gene, *optix*, which is repressed by the trunk *Hox* genes and activated by the more anterior (head) *Hox* genes (Coiffier *et al.*, 2008). For *hth* and *optix*, a limitation in this conclusion is that the Hox-binding sites in these genes (assuming the regulation is direct) have not yet been identified. Therefore, although these genes are clearly shared *Hox* targets by genetic criteria, it is possible that different Hox proteins use different binding sites within these genes to regulate their expression. Nevertheless, these and other examples (Greig and Akam, 1995; Hirth *et al.*, 2001) support the idea that not all Hox functions need to be paralog-specific. If true, it follows that many bona fide Hox-binding sites may not need to discriminate between different Hox proteins.

How many Hox targets are shared and how many are paralog-specific? Although the field may be getting closer to a definitive answer to this question, by applying ChIP-chip and/or ChIP-seq methodologies to Hox proteins, the currently available data provide an interesting estimate. Using overexpression of Hoxc8 in mouse embryo fibroblasts, the expression levels of 34 genes were found to change by twofold or more (Lei *et al.*, 2005). This relatively small number of potential Hoxc8 targets contrasts from the much larger numbers of regulated genes identified in whole embryo expression profiling experiments following the uniform (and ectopic) expression of individual Hox proteins during *Drosophila* embryogenesis (Hueber *et al.*, 2007). An important strength of these experiments is that the global transcriptional response to multiple Hox proteins was analyzed in parallel, using the same experimental conditions. Remarkably, of the ~1500 genes (about 10% of all *Drosophila* genes) whose expression changed significantly in response to ectopic Deformed (Dfd), Scr, Antp, Ubx, Abd-A, or Abd-B expression, more than two-thirds (~69%) were regulated by only one of these six Hox proteins. About one-third (~30%) of all Hox-responsive target genes responded to multiple Hox proteins, while only ~1% responded to all six of these Hox proteins. There are, however, a few caveats to these experiments. For one, they measured responses to ectopic

---

The number of binding sites used to generate each LOGO was as follows: Labial: 31 (B1H), 17 (*in vivo*); Dfd: 24 (B1H), 17 (*in vivo*); Scr: 34 (B1H), 12 (*in vivo*); Antp: 19 (B1H), 16 (*in vivo*); Ubx: 20 (B1H), 57 (*in vivo*; the resulting LOGO was only subtly affected if the 30 sites from the Antp-P2 element were omitted); Abd-A: 23 (B1H), 39 (*in vivo*); and AbdB: 21 (B1H), 49 (*in vivo*). (See Color Insert.)

Hox expression, which may not always reflect accurate gene regulation in their native expression domains. Second, these experiments did not distinguish between tissue-specific responses and third, they could not unambiguously discriminate between direct and indirect effects. Nevertheless, the results from this study are remarkable because they suggest that many, and perhaps the majority of Hox target genes are paralog-specific. However, the results also support the view that a significant number of genes are targeted by multiple Hox proteins, again raising the possibility that not all Hox-binding sites need to discriminate between Hox proteins.

Developmental context is another issue that should be considered when thinking about Hox specificity. In *Drosophila*, for example, each embryonic segment is built using the same reiterated set of signaling pathways that provide them with a common coordinate system, known as positional information. Once this "ground plan" is established, the non-*Hox* regulatory inputs into a gene are largely the same from segment to segment. One way to think about the *Hox* genes is that they impose identity information on top of this developmental ground plan, thus providing each segment its unique characteristics. Because the other regulatory inputs are largely equivalent, a gene that is specifically expressed (or repressed) in a small subset of embryonic segments is likely to be a paralog-specific Hox target gene. For example, for salivary glands to form only in the first thoracic (T1) segment, the *Hox* gene *Scr* must activate the salivary gland program, including its directly activated target gene *forkhead* (*fkh*), in a paralog-specific manner; other Hox proteins cannot activate this target in other segments (Bradley *et al.*, 2001). Similarly, while *Dll* is repressed by abdominal Hox proteins, the thoracic Hox proteins Scr and Antp must be permissive for *Dll* expression (Gebelein *et al.*, 2002). Repression by the abdominal Hox proteins therefore must have specificity. However, the same is not true for tissues where the ground plan, and thus the other regulatory inputs, is different. For example, the *Hox* gene *Ubx* is expressed in all cells of the developing haltere (a balancing organ used during flight) of the fly, where it regulates the expression of many genes (Crickmore and Mann, 2007, 2008; Lewis, 1978; Weatherbee and Carroll, 1999). Other than the developing wing, where *Hox* genes are not expressed for most of development, no other tissue in the fly has the same ground plan as the haltere. Therefore, the *cis*-regulatory elements used by *Ubx* to regulate genes in the haltere may not need to be highly selective for Ubx: other Hox proteins never have the opportunity to regulate these genes in the haltere/wing tissue simply because they are not expressed there. Confirmation of this idea comes from the finding that other Hox proteins, when expressed in the wing, can result in haltere-like phenotypes and mimic Ubx-like regulation (Casares *et al.*, 1996; R. S. Mann and M. Crickmore, unpublished observations). Similarly, although both Ubx and Abd-A have the potential to induce gonad development, the job is normally carried out by Abd-A simply because it is the only Hox protein that is expressed in the correct set of mesodermal progenitor cells (Greig and Akam, 1995). Finally, another remarkable example of functional redundancy among different Hox paralogs is that all of the *Drosophila* Hox proteins, except for Abd-B, have the ability to replace Labial in the specification of the tritocerebral neuromere in the fly's brain (Hirth *et al.*, 2001). Like Ubx in the haltere, the ground plan for this portion of the brain may be sufficiently unique so that it does not require exquisite Hox specificity, accounting for why nearly all Hox paralogs can, at least to some degree, carry out the same functions as Labial.

In summary, when thinking about what Hox-bindings sites may look like based on these considerations, we suggest that it is useful to distinguish between three types of Hox target genes (Fig. 3.3) (1) those that must be highly specific for one Hox paralog ("paralog-specific"; e.g., Scr → *fkh*), (2) those that are shared by a subset of Hox proteins ("semi-paralog-specific"; e.g., Ubx, Abd-A, and Abd-B ⊣ *Dll*), and (3) those that are regulated by most or all *Hox* genes ("general"; e.g., all of the trunk *Hox* genes ⊣ *optix*). In addition, we argue that for some Hox functions that have the appearance of paralog specificity (e.g., Ubx dictating haltere instead of wing fates), the Hox-binding sites, themselves, do not need to be paralog-specific as long as the developmental context is sufficient to specify a unique regulatory environment.

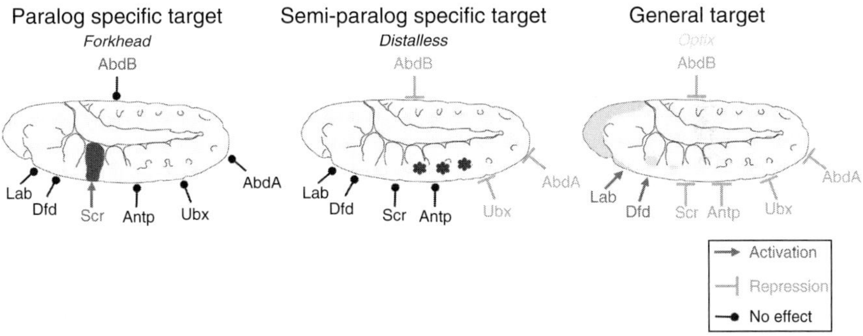

Figure 3.3 Three types of Hox target genes. "Paralog-specific" Hox target genes are those that are uniquely regulated by only a single Hox paralog, such as the activation of *fkh* by Scr. "Semi-paralog-specific" Hox target genes are those that are shared by a small subset of Hox paralogs, such as the repression of *Dll* by the abdominal Hox proteins Ubx, Abd-A, and Abd-B (schematized here is the *Dll304* embryonic enhancer element). "General" Hox target genes are those that are regulated by most, or perhaps all, Hox paralogs, such as the control of *optix* in *Drosophila*. Ideally, this classification should apply to individual *cis*-regulatory elements, not entire genes, to allow for the scenario that the same gene may fall into more than one of these categories (in two different tissues or times of development). For *fkh* and *Dll*, specific *cis*-regulatory elements that fit the "paralog-specific" and "semi-paralog-specific" criteria have been identified. In contrast, a single *cis*-regulatory element that is a "general" Hox target has not yet been identified and therefore remains hypothetical. (See Color Insert.)

4. Hox Cofactors

Given that some Hox functions truly require a high degree of specificity, and that Hox homeodomains, themselves, are not sufficiently discriminating to account for this specificity, how is specificity achieved? One well-established way in which Hox proteins achieve specificity *in vivo* is to bind DNA cooperatively with other DNA-binding cofactors. To date, the best-characterized cofactors are all TALE (three amino acid loop extension) homeodomain proteins (Mann and Chan, 1996; Moens and Selleri, 2006). In *Drosophila*, the known TALE Hox cofactors are Extradenticle (Exd) and Homothorax (Hth). In the mouse, there are four Exd-related proteins (Pbx1, Pbx2, Pbx3, Pbx4) and five Hth-related proteins (Meis1, Meis2, Meis3, Prep1, and Prep2). In *Caenorhabditis elegans*, there are three genes encoding Exd-like proteins (*ceh-20*, *ceh-40*, *ceh-60*) and two that encode Hth-like proteins, *unc-62* and *psa-2*, which encodes for a truncated form that has no homeodomain (see Mukherjee and Bürglin, 2007 for a thorough description of the TALE family genes). Here, we collectively refer to Exd/Pbx/Ceh-20 as PBC proteins. These proteins all have the ability, at least on some DNA sequences, to bind with Hox proteins in a highly cooperative manner. In addition, it is important to stress that where it has been analyzed, TALE family homeodomain proteins also carry out many Hox-independent functions *in vivo* (Bessa *et al.*, 2008; Casares and Mann, 1998, 2001; Jiang *et al.*, 2008; Laurent *et al.*, 2007; Moens and Selleri, 2006). Because they have both Hox-dependent and Hox-independent functions the genetic analysis of TALE family genes needs to be interpreted with caution, since only a subset of the observed phenotypes is due to their role as Hox cofactors.

Protein interaction domains characterized in Hox, PBC, and Hth/Meis/Prep proteins have provided many insights into how these three factors assemble when bound to DNA. PBC proteins interact with Hth/Meis/Prep family members in a DNA-independent manner via highly conserved domains present in the N-terminal regions (PBC-A of PBC and HM of Hth/Meis) of these proteins (Mann and Affolter, 1998). In several cases, the nuclear localization and/or stability of these proteins has been shown to depend on this protein-protein interaction (Arata *et al.*, 2006; Berthelsen *et al.*, 1999; Haller *et al.*, 2004; Huang *et al.*, 2003; Mann and Abu-Shaar, 1996; Ryoo and Mann, 1999; Saleh *et al.*, 2000; Stevens and Mann, 2007). In contrast, PBC-Hox interactions appear to be more complicated—and potentially more interesting. The traditional view, which has been supported by biochemical, *in vivo*, and structural studies, is that a motif common to most Hox proteins—YPWM—makes direct contacts with the TALE motif in PBC homeodomains, which creates a hydrophobic pocket that binds the W in YPWM (Chan and Mann, 1996; Chang *et al.*, 1995; Joshi *et al.*, 2007; LaRonde-LeBlanc and Wolberger, 2003; Lu and Kamps, 1996;

Neuteboom *et al.*, 1995; Passner *et al.*, 1999; Phelan *et al.*, 1995; Piper *et al.*, 1999). For those Hox proteins that do not have an obvious YPWM motif, in particular the Abd-B paralogs, there is a conserved W residue that, at least for a subset of Abd-B paralogs, plays an important role in this protein-protein interaction (Shen *et al.*, 1997). However, more recent studies make it clear that there is more to PBC-Hox-DNA complex formation than the YPWM-TALE interaction. On the one hand, two studies have shown that mutation of the YPWM motif in Ubx fail to abolish cooperative binding with PBC proteins *in vitro* and some Ubx functions *in vivo* (Galant *et al.*, 2002; Merabet *et al.*, 2003; Shen *et al.*, 1997). On the other hand, a peptide immediately following the Ubx homeodomain, termed the UbdA motif because of its similarity between Ubx and Abd-A, is playing an important, and perhaps dominant, role in the interaction between these Hox proteins and Exd on some binding sites (Merabet *et al.*, 2007). Interestingly, consistent with the idea that the UbdA motif is playing a role in PBC-Ubx (and likely Abd-A) interactions, it contributes to Ubx functional specificity *in vivo* (Chan and Mann, 1993; Gebelein *et al.*, 2002). Although the UbdA motif is not found outside of arthropods, these findings suggest the more general idea that PBC proteins may have modes of interaction with other Hox proteins that are in addition to the classical YPWM-TALE interaction. The existence of multiple PBC interaction domains in a single Hox protein such as Ubx suggest that the way in which PBC-Hox complexes assemble onto *cis*-regulatory elements may vary from target to target, potentially expanding their ability to recruit additional transcription factors.

In addition to TALE family homeodomain proteins, the *Drosophila* homeodomain protein Engrailed (En) has also been shown to be a Hox cofactor (Gebelein *et al.*, 2004). In this case, En bound cooperatively with both Ubx and Abd-A to a regulatory element from the *Dll* gene, and En input is required for *Dll* repression in the posterior compartments of the abdominal segments (Gebelein and Mann, 2007; Gebelein *et al.*, 2004). Unlike the TALE cofactors, which can function with Hox proteins to both activate and repress target genes, it is likely that En-Hox dimers are more typically involved in gene repression due to En's ability to directly bind the corepressor Groucho (Jiménez *et al.*, 1997).

A subset of Zn finger-containing transcription factors, most prominently *Drosophila* Teashirt (Tsh), has also been suggested to be Hox cofactors (Robertson *et al.*, 2004; Taghli-Lamallem *et al.*, 2007). Although a very appealing idea, these factors do not seem to exhibit the same robust cooperativity that is typically observed between TALE factors and Hox proteins. And, at least in the one target where Tsh-binding sites were identified (*modulo*), they are not adjacent to the Hox-binding sites (Taghli-Lamallem *et al.*, 2007). Thus, at this time we prefer to classify these Zn finger factors as Hox "collaborators," which provide additional, essential inputs into a subset of Hox-targeted *cis*-regulatory elements (discussed in more detail below) (Table 3.1).

Table 3.1 Direct Hox-DNA-binding sites

Name (gene)	Hox	Reporter Validation	DNA Binding Site (Hox, PBC, Meis/Hth, Other)	Cooperative Cofactors	Other Factors	Target Origin; Notes	References
EphA2-repeat E (*EphA2/eck*)	Hoxa1 Hoxb1	*Ex vivo*	gcaTGATGGATGGct	Pbx‡	ND	Mouse	(Chen and Ruley, 1998)
b1-ARE (*Hoxb1*)	Hoxb1 Lab	*In vivo*	ctcAGATGGATGGctgcgggTGATTGAAGTgTCTTTGTCATGCTAAT gcttggggggTGATGGATGGgcgctggggCTGCCAaac	Pbx1, Prep1	Sox2, Oct1	Mouse; Autoregulatory element, R3 site (also known as Repeat3) functions as a Labial target in *Drosophila*	(Chan and Mann, 1996; Di Rocco et al., 2001; Ferretti et al., 2005; Pöpperl et al., 1995)
r4-Hoxa2 (*Hoxa2*)	Hoxb1	*In vivo*	ggaTGATTTATTTgag; attTGACAGtaatgaagagTGATAGATTGctc; ggcTGATGCATTAatt	Pbx1, Prep1	ND	Chicken and Mouse	(Tümpel et al., 2007)
b2-PP2 (*Hoxb2*)	Hoxb1	*In vivo*	gagCTGTCAgggggctaAGATTGATCGccc	Pbx1†, Prep1	ND	Mouse	(Ferretti et al., 2000)
A3-PP2; A3-PhP1 (*Hoxa3*)	Hoxb1	*In vivo*	ggtTGATTATTGacc; gagTCATAAATCTtgcccagccataaaTGACAAaaa	Pbx1a†, Prep1	ND	Mouse and Chicken; Mutation of Prep site did not affect reporter expression in vivo	(Manzanares et al., 2001)
P2b_0.38 (*Phox2b*)	Hoxb1	*In vivo*	gcgTGATTGAATtaa; ttaTTGTCAtgt	Pbx1a†, Prep1	ND	Mouse	(Samad et al., 2004)
FPB (*COL5A2*)	Hoxb1	*Ex vivo*	gtcTGATTGATGGtaa	Pbx1a*	Prep1*	Human; Cotransfection of Prep1 increased expression ex vivo	(Penkov et al., 2000)
125bpHb9 (*Hb9*)	Hoxb1, Hoxb3	*In vivo*	agcTGATGAATTGacaaaaACTAATCA	Pbx1†	ND	Mouse	(Nakano et al., 2005)
EVIII (*CG11339*)	Lab	*In vivo*	tcgTGATCAATTAcagCTGACTggg	Exd, Hth	ND	*Drosophila*; Although the Exd half-site is required in vivo, cooperative binding of Lab/Exd dimers was not observed in vitro	(Ebner et al., 2005)
lab48/95 (*lab*)	Lab	*In vivo*	aatTGATGGATTGcccggcgccgaCTGTCAccg	Exd, Hth	ND	*Drosophila*; Autoregulatory element	(Ryoo et al., 1999)
enh450 (*ceh-13*)	CEH-13	*In vivo*	aaaTGATGGATGGttc	CEH-20	ND	*C. elegans*; Autoregulatory element	(Streit et al., 2002)
E2-A2RE (*Hoxa2*)	Hoxa2	*Ex vivo*	cggTGATTGATGGaag	Pbx1†	ND	Mouse, Autoregulatory element	(Lampe et al., 2004)
HRE1.25 (*Hoxa2*)	Hoxa2	*In vivo*	tctTGATTGATGAact	Pbx1a*	Prep1*	Mouse; Autoregulatory element, cotransfection of Prep1 increased expression ex vivo	(Lampe et al., 2008)
TTF-1-BS1&2 (*TTF-1*)	Hoxb3	*Ex vivo*	cctTAATTGgct; agaTAATTAgct	ND	ND	Rat; Both sites are necessary for expression ex vivo	(Guazzi et al., 1994)
CR3-HS1&2 (*Hoxb3, Hoxb4*)	Hoxb4, Hoxd4, Hoxb5	*In vivo*	agTGATTAATGGttttctgtaTAATTCtc	Exd†	ND	Mouse; Autoregulatory element; also responds to Dfd, Hoxb5, Scr, Antp, and Ubx in *Drosophila*	(Gould et al., 1997)
Rarb (*Rarb*)	Hoxb4, Hoxd4	*In vivo*	gggTGATAAATAAtgg	Pbx#	ND	Mouse; Role of Pbx not tested	(Serpente et al., 2005)
Repeat 3-TA (synthetic)	Dfd	*In vivo*	gggTGATTAATGGgcg	Exd	ND	Synthetic; (GG->TA) mutation of Repeat3 of b1-ARE	(Chan et al., 1997)
DfdEAE-module E (*Dfd*)	Dfd	*In vivo*	cccTAATTGccacgCATTAGctc	ND	Exd‡	*Drosophila*; Autoregulatory element, atypical Exd binding site, *exd* required in vivo	(Pinsonneault et al., 1997; Zeng et al., 1994)

Name (gene)	Hox	Reporter Validation	DNA Binding Site (**Hox**, PBC, **Meis/Hth**, **Other**)	Cooperative Cofactors	Other Factors	Target Origin; Notes	References
rprDfd (*rpr*)	Dfd	*In vivo*	caaTAATTAccc; ctcTAATTGccc; aacTAATTGaca; tcaTAATTGagg	Not Exd‡	ND	*Drosophila*; Cooperative binding with Exd not observed	(Lohmann et al., 2002)
1.28DRE (*1.28*)	Dfd	*In vivo*	gttTAATTGgtt; tctTAATAGccg; ccaCTACATTAATTATgaa; ccgCGATAATAAatc	Not Exd‡	ND	*Drosophila*; Sites not tested individually, cooperative binding with Exd not observed	(Pederson et al., 2000)
jkl-216-site1 (*hlh-8*)	LIN-39	*In vivo*	agtTGAAaaATTACCgcg	CEH-20	ND	*C. elegans*	(Liu and Fire, 2000)
egl-17 (*egl-17*)	LIN-39	*In vivo*	ctcTGATTAATCActg	CEH-20	ND	*C. elegans*	(Cui and Han, 2003)
egl-18/elt-6-site 1 (*elg-18, elt-6*)	LIN-39	*In vivo*	gggTGATATATATgtt	CEH-20†	ND	*C. elegans*	(Koh et al., 2002)
Fkh250 (*fkh*)	Scr	*In vivo*	tcaAGATTAATCGcca	Exd, Hth	ND	*Drosophila*; Hth binding site not found	(Ryoo and Mann, 1999)
Lin-39-site2 (*lin-39*)	LIN-39	*In vivo*	catTGATTTATTTttg	CEH-20†	ND	*C. elegans*; Autoregulatory element	(Wagmaister et al., 2006)
mod-84 (*mod*)	Scr	*In vivo*	acaTAATTTgttagtatgTAATATtccctgGTATTTTTCCAATGactgtcaa gcagtgctgGTAAATGGCGGAACaag	ND	Tsh	*Drosophila*; Although both Scr and Tsh regulate *mod*, a cooperative interaction was not observed	(Taghli-Lamallem et al., 2007; Taghli-Lamallem et al., 2008)
p53 (*p53*)	Hoxa5	*Ex vivo*	tctTAATTCaaa	ND	ND	Human and Mouse	(Raman et al., 2000a)
PRP-62 (*PR*)	Hoxa5	*Ex vivo*	cggTAATTGggg	ND	ND	Human	(Raman et al., 2000b)
PTN-106 (*PTN*)	Hoxa5	*Ex vivo*	tttTAATAAgct	ND	ND	Human	(Chen et al., 2005)
HBE (*flk1*)	Hoxb5	*In vivo*	tgcTGATTAATGAaaa	Pbx#	ND	Mouse; Mutation affecting reporter activity located in putative Pbx site	(Wu et al., 2003)
bFGF (*bFGF*)	Hoxb7	*Ex vivo*	cgcTAATCTgg	ND	ND	Human	(Caré et al., 1996)
egl-1+5995 (*egl-1*)	MAB-5	*In vivo*	cgtTGATTTATTTtta	CEH-20	ND	*C. elegans*	(Liu et al., 2006)
apME680 (*ap*)	Antp	*In vivo*	ccaATTAtttTGATTAATGCcaa; cccATAAATAATATtaa; tttTTATGAgtt; ttgAAATGAact; tgaATTATGATTATTGCcat	Exd†	ND	*Drosophila*; *exd* required in vivo	(Capovilla et al., 2001)
Tsh (*tsh*)	Antp, Ubx	*In vivo*	aacTAATGTAATTACaaa; tgaTAATTGact; cacATAAATctt; tttTAATATttt	ND	ND	*Drosophila*; Cluster of Hox binding sites tested by deletion	(McCormick et al., 1995)
sal1.1 (*sal*)	Ubx	*In vivo*	ttaTAATgTGCCCGTCTTAATATgat	Not Exd or Hth	Mad/Med	*Drosophila*; *exd* and *hth* not required in vivo	(Galant et al., 2002; Walsh and Carroll, 2007)
knot (*knot*)	Ubx	*In vivo*	gctTAATTTg; gctTAATTCt; cacTAATTAt; gccTAATTGTAATTGTTATTATTAATTAa	Not Exd or Hth	ND	*Drosophila*; Hox sites are additive, *exd* and *hth* not required in vivo	(Hersh and Carroll, 2005)
ß3 (*ßTub60D*)	Ubx	*In vivo*	ttcATAATTGcagcggccacactcCAATTAAatt	ND	ND	*Drosophila*	(Kremser et al., 1999)
CG13222-site1 (*CG13222*)	Ubx	*In vivo*	gtgTAATTTatc	Not Exd or Hth	ND	*Drosophila*; *exd* and *hth* not required in vivo	(Hersh et al., 2007)
Antp-P2 (*Antp*)	Ubx, AbdA	*In vivo*	30 Hox binding sites	ND	ND	*Drosophila*; Activity lost when all 30 sites are mutated.	(Appel and Sakonju, 1993)

(*Continued*)

Table 3.1 *(Continued)*

Name (gene)	Hox	Reporter Validation	DNA Binding Site (**Hox**, PBC, Meis/Hth, **Other**)	Cooperative Cofactors	Other Factors	Target Origin; Notes	References
DMX-R (*Dll*)	Ubx, AbdA	*In vivo*	gcaCTATAAAaCTGTCCgcggGAATGATTTAATTTcccaAATATTgtc	Exd, Hth, En	Slp	*Drosophila*; Slp and En function as repressors	(Gebelein et al., 2004)
dpp-midgut enhancer (*dpp*)	Ubx, AbdA	*In vivo*	caaaTTTATTACTAATTGGGtgTGAATTGCaggcagtgcaagtgctgca tatgcagcatgcagcATAATCGAAatgggtgCTAATTGATag	Exd	ND	*Drosophila*; Atypical Exd binding site, additional Hox and Exd sites required	(Capovilla and Botas, 1998; Capovilla et al., 1994; Chan et al., 1994; Manak et al., 1994; Stultz et al., 2006; Sun et al., 1995)
RhoA (*rho*)	AbdA	*In vivo*	agtTCATTGATTGACATTTTTATTAtgc	Exd, Hth	Sens	*Drosophila*; Sens competes for binding with AbdA-Exd-Hth complexes	(Li-Kroeger et al., 2008)
XC-Box2 (*wg*)	AbdA	*In vivo*	gcaTAATCTAATTGcgg	Not Exd or Hth‡	Mad, Medea, Creb	*Drosophila*; Exd and Hth do not form complexes with AbdA in vitro	(Grienenberger et al., 2003)
AH5 (*Hoxa4*)	Dfd, Scr, Antp, Ubx & AbdA	*In vivo*	ggCAATTAAATTTATGGgggcTATAATTActg	ND	ND	Mouse; Possible role for Exd, repressed by AbdA; activated by other Hox	(Haerry and Gehring, 1997)
FkhCON (synthetic)	Scr, Antp, Ubx, AbdA	*In vivo*	tcaAGATTTATGGcca	Exd, Hth	ND	Synthetic; Fkh250 binding site mutated to consensus Hox-Exd site, repressed by AbdA, activated by other Hox	(Ryoo and Mann, 1999)
MLC (*MLC1/3*)	Hoxc8, Hoxa10	*Ex vivo*	cttATTAAATTAcCATGTGtga	ND	ND	Rat; Factor binding to CATGTG (E-box) not identified, Hoxc8 activates and Hoxa10 represses	(Ceccarelli et al., 1999; Houghton and Rosenthal, 1999; Rao et al., 1996)
SBE (*OPN*)	Hoxc8, Hoxa9	*Ex vivo*	atgCAGTCtataaatgaaaagggtagtTAATGAcat	ND	Smad1, Smad3, Smad4	Mouse; Smad3 directly binds DNA site while Smad1 and Smad4 antagonize Hox binding	(Shi et al., 2001; Shi et al., 1999)
EphB4-1365 (*EphB4*)	Hoxa9	*Ex vivo*	agcTTATT	ND	ND	Human	(Bruhl et al., 2004)
E-510 (*E-selectin*)	Hoxa9	*Ex vivo*	atgCAATTTTATTAAtat	ND	ND	Human	(Bandyopadhyay et al., 2007)
N-CAM (*N-CAM*)	Hoxc6, Hoxb8, Hoxb9	*Ex vivo*	tTAATAATtac; cctTAATCAg	ND	ND	Mouse; Hoxc6 and Hoxb9 activate while Hoxb8 represses reporter expression ex vivo	(Edelman and Jones, 1995; Jones et al., 1993; Jones et al., 1992)
p21A10RE (*p21*)	Hoxa10	*Ex vivo*	tttTTATAAttt	Pbx1a*, Meis1b*	ND	Human; Cotransfection of Pbx1a and Meis1b increased reporter expression ex vivo, no direct binding sites identified	(Bromleigh and Freedman, 2000)
B3A (*β_3-integrin*)	Hoxa10	*Ex vivo*	aatGTTATTTtta	ND	ND	Human; Hox binding sites not confirmed by mutagenesis	(Daftary et al., 2002)
EMXC (*Emx2*)	Hoxa10	*Ex vivo*	tgtTTATGTgat	Not Pbx‡	ND	Human; Pbx did not form complexes with Hoxa10 in vitro	(Troy et al., 2003)
Runx2-site1 (*Runx2*)	Hoxa10	*Ex vivo*	aggTTATAGctt	ND	ND	Rat	(Hassan et al., 2007)

Name (gene)	Hox	Reporter Validation	DNA Binding Site (**Hox**, PBC, Meis/Hth, **Other**)	Cooperative Cofactors	Other Factors	Target Origin; Notes	References
PPE (*Rea-1*[c])	Hoxd10	*In vivo*	tgtTTCCACACt; cgcTTCCggc; cttTGATTTATTAccc	Pbx1b, Prep1*	N1IC, Ets-1	Mouse; Homeodomain of Prep1 not necessary for activation ex vivo, Prep1 binding site not identified	(Glenn et al., 2008; Pan et al., 2005; Pan et al., 2001)
Six2 (*Six2*)	Hox11	*In vivo*	gTTATCTgacccggggcctgcccgcgccaGacaatagTCGaGTCA aattaTTc	ND	Pax2, Eya1	Mouse	(Gong et al., 2007)
Enpp2 (*Enpp2*)	Hoxa13	*Ex vivo*	TTAATTG; TTAACAT; TTTATAT	ND	ND	Mouse	(McCabe and Innis, 2005)
Sostdc1 (*Sostdc1*)	Hoxa13	*Ex vivo*	15 sites tested	ND	ND	Mouse	(Knosp et al., 2007)
EphA7-site 3 (*EphA7*)	Hoxa13 Hoxd13	*Ex vivo*	ataTTATTGgag	ND	ND	Mouse	(Salsi and Zappavigna, 2006)
hHa2-motifs: 1, 3, 6, 11 (*hHa2*)	Hoxc13	*Ex vivo*	aaaTTAATTAgcag; tctTTTATTGgaa; cttTTAATGAaag; catTTAATATgttg	Not Meis/Prep	ND	Human; Meis/Prep localization cytoplasmic	(Jave-Suárez and Schweizer, 2006; Jave-Suárez et al., 2002)
hHa5-motifs: 2, 3, 4, 6, 9, and 13 (*hHa5*)	Hoxc13	*Ex vivo*	actTTAATGAgga; gttTTAATAGaaa; aggTTAATGAggg; tatTTTATGAgact; tctTTTATTGgcct; cagTTAATTGgac	Not Meis/Prep	ND	Human; Meis/Prep localization cytoplasmic	
hHa7-motif 2 (*hHa7*)	Hoxc13	*Ex vivo*	gctTTAATGAgct	ND	ND	Human	
Foxq1 (*Foxq1*)	Hoxc13	*Ex vivo*	accTTCATTAcaa; tccTCCATAAAaca; agcTTAATAGGgac	ND	ND	Mouse; Other Hoxc13 sites may be required	(Potter et al., 2006)
BE3-6 (*yellow*)	AbdB	*In vivo*	aggTCGTAAAAcgtatttttacccatttgcatgtTTATTATGcgt	ND	ND	*Drosophila*	(Jeong et al., 2006)
BabDE (*bab*)	AbdB	*In vivo*	7 TTTAT, 7 TTTAC, 8 TTAT, and 2 ACAATGT binding sites	ND	Dsx	*Drosophila*	(Williams et al., 2008)
i4c-S2 (*psa-3*)	NOB-1	*In vivo*	tttTGATAGTAATttt	CEH-20	ND	*C. elegans*; CEH-20 facilitates NOB-1 DNA binding independent of its own binding site	(Arata et al., 2006)

[a] Tested only *in vitro*.
[b] Half-site not directly tested.
[c] Tested *ex vivo*.
[d] PBC-binding site recognized in sequence but untested.

Targets are ordered according to Hox paralog and are derived from human, mouse, chicken, *Drosophila*, and *C. elegans*. All of these elements have been shown to require direct Hox input for activity either *in vivo* or *ex vivo* (cell culture). Colored shading in the first column indicates if the element is repressed, activated or both (red, green, and blue, respectively) by Hox input. Nonadjacent binding sites within the same enhancer are separated by semicolons. Underlined sites are shared between two proteins using the colors defined in the DNA-binding site column. ND, not determined. (See Color Insert.)

Given the high degree of specificity required for some Hox functions, and that there are dozens of *Hox* genes in vertebrates, it is perhaps surprising how few bona fide Hox cofactors have been identified. One answer to this paradox, discussed later in this chapter, comes from a recent atomic-level resolution view of how TALE cofactors bind to DNA with Hox proteins. Another possible answer, especially in vertebrates, is that multiple *Pbx* and *Meis/Prep* genes encode for proteins with different biochemical properties and thus expand the number of Hox cofactors. Perhaps analogously, although *Drosophila* has a single *exd*-like gene and a single *hth*-like gene, alternative splicing of *Drosophila hth* also adds to the repertoire of Hox cofactors present in the fly (Noro *et al.*, 2006). Specifically, *hth* encodes both homeodomain-containing and homeodomain-less (HDless) isoforms. Not only do both of these isoforms contribute to *hth* functions, there is a clear division of labor for these isoforms. For example, the homeodomain-containing form of Hth is not required for a large number of embryonic functions, including many (but not all) Hox-dependent functions. In contrast, the homeodomain-containing form is essential for *hth* to specify antennal development, which is one of its Hox-independent functions. The fact that these two Hth isoforms exist suggests the possibility that they may also be used in different ways to achieve Hox specificity. Strikingly, in *C. elegans*, the gene *psa-3* encodes an HDless Hth/Meis ortholog (Arata *et al.*, 2006). Thus, *C. elegans* produces a very similar HDless isoform, but via gene duplication and truncation, instead of by alternative splicing as in *Drosophila* and vertebrates. The presence of HDless isoforms of Hth/Meis in *C. elegans*, *Drosophila*, and vertebrates reinforces the idea that it carries out critical functions that are distinct from those executed by homeodomain-containing isoforms.

Although *exd* appears to produce only a single isoform, some of the vertebrate *Pbx* genes produce multiple isoforms via alternative splicing (Milech *et al.*, 2001; Monica *et al.*, 1991; Wagner *et al.*, 2001). Using the yeast two-hybrid assay, some evidence exists that a subset of isoforms have distinct abilities to interact with Meis 1, Meis 2a, and Prep1 (Milech *et al.*, 2001). Such differences may also be important for Hox specificity, and may be reflected in the arrangement of binding sites for Hox and TALE proteins in Hox-targeted *cis*-regulatory elements.

5. What Do *In Vivo* Hox-Binding Sites Look Like?

An important approach to understand how Hox proteins regulate target gene expression, and to reveal potential generalizations, is to examine the *cis*-regulatory elements they directly bind to *in vivo*. Once a set of *in vivo*-validated Hox-targeted *cis*-regulatory elements are in hand, several questions can be asked. These include: How many also require input from known cofactors? How many Hox-binding sites are present in each element?, and What other regulatory inputs are there? To provide initial answers to these questions, we have surveyed the literature with the goal of cataloging the majority of the direct Hox-binding sites that have been examined to date, in both vertebrates and invertebrates. By "direct," we included in this survey only Hox-binding sites that have been shown by a reporter gene assay (in cell culture (*ex vivo*) or *in vivo*) to be required for the activity of a *cis*-regulatory element (Table 3.1). Therefore, some recent genome-wide studies fell short of these stringent criteria for validation (Ebner *et al.*, 2005; Hueber and Lohmann, 2008; Hueber *et al.*, 2007; McCabe and Innis, 2005). Below, we discuss the results of this survey and their implications for Hox specificity.

We found 66 *cis*-regulatory elements for which there is strong experimental evidence for direct and essential Hox input (Table 3.1). Of these, 29 have been shown to use PBC cofactors (Exd/Pbx/Ceh-20). Two additional elements appear, by sequence, to have PBC-Hox composite sites, making a total of 31 elements with PBC-Hox sites in this data set (Table 3.1). For seven elements, there is experimental evidence that they do not use these cofactors. The remaining 30 targets have not been directly tested for PBC input, although two of these have been shown not to use Hth/Meis or Prep proteins (Table 3.1). Finally, among the 66 targets in this list, there are 11 examples in which other direct inputs have been shown to be required for the activity of the *cis*-regulatory element.

Although we need to cautiously interpret this relatively small number of elements, several interesting features emerge by analyzing these examples. First, it is noteworthy that for a large fraction of the elements (29 of the 36 elements where it was examined) Hox proteins bind their binding sites cooperatively with a PBC factor (Table 3.1). This is likely an overestimate of the frequency of PBC-Hox-binding sites, because sequence gazing of the 30 elements that were untested for PBC input suggests that many do not have an obvious PBC-binding site. Nevertheless, the abundance of PBC-Hox composite-binding sites in this list underscores the widespread contribution of these cofactors to Hox-binding *in vivo*. Strikingly, with only two exceptions (*Dll* DMX and *fkh250*^{con}), PBC-Hox-binding sites are used for gene activation, not repression. In contrast, the Hox sites that do not have clear PBC input are used for both repression and activation. If this overall correlation continues to hold up, it suggests that PBC-Hox complexes are, in general, more likely to recruit transcriptional coactivators rather than corepressors.

Second, there is a trend for the anterior Hox proteins (paralogs 1-5) to use PBC cofactors more than the posterior Hox proteins (paralogs 6-13) (Table 3.1). Of the 30 elements targeted by a Hox 1-5 paralog, 20 have a required PBC-Hox-binding site. In contrast, of the 36 elements targeted by a Hox 6-13 paralog, only nine have been shown to have an essential PBC-Hox-binding site.

Third, elements that do not have PBC input are more likely to have multiple Hox-binding sites than elements that have PBC input (Table 3.1). For those elements that use Hox-PBC sites, the average number of Hox-binding sites is 1.2 (ranging from 1 to 3), whereas for those elements that do not appear to use Hox-PBC sites, the average number of sites is 2.8 (ranging from 1 to 30). Thus, from this data set, it appears that Hox sites without PBC input often function in groups. If this trend holds up, it may reflect a lower affinity for non-PBC Hox sites when compared to PBC-Hox sites. Perhaps multiple non-PBC Hox sites are therefore required in an additive manner to elicit a transcriptional response.

Although many of the targets listed in Table 3.1 have Hox-binding sites that do not appear to have direct PBC input, we avoid calling them Hox "monomer" sites because it is plausible that currently unidentified factors bind with Hox proteins (cooperatively or noncooperatively) to these sites. In fact, when true Hox "monomer"-binding sites—synthetic, but high-affinity-binding sites—were used to drive a reporter gene in *Drosophila* embryos, they did not produce expression patterns consistent with their ability to bind dozens of homeodomain proteins (Vincent *et al.*, 1990). This experiment, together with the analysis of *in vivo* Hox targets listed in Table 3.1, suggest that Hox proteins never work as monomers.

The arrangement of binding sites also varies in interesting ways within this data set. Of the 40 PBC-Hox sites (distributed among 31 elements), 33 have the PBC half-site adjacent to the Hox half-site, and the majority of these (26) have the structure nnATnnATnn (where the first and second ATs form the core of the PBC and Hox half-sites, respectively). In one case (DMX-R from *Dll*) the PBC-Hox site has the structure nnATnnnATnn and there is one example (in the *dpp* midgut element) where the Hox site precedes the PBC site (Table 3.1). It is possible that these atypical arrangements help these sites be more selective for a subset of Hox proteins. It is also possible that the unique three-dimensional architectures of the protein complexes assembled by these atypical PBC-Hox-binding sites are important for recruiting additional, element-specific transcriptional effectors. Consistent with these ideas, both *in vitro* Hox-binding specificity and *in vivo* activity of DMX-R were reduced when the spacing between the two half-sites was changed from 3 to 2 bp (Gebelein *et al.*, 2002). Of the 40 PBC-Hox sites, eight have been shown to have a nearby Hth/Meis or Prep-binding site. Although the low number of identified Hth/Meis/Prep-binding sites may in part be because they are not always looked for, it may also reflect the fact that there are isoforms of Hth and Meis that do not have a homeodomain and thus would not be expected to make DNA contacts.

Because both PBC-Hox and Hth/Meis/Prep-binding sites have a clear orientation, four possible arrangements of these two binding sites are possible. Interestingly, of these four, only one is not observed (TGACAG...PBC-Hox, where TGACAG represents one orientation of a Hth/Meis site) (Table 3.1). That three of the potential orientations have been observed suggests an inherent flexibility in how these complexes can bind DNA. Further, because the different orientations are expected to orchestrate the assembly of protein-DNA complexes that have unique three-dimensional architectures, these observations also suggest the possibility that they have unique biochemical properties, such as their ability to recruit additional transcriptional coactivators or corepressors.

An unusual arrangement of binding sites is found in an Abd-A-targeted element from the *Drosophila rhomboid* (*rho*) gene (Table 3.1) (Li-Kroeger *et al.*, 2008). In this element, the order of binding sites is PBC-Hth-Hox. Robust cooperative DNA binding to this element was observed between a preformed Exd-Hth dimer and Abd-A. In principle, because both Exd and Hth are TALE homeodomain proteins, they both have the ability to bind the "YPWM" motif of Abd-A, raising the question of which, if either, TALE motif Abd-A is interacting with. Interestingly, mutation of the Hth site, but not the Exd site, dramatically reduced complex formation (Li-Kroeger *et al.*, 2008), suggesting that the Hth-binding site and, perhaps, the Hth TALE motif are more critical for complex formation. Perhaps analogously, a direct Hox-Hth interaction was also proposed to exist in the repressor element from the *Dll* gene (Table 3.1) (Gebelein *et al.*, 2004). The existence of these atypical arrangements suggests that there may be additional flexibility in how Hox, PBC, and Meis/Hth/Prep proteins assemble onto target DNAs. The dissection of the *rho* element in particular emphasizes that carrying out careful mutagenesis and follow-up *in vivo* studies will be critical for identifying additional novel architectures that are used by these factors *in vivo*.

As discussed above, two recent reports described the *in vitro* binding site preferences for nearly all mouse and fly homeodomains, including all Hox homeodomains (Berger *et al.*, 2008; Noyes *et al.*, 2008). How do these results compare with the *in vivo* Hox-binding sites listed in Table 3.1? To answer this question, we generated binding site logo diagrams using the B1H-derived binding sites for the *Drosophila* Hox homeodomains (Noyes *et al.*, 2008) and the *in vivo* Hox-binding sites listed in Table 3.1 (Fig. 3.2). The B1H-derived logos are all based on at least 19 individual binding sites, while the number of individual *in vivo* binding sites for each Hox protein that went into this analysis ranged from 12 (for Scr/Hox5) to 57 (for Ubx/Hox7). Side-by-side analysis of these two sets of sequences reveals some noteworthy differences (Fig. 3.2). First, consistent with the high proportion of PBC-Hox sites in the Hox1/Labial targets, the *in vivo* consensus sequence readily identified a PBC half-site (TGAT). In addition, while the B1H selection tended to identify TAATTA for Hox1/Labial, GGATGG is commonly observed in the *in vivo* data set for these Hox proteins. Other, though less dramatic, differences are also observed between the B1H and *in vivo* data sets for nearly all of the Hox paralogs (Fig. 3.2). These comparisons reinforce the view that the *in vivo* environment, due to the presence of other cofactors, collaborators, or differences in DNA and/or chromatin structure, influences Hox-binding site preferences.

6. Insights into Hox Specificity from Structural Studies

Several monomeric homeodomain-DNA structures have been solved, and all reveal a very similar mode of DNA recognition by this DNA-binding domain (Gehring *et al.*, 1994). Briefly, the third alpha-helix, also called the recognition helix, lies in the major groove of the DNA, where it makes several direct and water-mediated contacts with specific bases and the phosphate backbone. Ile47, Gln50, Asn51, and Met54, residues that are present in all Hox homeodomains, are primarily responsible for making these contacts. In addition, the so-called N-terminal arm, which precedes the first alpha-helix, is typically observed in the minor groove. Arg5, an N-terminal arm residue present in nearly all homeodomains, is the most commonly observed residue in the minor groove.

In addition to these monomeric homeodomain-DNA structures, we now have X-ray structures of five different PBC-Hox-DNA complexes (Fig. 3.4). The Hox homeodomains in these structures recognize the DNA using the same contacts that were observed in the monomeric structures, demonstrating that the presence of PBC does not grossly alter the way in which Hox homeodomains bind DNA. In all five of the PBC-Hox structures (with four different Hox proteins: Hoxb1, Scr, Ubx, and Hoxa9), the PBC and Hox homeodomains bind DNA in a head-to-tail orientation, with very similar overall arrangements. In all five structures the Hox YPWM motif binds the hydrophobic TALE pocket in the PBC homeodomain. Thus, these four Hox proteins have the capacity to bind DNA cooperatively with PBC proteins using very similar protein-protein and protein-DNA contacts. However, we note that the currently available structures provide an incomplete picture because, for some binding sites, other protein motifs, such as the UbdA motif of Ubx and Abd-A, play an important role in forming PBC-Hox-DNA complexes (Merabet *et al.*, 2007). Currently, no structural information exists about these domains or how they contact PBC proteins.

PBC proteins not only bind cooperatively to DNA with Hox proteins, they also increase Hox-DNA-binding selectivity. This phenomenon is best illustrated with a few examples. In the absence of cofactors, the Hox1/Labial paralog shows a preference for binding the sequence TAATTA

(Fig. 3.2) (Berger *et al.*, 2008; Noyes *et al.*, 2008). In the presence of a PBC protein, a PBC-Hox1/Lab heterodimer prefers to bind the sequence TGAT[t/g]GATgg, where [t/g]GATgg is the Hox-binding site (base pairs with brackets indicate multiple possibilities at the same position and lowercase letters indicate only partially preferred base pairs) (Fig. 3.2). In contrast, while Ubx also prefers to bind TAATTA as a monomer, it will readily bind TGATTTATTT as a PBC-Ubx heterodimer, where TTATTT is the Hox-binding site (Berger *et al.*, 2008; Noyes *et al.*, 2008) (Fig. 3.2). Thus, the presence of PBC changes the DNA-binding preferences of both Labial/Hox1 and Ubx, but toward different sequences.

These observations raise the question of how the same cofactor can generate two different outcomes for these two Hox paralogs. The likely answer is that the specificity information is in the Hox protein, but is only revealed in the presence of the cofactor. Two recent crystal structures of PBC-Hox-DNA complexes support this idea (Joshi *et al.*, 2007). In one, an Exd-Scr heterodimer is bound to a paralog-specific PBC-Hox-binding site (*fkh250*) from the *fkh* gene, an *in vivo* Scr target gene. A second crystal structure shows the same two proteins bound to a consensus PBC-Hox sequence (*fkh250^{con}*). Importantly, additional protein-DNA contacts were observed in the Exd-Scr-*fkh250* complex, but not in the Exd-Scr-*fkh250^{con}* complex. These contacts are derived from the N-terminal arm of the Scr homeodomain and, surprisingly, a nonhomeodomain residue in the linker region between Scr's YPWM motif and homeodomain. Both of these side chains are inserted into the minor groove of the *fkh250*-binding site (Figs. 3.4 and 3.5) (Joshi *et al.*, 2007). Two additional observations are of interest. First, the DNA minor groove where these two side chains insert is unusually narrow, and significantly narrower than the analogous region of the *fkh250^{con}*-binding site (Fig. 3.5). This suggests that subtle differences in DNA structure are likely to contribute to Hox-binding specificity. Second, the residues making these minor groove contacts, which are part of a normally unstructured region of Scr, require the YPWM-Exd interaction to be positioned in close proximity to the minor groove. Thus, the paralog-specific binding of Scr to its binding site in *fkh* depends on three contributing features (1) an unusual DNA structure; (2) paralog-specific residues in the Scr homeodomain and linker that insert into this DNA structure; and (3) the YPWM-Exd interaction, which positions the N-terminal arm and linker region so that this normally unstructured peptide can make these contacts.

Although these structures provide insights into why Exd-Scr specifically binds *fkh250*, they raise the question of how general these findings are. Two additional observations suggest that the underlying principles revealed by these structures may, in fact, be a general feature of PBC-Hox-DNA interactions, at least for paralog-specific and semi-paralog-specific target sites. For one, Hox N-terminal arms and linker regions are evolutionarily conserved in a paralog-specific manner (Joshi *et al.*, 2007; Mann, 1995;

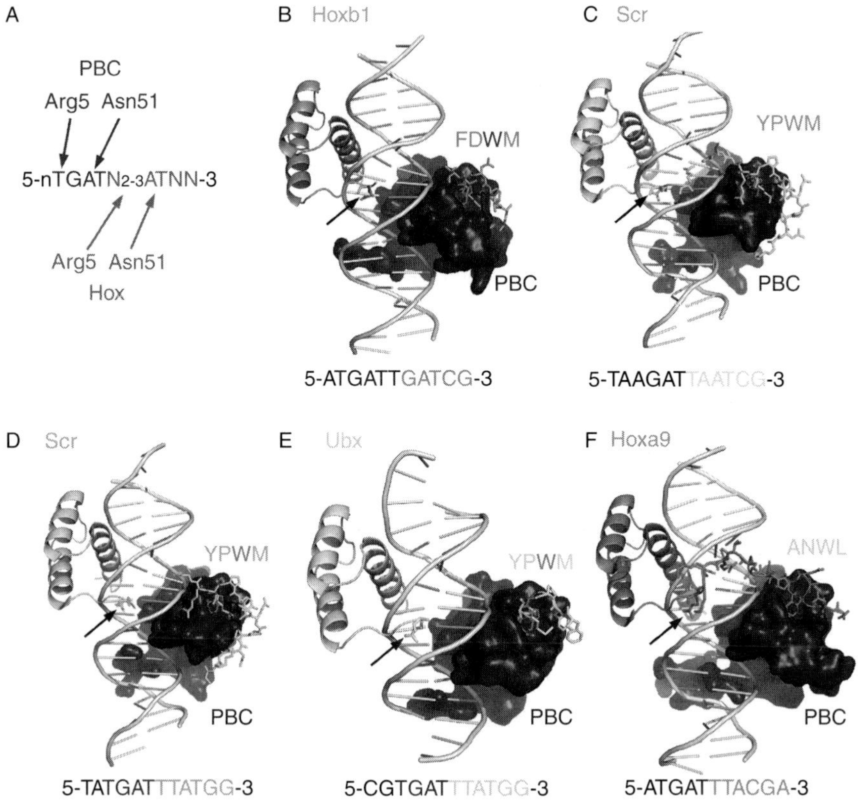

Figure 3.4 Common and unique features of PBC-Hox-DNA complexes. (A) Consensus PBC-Hox-binding sites have a PBC half-site (typically TGAT or AGAT, blue) and a Hox half-site (typically NNATNN, red). Minor groove (Arg5) and major groove (Asn51) contacts observed in all five of the PBC-Hox crystal structures are indicated. $N_{2\text{-}3}$ reflects the observation that the PBC and Hox Asn51-contacted "AT" are usually separated by 2 bp, but 3 bp spacings have also been observed. (B)-(F) Overviews of the five existing PBC-Hox-DNA crystal structures. In all examples, the PBC protein (for most examples, just its homeodomain) is shown as a blue surface. The Hox proteins, which include the YPWM motif (which is FDWM in Hoxb1 and ANWL in Hoxa9), linker, and homeodomain, are color-coded as indicated. Only side chains around the YPWM, linker, and N-terminal arm are shown; homeodomain helices and loops are shown in cartoon format. The Trp (W) in the YPWM motif is colored red in all cases to indicate its conserved interaction with the TALE motif in the PBC homeodomain. In all cases, Arg5 of the Hox N-terminal arm is observed in the minor groove (black arrows). In only two cases (C; Exd-Scr bound to *fkh250* and F; Pbx-Hoxa9 bound to a consensus sequence) are additional N-terminal arm and linker regions observed; these regions are disordered in the other three structures. The DNA sequences present in these structures are shown below the structure, with the PBC and Hox half-sites color-coded. These images were generated using PyMol; the PDB accession numbers for these structures are (B) 1B72, (C) 2R5Z, (D) 2R5Y, (E) 1B8I, and (F) 1PUF. (See Color Insert.)

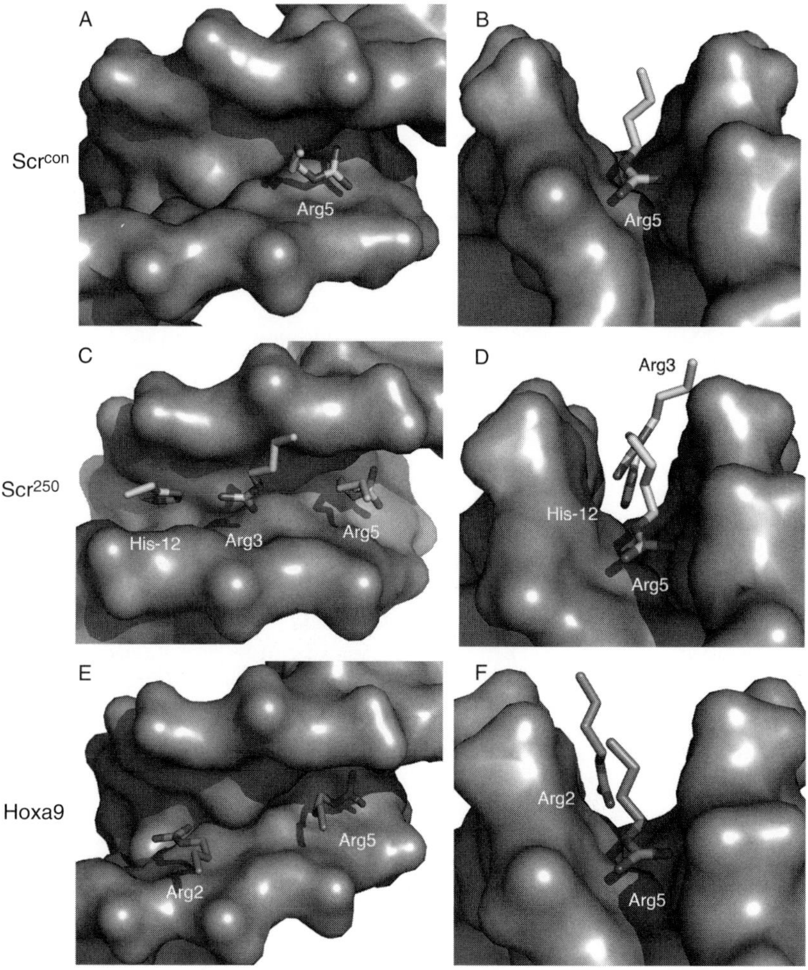

Figure 3.5 Interactions between Hox proteins and the DNA minor groove. Shown are images from X-ray crystal structures of PBC-Hox-DNA complexes, focused only on the interaction between the minor groove (shown as the gray surfaces) and the amino acid side chains of N-terminal arm/linker residues. The left-hand images (A, C, E) look into the minor groove from the top; the right-hand images (B, D, F) look along the axis of the minor groove. (A, B) Exd-Scr bound to the *fkh250^con* consensus-binding site. Only Arg5 from the N-terminal is observed in the minor groove. (C, D) Exd-Scr bound to the *fkh250 in vivo* binding site. In contrast to the *fkh250^con* structure, Arg3 (from the N-terminal arm) and His-12 (from the linker) are observed in the minor groove, in addition to Arg5. Note also that the minor groove in the *fkh250* structure appears narrower than in the *fkh250^con* structure (compare B with D). See Joshi *et al.* (2007) for details. (E, F). In the Pbx-Hoxa9 structure, one additional N-terminal arm residue, Arg2, is observed, together with Arg5. The Hoxa9 linker is unusually short (four residues), and none of them are seen inserting into the minor groove. See LeRonde-LeBlanc and Wolberger (2003) for details. (See Color Insert.)

Morgan *et al.*, 2000). The two minor groove inserting residues in Scr are conserved in all Hox5 paralogs, while other Hox paralogs have different N-terminal arm and linker sequences that are also equally conserved in a paralog-specific manner. Second, another PBC-Hox structure, that of Pbx-Hoxa9, also has significant N-terminal arm-minor groove contacts (LaRonde-LeBlanc and Wolberger, 2003) (Figs. 3.4 and 3.5). Although the binding site used in the Pbx-HoxA9 structure is not from an *in vivo* HoxA9 target, the N-terminal arm-minor groove contacts are also dependent on the YPWM-Pbx interaction, illustrating the potential generality of DNA minor groove recognition by PBC-Hox complexes. The Pbx-Hoxa9 structure also reveals significantly more contacts with the phosphate backbone of the DNA, perhaps accounting for its higher affinity compared to other PBC-Hox complexes (LaRonde-LeBlanc and Wolberger, 2003).

Along the same lines, it is also noteworthy that linker length—and, consequently, the distance between the YPWM motif and the N-terminal arm—varies significantly among Hox proteins. Not only are there huge differences in linker lengths (ranging from >50 in Labial to <5 in Hoxa9), linker length roughly correlates with Hox paralog: anterior (3′) Hox paralogs have a much greater tendency for long linkers than more posterior (5′) Hox paralogs. In addition, Hox linkers also vary for individual paralogs due to alternative splicing; the Ubx linker, for example, ranges from 8 to 51 depending on the Ubx isoform (Kornfeld *et al.*, 1989; O'Connor *et al.*, 1988). Consequently, the distance between the YPWM motif and the homeodomain varies and would therefore be expected to affect how the N-terminal arm and/or linker region interacts with the DNA. These intriguing observations contribute to the idea that DNA contacts made by linker and N-terminal arm residues may be generally critical for paralog-specific and cofactor-dependent DNA binding. These observations are also consistent with the alternative idea that linker residues interact with additional proteins, although such factors have not yet been identified (Merabet *et al.*, 2003).

In summary, the common and unique features revealed in these five PBC-Hox-DNA structures, together with previously solved monomeric homeodomain-DNA structures, suggest that DNA recognition by Hox proteins uses two tiers of information that provide different degrees of specificity (Fig. 3.6). The first tier uses DNA-contacting residues that are common to all Hox proteins (Arg5, Ile47, Gln50, Asn51, and Met54) to promote Hox binding to "AT"-rich sequences, such as TAAT[gt][ga]. The second tier uses additional DNA-contacting side chains that come from the N-terminal arm and linker regions; these contacts are cofactor-dependent and paralog-specific (Fig. 3.6). Further, at least for the case of Exd-Scr bound to *fkh250*, these side chains recognize a DNA structure, rather than a specific DNA sequence. The fact that N-terminal arm and linker residues are conserved in a paralog-specific manner suggests that the two

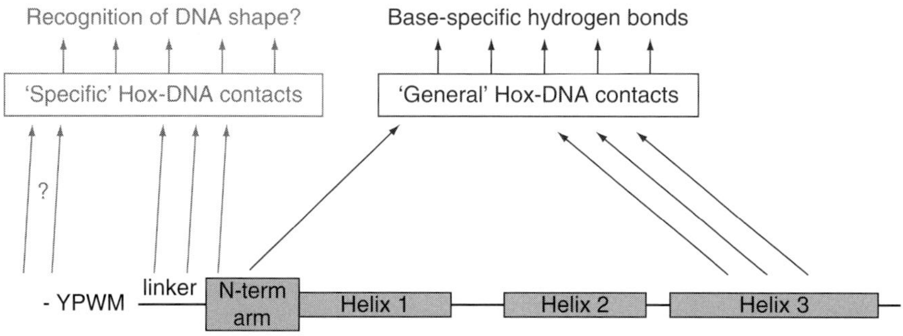

Figure 3.6 Two tiers of Hox-DNA-binding specificity. Hox proteins bind DNA using two levels of protein-DNA contacts. DNA contacts made by Arg5 (in the N-terminal arm) and Ile47, Gln50, Asn51, Met54 (in the third helix) are used by all Hox proteins to bind "AT"-rich DNA sequences ("general" Hox-DNA contacts), but are not good at distinguishing between Hox paralogs. With the help of cofactors (such as PBC proteins), paralog-specific DNA contacts are mediated by linker and N-terminal arm residues. "General" DNA contacts make hydrogen bonds in the DNA major groove. "Paralog-specific" DNA contacts may read a DNA structure, such as the narrow minor groove seen in the Exd-Scr-*fkh250* structure.

tiers of recognition may underlie the recognition of paralog-specific DNA sequences by other Hox-cofactor complexes.

7. Activity Regulation of Hox Proteins: The Role of Hox Collaborators

Although TALE family proteins clearly play an important role in DNA-binding site recognition, Hox proteins use these cofactors to both activate and repress target genes, raising the question of how gene activation versus repression is determined. Although there is currently only one example, one answer is that Hox proteins may use dedicated repressors, such as En, as Hox cofactors in gene repression (Gebelein *et al.*, 2004). Another possibility, which will be no surprise to people used to thinking about *cis*-regulatory elements, is that additional factors bind to Hox-targeted elements and contribute to their activities. Given the increasing number of directly regulated Hox targets that have been characterized, several such accessory factors, which we refer to here as Hox collaborators, have been identified. We have classified a factor as a Hox collaborator if it provides a direct and essential input into a Hox-regulated element, but has not been definitively shown to bind DNA cooperatively with Hox proteins (Table 3.1). These factors include the *Drosophila* Forkhead domain protein Sloppy paired (Slp), which collaborates with Ubx and Abd-A to repress *Dll* in the *Drosophila* abdomen (Gebelein *et al.*, 2004).

Interestingly, recent results suggest that a vertebrate ortholog of Slp, FoxP1, collaborates with Hox proteins during the establishment of motor neuron identities in the mouse (see Chapter 1) (Dasen *et al.*, 2008; Rousso *et al.*, 2008). Thus, the collaboration of FoxP1/Slp (and perhaps other Forkhead domain factors) with Hox proteins appears to be evolutionarily conserved. Like En, Slp may be a dedicated transcriptional repressor due to its Groucho interaction domain, thus providing a mechanism to explain the regulatory sign of *Hox-Slp*-targeted genes. In addition to these cases, protein-protein interaction and genetic studies suggest that the range of potential Hox collaborators is extensive (Kataoka *et al.*, 2001; Luo *et al.*, 2004; Plaza *et al.*, 2008; Prévôt *et al.*, 2000).

Transcription factors that provide cells with identity information, like the Hox factors, have been generally referred to as Selector transcription factors, a term originally coined 40 years ago to describe common properties of the genes *Ubx* and *en* (García-Bellido, 1975; Mann and Carroll, 2002; Mann and Morata, 2000). More recently, selector proteins have been proposed to frequently, if not always, function together with effector transcription factors that are downstream of cell-cell signaling pathways (reviewed previously by Bondos and Tan, 2001; Curtiss *et al.*, 2002; Mann and Affolter, 1998). Thus, not surprisingly, as more elements that are directly targeted by Hox proteins are dissected, signal effector transcription factors are being identified as Hox collaborators. In particular, vertebrate and *Drosophila* SMADs, effectors of the TGF-beta and Decapentaplegic (Dpp) pathways, have been identified as Hox collaborators in several *cis*-regulatory elements (Galant *et al.*, 2002; Grienenberger *et al.*, 2003; Shi *et al.*, 1999, 2001; Walsh and Carroll, 2007) (Table 3.1). Although SMAD-Hox-DNA-binding cooperativity has not been described, there are several reports suggesting that SMADs and Hox proteins can directly interact with each other (Wang *et al.*, 2006; Williams *et al.*, 2005; Zhou *et al.*, 2008). Such interactions may be critical for building an enhanceosome-like structure on Hox-targeted *cis*-regulatory elements. Although the number of examples shown to have direct inputs by signaling effectors is currently low, genetic analyses suggest that this phenomenon is likely to be a general feature of Hox-targeted *cis*-regulatory elements, and will probably extend to other signaling pathways, including Hedgehog (Hh), Wnts, and Notch (Arata *et al.*, 2006; Crickmore and Mann, 2007, 2008; Hersh *et al.*, 2007; Joulia *et al.*, 2006; Marty *et al.*, 2001; Merabet *et al.*, 2005; Weatherbee and Carroll, 1999).

Like Hox cofactors, the presence of a particular Hox collaborator does not guarantee the sign of the transcriptional regulation. In the *sal1.1* element, for example, Ubx collaborates with Mad/Medea to repress transcription, while in the *XC* midgut element from the *wg* gene, Mad/Medea collaborates with Abd-A to activate transcription (Grienenberger *et al.*, 2003; Walsh and Carroll, 2007). This difference is not simply due to different Hox paralogs, because both Ubx and Abd-A can both directly repress and directly activate transcription (Table 3.1). Instead, these observations imply that additional, currently unknown, factors are being recruited to these elements to determine the sign of the transcriptional regulation.

Based on these direct examples, together with the larger number of genetically defined examples of Hox—signaling collaborations, we suggest that it may be a general feature of the multiprotein complexes that are built on Hox-targeted *cis*-regulatory elements. In fact, because Hox proteins work in so many different developmental contexts, it is likely that Hox collaborators will ultimately include a very large number of different types of transcription factors. Perhaps the ability of Hox proteins and PBC-Hox dimers to interact with a large number of different collaborators makes these proteins such ideal regulators of cell type and tissue identities.

Yet, despite this flexibility, it is critical to stress that the Hox factors play the central role in the function (and/or the assembly) of these multiprotein complexes, because without them, these complexes cannot function. Moreover, for paralog-specific functions, the activity and/or assembly of these complexes must depend on the correct Hox paralog and cofactors. Because of their central role, we would therefore like to coin the term "Hoxasome" to describe these multiprotein complexes, which include the Hox proteins, their cofactors, and their collaborators.

8. Insights into Hoxasome Function from *cis*-Regulatory Element Architecture

One straightforward view for how Hoxasomes function is that, once assembled, they recruit coactivators, corepressors, and/or chromatin remodeling complexes that ultimately carry out transcriptional regulation much like any other enhanceosome. Indeed, consistent with this view, there have been numerous reports describing direct interactions between Hox proteins and/or TALE cofactors with these more general components of the transcriptional machinery (Chariot *et al.*, 1999; Prince *et al.*, 2008; Saleh *et al.*, 2000; Shen *et al.*, 2004) and, in some cases, activation and repression domains have been mapped in Hox proteins (Rambaldi *et al.*, 1994; Tour, 2005; Viganò *et al.*, 1998; Zhao *et al.*, 1996). Covalent modifications, such as phosphorylation, have also been shown to influence Hox activities in interesting ways (Berry and Gehring, 2000; Galant and Carroll, 2002; Jaffe *et al.*, 1997; Ronshaugen *et al.*, 2002; Taghli-Lamallem *et al.*, 2008; reviewed elsewhere by Pearson *et al.*, 2005).

In addition, to these mechanisms, there are some recent examples suggesting that Hoxasomes may regulate transcription—and be regulated themselves—in other, mechanistically distinct ways. One example concerns the way in which *Drosophila* Abd-A activates the expression of the gene *rho* in the peripheral nervous system (PNS). Abd-A-dependent expression of *rho*, which encodes a protease required for the processing of a ligand for the epidermal growth factor (EGF) receptor pathway, is necessary for abdominal-specific cell types (Brodu *et al.*, 2002). Abd-A carries out this paralog-specific function by assembling a multiprotein complex that includes both Exd and Hth cofactors (Gebelein and Mann, 2007). However, instead of the Abd-A Hoxasome directly activating *rho* transcription, it appears that it functions by competing with the binding of another transcription factor, Senseless, which is a *rho* repressor. Thus, the architecture of the *rho cis*-regulatory element is organized in a manner whereby the sequence-specific binding of an Abd-A Hoxasome permits *rho* expression by blocking binding of the Senseless repressor. Perhaps analogously, a zinc finger protein called ZFPIP has been shown to bind to Pbx1 and inhibit the binding of Hoxa9-Pbx1 complexes to a Hox-PBC consensus site (Laurent *et al.*, 2007). Thus, competition in DNA binding, rather than a direct influence on transcription, may underlie other examples of gene regulation by Hox factors.

A second interesting example of the importance of *cis*-regulatory element architecture comes from the analysis of an element from the *Drosophila bric-a-brac* (*bab*) gene, which is a direct target of Abd-B (Williams *et al.*, 2008). The activity of this element in the fifth and sixth abdominal segments of females—but not males—is critical for the dimorphic nature of abdominal pigmentation in male and female flies (Williams *et al.*, 2008). As Abd-B expression is the same in male and female flies, the sex-specific activities of this element stem from the Hox collaborator, *doublesex* (*dsx*), which is a downstream effector in the sex-determination pathway (Christiansen *et al.*, 2002). *dsx* encodes both male and female-specific isoforms. In males, the Dsx-M isoform collaborates with Abd-B to repress *bab*, while in females, the Dsx-F isoform collaborates with Abd-B to activate this *bab* element. In this element, there are two required Bab-binding sites, and more than 15 Abd-B-binding sites, suggesting the existence an unusually Hox-dense Hoxasome (Table 3.1). Moreover, the analysis of the same *cis*-regulatory element from other *Drosophila* species in which abdominal pigmentation pattern is the same in males and females suggests that the dimorphic activity of the *D. melanogaster bab* element is due to the relative orientation and specific spacing of the Dsx and Abd-B-binding sites. Mechanistically, it is currently unclear if these changes affect the stable assembly of this Hoxasome or, alternatively, its ability to recruit transcriptional coactivators.

Both of the studies highlighted above emphasize the value in characterizing bona fide Hox-targeted *cis*-regulatory elements at high resolution. Moreover, they also make it clear that a complete understanding of gene regulation by Hox proteins not only depends on understanding how these transcription factors bind DNA, but also how the bound factors, together with their cofactors and collaborators, assemble and regulate transcription. We suspect that the discovery of additional regulatory mechanisms will depend on similar fine-scale analysis of other Hox-targeted *cis*-regulatory elements.

9. Conclusions

In this review, we have summarized a wide range of mechanisms that Hox proteins employ to regulate their target genes. For one, Hox proteins often require cofactors to bind to their binding sites in paralog-specific and semi-paralog-specific target genes. Cofactors may not be as essential, however, for shared Hox functions or those executed by Hox proteins in a unique regulatory environment, such as the *Drosophila* haltere. Structural studies have suggested that TALE family cofactors not only increase the size of the binding site, they help to impose additional structure onto otherwise unstructured homeodomain and nonhomeodomain residues, allowing them to read additional features present in Hox-cofactor-binding sites. It will be interesting to see how generally applicable this model for Hox-DNA binding (and perhaps other homeodomain proteins) will be as more PBC-Hox-DNA complexes are characterized at high resolution. Finally, we have also seen that Hox-regulated *cis*-regulatory elements utilize a potentially large number of protein collaborators, such as effector transcription factors that are downstream of cell-cell signaling pathways. The assembly of these multiprotein-DNA complexes, which we have called Hoxasomes to emphasize the central importance of the Hox input, is essential for dictating the sign (repression or activation) of the transcriptional regulation.

Acknowledgments

We thank Matthew Slattery for comments on the manuscript and Barry Honig for discussions related to this review. This work was supported by an NIH RO1 grant awarded to R.S.M.

References

Affolter, M., Percival-Smith, A., Müller, M., Leupin, W., and Gehring, W. J. (1990). DNA binding properties of the purified Antennapedia homeodomain. *Proc. Natl. Acad. Sci. USA* **87,** 4093–4097.

Affolter, M., Slattery, M., and Mann, R. S. (2008). A lexicon for homeodomain-DNA recognition. *Cell* **133,** 1133–1135.

Akam, M. (1989). Hox and HOM: homologous gene clusters in insects and vertebrates. *Cell* **57,** 347–349.

Appel, B., and Sakonju, S. (1993). Cell-type-specific mechanisms of transcriptional repression by the homeotic gene products UBX and ABD-A in Drosophila embryos. *EMBO J.* **12,** 1099–1109.

Arata, Y., Kouike, H., Zhang, Y., Herman, M. A., Okano, H., and Sawa, H. (2006). Wnt signaling and a Hox protein cooperatively regulate psa-3/Meis to determine daughter cell fate after asymmetric cell division in C. *elegans*. *Dev. Cell* **11,** 105–115.

Bandyopadhyay, S., Ashraf, M. Z., Daher, P., Howe, P. H., and DiCorleto, P. E. (2007). HOXA9 participates in the transcriptional activation of E-selectin in endothelial cells. *Mol. Cell Biol.* **27,** 4207–4216.

Berger, M. F., Badis, G., Gehrke, A. R., Talukder, S., Philippakis, A. A., Peña-Castillo, L., Alleyne, T. M., Mnaimneh, S., Botvinnik, O. B., Chan, E. T., Khalid, F., Zhang, W., *et al.* (2008). Variation in homeodomain DNA binding revealed by high-resolution analysis of sequence preferences. *Cell* **133,** 1266–1276.

Berry, M., and Gehring, W. (2000). Phosphorylation status of the SCR homeodomain determines its functional activity: essential role for protein phosphatase 2A,B'. *EMBO J.* **19,** 2946–2957.

Berthelsen, J., Kilstrup-Nielsen, C., Blasi, F., Mavilio, F., and Zappavigna, V. (1999). The subcellular localization of PBX1 and EXD proteins depends on nuclear import and export signals and is modulated by association with PREP1 and HTH. *Genes. Dev.* **13,** 946–953.

Bessa, J., Tavares, M. J., Santos, J., Kikuta, H., Laplante, M., Becker, T. S., Gómez-Skarmeta, J. L., and Casares, F. (2008). meis1 regulates cyclin D1 and c-myc expression, and controls the proliferation of the multipotent cells in the early developing zebrafish eye. *Development* **135,** 799–803.

Bondos, S. E., and Tan, X. X. (2001). Combinatorial transcriptional regulation: the interaction of transcription factors and cell signaling molecules with homeodomain proteins in Drosophila development. *Crit. Rev. Eukaryot. Gene. Expr.* **11,** 145–171.

Bradley, P. L., Haberman, A. S., and Andrew, D. J. (2001). Organ formation in Drosophila: specification and morphogenesis of the salivary gland. *Bioessays* **23,** 901–911.

Brodu, V., Elstob, P. R., and Gould, A. P. (2002). abdominal A specifies one cell type in Drosophila by regulating one principal target gene. *Development* **129,** 2957–2963.

Bromleigh, V. C., and Freedman, L. P. (2000). p21 is a transcriptional target of HOXA10 in differentiating myelomonocytic cells. *Genes. Dev.* **14,** 2581–2586.

Bruhl, T., Urbich, C., Aicher, D., Acker-Palmer, A., Zeiher, A. M., and Dimmeler, S. (2004). Homeobox A9 transcriptionally regulates the EphB4 receptor to modulate endothelial cell migration and tube formation. *Circ. Res.* **94,** 743–751.

Capovilla, M., and Botas, J. (1998). Functional dominance among Hox genes: repression dominates activation in the regulation of Dpp. *Development* **125,** 4949–4957.

Capovilla, M., Brandt, M., and Botas, J. (1994). Direct regulation of decapentaplegic by Ultrabithorax and its role in Drosophila midgut morphogenesis. *Cell* **76,** 461–475.

Capovilla, M., Kambris, Z., and Botas, J. (2001). Direct regulation of the muscle-identity gene apterous by a Hox protein in the somatic mesoderm. *Development* **128,** 1221–1230.

Caré, A., Silvani, A., Meccia, E., Mattia, G., Stoppacciaro, A., Parmiani, G., Peschle, C., and Colombo, M. P. (1996). HOXB7 constitutively activates basic fibroblast growth factor in melanomas. *Mol. Cell Biol.* **16,** 4842–4851.

Casares, F., Calleja, M., and Sánchez-Herrero, E. (1996). Functional similarity in appendage specification by the Ultrabithorax and abdominal-A Drosophila HOX genes. *EMBO J.* **15,** 3934–3942.

Casares, F., and Mann, R. S. (1998). Control of antennal versus leg development in Drosophila. *Nature* **392,** 723–726.

Casares, F., and Mann, R. S. (2001). The ground state of the ventral appendage in Drosophila. *Science* **293,** 1477–1480.

Ceccarelli, E., McGrew, M. J., Nguyen, T., Grieshammer, U., Horgan, D., Hughes, S. H., and Rosenthal, N. (1999). An E box comprises a positional sensor for regional differences in skeletal muscle gene expression and methylation. *Dev. Biol.* **213,** 217–229.

Chan, S. K., Jaffe, L., Capovilla, M., Botas, J., and Mann, R. S. (1994). The DNA binding specificity of Ultrabithorax is modulated by cooperative interactions with extradenticle, another homeoprotein. *Cell* **78,** 603–615.

Chan, S. K., and Mann, R. S. (1993). The segment identity functions of Ultrabithorax are contained within its homeo domain and carboxy-terminal sequences. *Genes Dev.* **7,** 796–811.

Chan, S. K., and Mann, R. S. (1996). A structural model for a homeotic protein-extradenticle-DNA complex accounts for the choice of HOX protein in the heterodimer. *Proc. Natl. Acad. Sci. USA* **93,** 5223–5228.

Chan, S. K., Ryoo, H. D., Gould, A., Krumlauf, R., and Mann, R. S. (1997). Switching the *in vivo* specificity of a minimal Hox-responsive element. *Development* **124,** 2007–2014.

Chang, C. P., Shen, W. F., Rozenfeld, S., Lawrence, H. J., Largman, C., and Cleary, M. L. (1995). Pbx proteins display hexapeptide-dependent cooperative DNA binding with a subset of Hox proteins. *Genes Dev.* **9,** 663–674.

Chariot, A., van Lint, C., Chapelier, M., Gielen, J., Merville, M. P., and Bours, V. (1999). CBP and histone deacetylase inhibition enhance the transactivation potential of the HOXB7 homeodomain-containing protein. *Oncogene* **18,** 4007–4014.

Chen, H., Rubin, E., Zhang, H., Chung, S., Jie, C. C., Garrett, E., Biswal, S., and Sukumar, S. (2005). Identification of transcriptional targets of HOXA5. *J. Biol. Chem.* **280,** 19373–19380.

Chen, J., and Ruley, H. E. (1998). An enhancer element in the EphA2 (Eck) gene sufficient for rhombomere-specific expression is activated by HOXA1 and HOXB1 homeobox proteins. *J. Biol. Chem.* **273,** 24670–24675.

Christiansen, A. E., Keisman, E. L., Ahmad, S. M., and Baker, B. S. (2002). Sex comes in from the cold: the integration of sex and pattern. *Trends Genet.* **18,** 510–516.

Coiffier, D., Charroux, B., and Kerridge, S. (2008). Common functions of central and posterior Hox genes for the repression of head in the trunk of Drosophila. *Development* **135,** 291–300.

Crickmore, M. A., and Mann, R. S. (2007). Hox control of morphogen mobility and organ development through regulation of glypican expression. *Development* **134,** 327–334.

Crickmore, M. A., and Mann, R. S. (2008). The control of size in animals: insights from selector genes. *Bioessays* **30,** 843–853.

Cui, M., and Han, M. (2003). Cis regulatory requirements for vulval cell-specific expression of the Caenorhabditis elegans fibroblast growth factor gene egl-17. *Dev. Biol.* **257,** 104–116.

Curtiss, J., Halder, G., and Mlodzik, M. (2002). Selector and signalling molecules cooperate in organ patterning. *Nat. Cell Biol.* **4,** E48–E51.

Daftary, G. S., Troy, P. J., Bagot, C. N., Young, S. L., and Taylor, H. S. (2002). Direct regulation of beta3-integrin subunit gene expression by HOXA10 in endometrial cells. *Mol. Endocrinol.* **16,** 571–579.

Dasen, J. S., De Camilli, A., Wang, B., Tucker, P. W., and Jessell, T. M. (2008). Hox repertoires for motor neuron diversity and connectivity gated by a single accessory factor, FoxP1. *Cell* **134,** 304–316.

Desplan, C., Theis, J., and O'Farrell, P. H. (1988). The sequence specificity of homeodomain-DNA interaction. *Cell* **54,** 1081–1090.

Dessain, S., Gross, C. T., Kuziora, M. A., and McGinnis, W. (1992). Antp-type homeodomains have distinct DNA binding specificities that correlate with their different regulatory functions in embryos. *EMBO J.* **11,** 991–1002.

Di Rocco, G., Gavalas, A., Popperl, H., Krumlauf, R., Mavilio, F., and Zappavigna, V. (2001). The recruitment of SOX/OCT complexes and the differential activity of HOXA1 and HOXB1 modulate the Hoxb1 auto-regulatory enhancer function. *J. Biol. Chem.* **276,** 20506–20515.

Ebner, A., Cabernard, C., Affolter, M., and Merabet, S. (2005). Recognition of distinct target sites by a unique Labial/Extradenticle/Homothorax complex. *Development* **132,** 1591–1600.

Edelman, G. M., and Jones, F. S. (1995). Developmental control of N-CAM expression by Hox and Pax gene products. *Philos. Trans. R. Soc. Lond. B. Biol. Sci.* **349,** 305–312.

Ekker, S. C., Jackson, D. G., von Kessler, D. P., Sun, B. I., Young, K. E., and Beachy, P. A. (1994). The degree of variation in DNA sequence recognition among four Drosophila homeotic proteins. *EMBO J.* **13,** 3551–3560.

Ekker, S. C., Young, K. E., von Kessler, D. P., and Beachy, P. A. (1991). Optimal DNA sequence recognition by the Ultrabithorax homeodomain of Drosophila. *EMBO J.* **10,** 1179–1186.

Estrada, B., and Sánchez-Herrero, E. (2001). The Hox gene Abdominal-B antagonizes appendage development in the genital disc of Drosophila. *Development* **128,** 331–339.

Ferretti, E., Cambronero, F., Tümpel, S., Longobardi, E., Wiedemann, L. M., Blasi, F., and Krumlauf, R. (2005). Hoxb1 enhancer and control of rhombomere 4 expression: complex interplay between PREP1-PBX1-HOXB1 binding sites. *Mol. Cell Biol.* **25,** 8541–8552.

Ferretti, E., Marshall, H., Pöpperl, H., Maconochie, M., Krumlauf, R., and Blasi, F. (2000). Segmental expression of Hoxb2 in r4 requires two separate sites that integrate cooperative interactions between Prep1, Pbx and Hox proteins. *Development* **127,** 155–166.

Furukubo-Tokunaga, K., Flister, S., and Gehring, W. J. (1993). Functional specificity of the Antennapedia homeodomain. *Proc. Natl. Acad. Sci. USA* **90,** 6360–6364.

Galant, R., and Carroll, S. B. (2002). Evolution of a transcriptional repression domain in an insect Hox protein. *Nature* **415,** 910–913.

Galant, R., Walsh, C. M., and Carroll, S. B. (2002). Hox repression of a target gene: extradenticle-independent, additive action through multiple monomer binding sites. *Development* **129,** 3115–3126.

García-Bellido, A. (1975). Genetic control of wing disc development in Drosophila. *Ciba Found Symp.* **0,** 161–182.

Gebelein, B., Culi, J., Ryoo, H. D., Zhang, W., and Mann, R. S. (2002). Specificity of Distalless repression and limb primordia development by abdominal Hox proteins. *Dev. Cell* **3,** 487–498.

Gebelein, B., and Mann, R. (2007). Compartmental modulation of abdominal Hox expression by engrailed and sloppy-paired patterns the fly ectoderm. *Dev. Biol.* **308,** 593–605.

Gebelein, B., McKay, D. J., and Mann, R. S. (2004). Direct integration of Hox and segmentation gene inputs during Drosophila development. *Nature* **431,** 653–659.

Gehring, W. J., Qian, Y. Q., Billeter, M., Furukubo-Tokunaga, K., Schier, A. F., Resendez-Perez, D., Affolter, M., Otting, G., and Wüthrich, K. (1994). Homeodomain-DNA recognition. *Cell* **78,** 211–223.

Gibson, G., Schier, A., LeMotte, P., and Gehring, W. J. (1990). The specificities of Sex combs reduced and Antennapedia are defined by a distinct portion of each protein that includes the homeodomain. *Cell* **62,** 1087–1103.

Glenn, S. T., Jones, C. A., Pan, L., and Gross, K. W. (2008). *In Vivo* Analysis of Key Elements within the Renin Regulatory Region. *Physiol. Genomics.* **35**(3), 243–253.

Gong, K. Q., Yallowitz, A. R., Sun, H., Dressler, G. R., and Wellik, D. M. (2007). A Hox-Eya-Pax complex regulates early kidney developmental gene expression. *Mol. Cell Biol.* **27,** 7661–7668.

Goodman, F. R. (2002). Limb malformations and the human HOX genes. *Am. J. Med. Genet.* **112,** 256–265.

Gould, A., Morrison, A., Sproat, G., White, R. A., and Krumlauf, R. (1997). Positive cross-regulation and enhancer sharing: two mechanisms for specifying overlapping Hox expression patterns. *Genes Dev.* **11,** 900–913.

Greig, S., and Akam, M. (1995). The role of homeotic genes in the specification of the Drosophila gonad. *Curr. Biol.* **5,** 1057–1062.

Grienenberger, A., Merabet, S., Manak, J., Iltis, I., Fabre, A., Bérenger, H., Scott, M. P., Pradel, J., and Graba, Y. (2003). Tgfbeta signaling acts on a Hox response element to confer specificity and diversity to Hox protein function. *Development* **130,** 5445–5455.

Guazzi, S., Lonigro, R., Pintonello, L., Boncinelli, E., Di Lauro, R., and Mavilio, F. (1994). The thyroid transcription factor-1 gene is a candidate target for regulation by Hox proteins. *EMBO J.* **13,** 3339–3347.

Haerry, T. E., and Gehring, W. J. (1997). A conserved cluster of homeodomain binding sites in the mouse Hoxa-4 intron functions in Drosophila embryos as an enhancer that is directly regulated by Ultrabithorax. *Dev. Biol.* **186,** 1–15.

Haller, K., Rambaldi, I., Daniels, E., and Featherstone, M. (2004). Subcellular localization of multiple PREP2 isoforms is regulated by actin, tubulin, and nuclear export. *J. Biol. Chem.* **279,** 49384–49394.

Hassan, M. Q., Tare, R., Lee, S. H., Mandeville, M., Weiner, B., Montecino, M., van Wijnen, A. J., Stein, J. L., Stein, G. S., and Lian, J. B. (2007). HOXA10 controls osteoblastogenesis by directly activating bone regulatory and phenotypic genes. *Mol. Cell Biol.* **27,** 3337–3352.

Hersh, B. M., and Carroll, S. B. (2005). Direct regulation of knot gene expression by Ultrabithorax and the evolution of cis-regulatory elements in Drosophila. *Development* **132,** 1567–1577.

Hersh, B. M., Nelson, C. E., Stoll, S. J., Norton, J. E., Albert, T. J., and Carroll, S. B. (2007). The UBX-regulated network in the haltere imaginal disc of D. melanogaster. *Dev. Biol.* **302,** 717–727.

Hirth, F., Loop, T., Egger, B., Miller, D. F., Kaufman, T. C., and Reichert, H. (2001). Functional equivalence of Hox gene products in the specification of the tritocerebrum during embryonic brain development of Drosophila. *Development* **128,** 4781–4788.

Hoey, T., and Levine, M. (1988). Divergent homeo box proteins recognize similar DNA sequences in Drosophila. *Nature* **332,** 858–861.

Houghton, L., and Rosenthal, N. (1999). Regulation of a muscle-specific transgene by persistent expression of Hox genes in postnatal murine limb muscle. *Dev. Dyn.* **216,** 385–397.

Huang, H., Paliouras, M., Rambaldi, I., Lasko, P., and Featherstone, M. (2003). Nonmuscle myosin promotes cytoplasmic localization of PBX. *Mol. Cell Biol.* **23,** 3636–3645.

Hueber, S. D., Bezdan, D., Henz, S. R., Blank, M., Wu, H., and Lohmann, I. (2007). Comparative analysis of Hox downstream genes in Drosophila. *Development* **134,** 381–392.

Hueber, S. D., and Lohmann, I. (2008). Shaping segments: Hox gene function in the genomic age. *Bioessays* **30,** 965–979.

Jaffe, L., Ryoo, H. D., and Mann, R. S. (1997). A role for phosphorylation by casein kinase II in modulating Antennapedia activity in Drosophila. *Genes Dev.* **11,** 1327–1340.

Jave-Suárez, L. F., and Schweizer, J. (2006). The HOXC13-controlled expression of early hair keratin genes in the human hair follicle does not involve TALE proteins MEIS and PREP as cofactors. *Arch. Dermatol. Res.* **297,** 372–376.

Jave-Suarez, L. F., Winter, H., Langbein, L., Rogers, M. A., and Schweizer, J. (2002). HOXC13 is involved in the regulation of human hair keratin gene expression. *J. Biol. Chem.* **277,** 3718–3726.

Jeong, S., Rokas, A., and Carroll, S. B. (2006). Regulation of body pigmentation by the Abdominal-B Hox protein and its gain and loss in Drosophila evolution. *Cell* **125,** 1387–1399.

Jiang, Y., Shi, H., Amin, N. M., Sultan, I., and Liu, J. (2008). Mesodermal expression of the *C. elegans* HMX homolog mls-2 requires the PBC homolog CEH-20. *Mech. Dev.* **125,** 451–461.

Jiménez, G., Paroush, Z., and Ish-Horowicz, D. (1997). Groucho acts as a corepressor for a subset of negative regulators, including Hairy and Engrailed. *Genes Dev.* **11,** 3072–3082.

Jones, F. S., Holst, B. D., Minowa, O., De Robertis, E. M., and Edelman, G. M. (1993). Binding and transcriptional activation of the promoter for the neural cell adhesion molecule by HoxC6 (Hox-3.3). *Proc. Natl. Acad. Sci. USA* **90,** 6557–6561.

Jones, F. S., Prediger, E. A., Bittner, D. A., De Robertis, E. M., and Edelman, G. M. (1992). Cell adhesion molecules as targets for Hox genes: neural cell adhesion molecule promoter activity is modulated by cotransfection with Hox-2.5 and -2.4. *Proc. Natl. Acad. Sci. USA* **89,** 2086–2090.

Joshi, R., Passner, J., Rohs, R., Jain, R., Sosinsky, A., Crickmore, M., Jacob, V., Aggarwal, A., Honig, B., and Mann, R. (2007). Functional Specificity of a Hox Protein Mediated by the Recognition of Minor Groove Structure. *Cell* **131,** 530–543.

Joulia, L., Deutsch, J., Bourbon, H. M., and Cribbs, D. L. (2006). The specification of a highly derived arthropod appendage, the Drosophila labial palps, requires the joint action of selectors and signaling pathways. *Dev. Genes Evol.* **216,** 431–442.

Kataoka, K., Yoshitomo-Nakagawa, K., Shioda, S., and Nishizawa, M. (2001). A set of Hox proteins interact with the Maf oncoprotein to inhibit its DNA binding, transactivation, and transforming activities. *J. Biol. Chem.* **276,** 819–826.

Knosp, W. M., Saneyoshi, C., Shou, S., Bächinger, H. P., and Stadler, H. S. (2007). Elucidation, quantitative refinement, and *in vivo* utilization of the HOXA13 DNA binding site. *J. Biol. Chem.* **282,** 6843–6853.

Koh, K., Peyrot, S. M., Wood, C. G., Wagmaister, J. A., Maduro, M. F., Eisenmann, D. M., and Rothman, J. H. (2002). Cell fates and fusion in the *C. elegans* vulval primordium are regulated by the EGL-18 and ELT-6 GATA factors -- apparent direct targets of the LIN-39 Hox protein. *Development* **129,** 5171–5180.

Kornfeld, K., Saint, R. B., Beachy, P. A., Harte, P. J., Peattie, D. A., and Hogness, D. S. (1989). Structure and expression of a family of Ultrabithorax mRNAs generated by alternative splicing and polyadenylation in Drosophila. *Genes Dev.* **3,** 243–258.

Kremser, T., Hasenpusch-Theil, K., Wagner, E., Buttgereit, D., and Renkawitz-Pohl, R. (1999). Expression of the beta3 tubulin gene (beta Tub60D) in the visceral mesoderm of Drosophila is dependent on a complex enhancer that binds Tinman and UBX. *Mol. Gen. Genet.* **262,** 643–658.

Kuziora, M. A., and McGinnis, W. (1989). A homeodomain substitution changes the regulatory specificity of the deformed protein in Drosophila embryos. *Cell* **59,** 563–571.

Kuziora, M. A., and McGinnis, W. (1990). Altering the regulatory targets of the Deformed protein in Drosophila embryos by substituting the Abdominal-B homeodomain. *Mech. Dev.* **33,** 83–93.

Lampe, X., Picard, J. J., and Rezsohazy, R. (2004). The Hoxa2 enhancer 2 contains a critical Hoxa2 responsive regulatory element. *Biochem. Biophys. Res. Commun.* **316,** 898–902.

Lampe, X., Samad, O. A., Guiguen, A., Matis, C., Remacle, S., Picard, J. J., Rijli, F. M., and Rezsohazy, R. (2008). An ultraconserved Hox-Pbx responsive element resides in the coding sequence of Hoxa2 and is active in rhombomere 4. *Nucleic Acids Res.* **36,** 3214–3225.

LaRonde-LeBlanc, N. A., and Wolberger, C. (2003). Structure of HoxA9 and Pbx1 bound to DNA: Hox hexapeptide and DNA recognition anterior to posterior. *Genes Dev.* **17,** 2060–2072.

Laurent, A., Bihan, R., Deschamps, S., Guerrier, D., Dupé, V., Omilli, F., Burel, A., and Pellerin, I. (2007). Identification of a new type of PBX1 partner that contains zinc finger motifs and inhibits the binding of HOXA9-PBX1 to DNA. *Mech. Dev.* **124,** 364–376.

Lei, H., Wang, H., Juan, A. H., and Ruddle, F. H. (2005). The identification of Hoxc8 target genes. *Proc. Natl. Acad. Sci. USA* **102,** 2420–2424.

Lewis, E. B. (1978). A gene complex controlling segmentation in Drosophila. *Nature* **276,** 565–570.

Li-Kroeger, D., Witt, L. M., Grimes, H. L., Cook, T. A., and Gebelein, B. (2008). Hox and senseless antagonism functions as a molecular switch to regulate EGF secretion in the Drosophila PNS. *Dev. Cell* **15,** 298–308.

Lin, L., and McGinnis, W. (1992). Mapping functional specificity in the Dfd and Ubx homeo domains. *Genes Dev.* **6,** 1071–1081.

Liu, H., Strauss, T. J., Potts, M. B., and Cameron, S. (2006). Direct regulation of egl-1 and of programmed cell death by the Hox protein MAB-5 and by CEH-20, a *C. elegans* homolog of Pbx1. *Development* **133,** 641–650.

Liu, J., and Fire, A. (2000). Overlapping roles of two Hox genes and the exd ortholog ceh-20 in diversification of the *C. elegans* postembryonic mesoderm. *Development* **127,** 5179–5190.

Lohmann, I., McGinnis, N., Bodmer, M., and McGinnis, W. (2002). The Drosophila Hox gene deformed sculpts head morphology via direct regulation of the apoptosis activator reaper. *Cell* **110,** 457–466.

Lu, Q., and Kamps, M. P. (1996). Structural determinants within Pbx1 that mediate cooperative DNA binding with pentapeptide-containing Hox proteins: proposal for a model of a Pbx1-Hox-DNA complex. *Mol. Cell Biol.* **16,** 1632–1640.

Luo, L., Yang, X., Takihara, Y., Knoetgen, H., and Kessel, M. (2004). The cell-cycle regulator geminin inhibits Hox function through direct and polycomb-mediated interactions. *Nature* **427,** 749–753.

Manak, J. R., Mathies, L. D., and Scott, M. P. (1994). Regulation of a decapentaplegic midgut enhancer by homeotic proteins. *Development* **120,** 3605–3619.

Mann, R. S. (1995). The specificity of homeotic gene function. *Bioessays* **17,** 855–863.

Mann, R. S., and Abu-Shaar, M. (1996). Nuclear import of the homeodomain protein extradenticle in response to Wg and Dpp signalling. *Nature* **383,** 630–633.

Mann, R. S., and Affolter, M. (1998). Hox proteins meet more partners. *Curr. Opin. Genet. Dev.* **8,** 423–429.

Mann, R. S., and Carroll, S. B. (2002). Molecular mechanisms of selector gene function and evolution. *Curr. Opin. Genet. Dev.* **12,** 592–600.

Mann, R. S., and Chan, S. K. (1996). Extra specificity from extradenticle: the partnership between HOX and PBX/EXD homeodomain proteins. *Trends Genet.* **12,** 258–262.

Mann, R. S., and Hogness, D. S. (1990). Functional dissection of Ultrabithorax proteins in D. melanogaster. *Cell* **60,** 597–610.

Mann, R. S., and Morata, G. (2000). The developmental and molecular biology of genes that subdivide the body of Drosophila. *Annu. Rev. Cell Dev. Biol.* **16,** 243–271.

Manzanares, M., Bel-Vialar, S., Ariza-McNaughton, L., Ferretti, E., Marshall, H., Maconochie, M. M., Blasi, F., and Krumlauf, R. (2001). Independent regulation of initiation and maintenance phases of Hoxa3 expression in the vertebrate hindbrain involve auto- and cross-regulatory mechanisms. *Development* **128,** 3595–3607.

Marty, T., Vigano, M. A., Ribeiro, C., Nussbaumer, U., Grieder, N. C., and Affolter, M. (2001). A HOX complex, a repressor element and a 50 bp sequence confer regional specificity to a DPP-responsive enhancer. *Development* **128,** 2833–2845.

McCabe, C. D., and Innis, J. W. (2005). A genomic approach to the identification and characterization of HOXA13 functional binding elements. *Nucleic Acids Res.* **33,** 6782–6794.

McCormick, A., Coré, N., Kerridge, S., and Scott, M. P. (1995). Homeotic response elements are tightly linked to tissue-specific elements in a transcriptional enhancer of the teashirt gene. *Development* **121,** 2799–2812.

Merabet, S., Hombria, J. C., Hu, N., Pradel, J., and Graba, Y. (2005). Hox-controlled reorganisation of intrasegmental patterning cues underlies Drosophila posterior spiracle organogenesis. *Development* **132,** 3093–3102.

Merabet, S., Kambris, Z., Capovilla, M., Bérenger, H., Pradel, J., and Graba, Y. (2003). The hexapeptide and linker regions of the AbdA Hox protein regulate its activating and repressive functions. *Dev. Cell* **4,** 761–768.

Merabet, S., Saadaoui, M., Sambrani, N., Hudry, B., Pradel, J., Affolter, M., and Graba, Y. (2007). A unique Extradenticle recruitment mode in the Drosophila Hox protein Ultrabithorax. *Proc. Natl. Acad. Sci. USA* **104,** 16946–16951.

Milech, N., Kees, U. R., and Watt, P. M. (2001). Novel alternative PBX3 isoforms in leukemia cells with distinct interaction specificities. *Genes Chromosomes Cancer* **32,** 275–780.

Moens, C. B., and Selleri, L. (2006). Hox cofactors in vertebrate development. *Dev. Biol.* **291,** 193–206.

Monica, K., Galili, N., Nourse, J., Saltman, D., and Cleary, M. L. (1991). PBX2 and PBX3, new homeobox genes with extensive homology to the human proto-oncogene PBX1. *Mol. Cell Biol.* **11,** 6149–6157.

Morgan, R., In der Rieden, P., Hooiveld, M. H., and Durston, A. J. (2000). Identifying HOX paralog groups by the PBX-binding region. *Trends Genet.* **16,** 66–67.

Mukherjee, K., and Bürglin, T. R. (2007). Comprehensive analysis of animal TALE homeobox genes: new conserved motifs and cases of accelerated evolution. *J. Mol. Evol.* **65,** 137–153.

Nakano, T., Windrem, M., Zappavigna, V., and Goldman, S. A. (2005). Identification of a conserved 125 base-pair Hb9 enhancer that specifies gene expression to spinal motor neurons. *Dev. Biol.* **283,** 474–485.

Neuteboom, S. T., Peltenburg, L. T., van Dijk, M. A., and Murre, C. (1995). The hexapeptide LFPWMR in Hoxb-8 is required for cooperative DNA binding with Pbx1 and Pbx2 proteins. *Proc. Natl. Acad. Sci. USA* **92,** 9166–9170.

Noro, B., Culi, J., McKay, D. J., Zhang, W., and Mann, R. S. (2006). Distinct functions of homeodomain-containing and homeodomain-less isoforms encoded by homothorax. *Genes Dev.* **20,** 1636–1650.

Noyes, M. B., Christensen, R. G., Wakabayashi, A., Stormo, G. D., Brodsky, M. H., and Wolfe, S. A. (2008). Analysis of homeodomain specificities allows the family-wide prediction of preferred recognition sites. *Cell* **133,** 1277–1289.

O'Connor, M. B., Binari, R., Perkins, L. A., and Bender, W. (1988). Alternative RNA products from the Ultrabithorax domain of the bithorax complex. *EMBO J.* **7,** 435–445.

Pan, L., Glenn, S. T., Jones, C. A., and Gross, K. W. (2005). Activation of the rat renin promoter by HOXD10.PBX1b.PREP1, Ets-1, and the intracellular domain of notch. *J. Biol. Chem.* **280,** 20860–20866.

Pan, L., Xie, Y., Black, T. A., Jones, C. A., Pruitt, S. C., and Gross, K. W. (2001). An Abd-B class HOX.PBX recognition sequence is required for expression from the mouse Ren-1c gene. *J. Biol. Chem.* **276,** 32489–32494.

Passner, J. M., Ryoo, H. D., Shen, L., Mann, R. S., and Aggarwal, A. K. (1999). Structure of a DNA-bound Ultrabithorax-Extradenticle homeodomain complex. *Nature* **397,** 714–719.

Pearson, J. C., Lemons, D., and McGinnis, W. (2005). Modulating Hox gene functions during animal body patterning. *Nat. Rev. Genet.* **6,** 893–904.

Pederson, J. A., LaFollette, J. W., Gross, C., Veraksa, A., McGinnis, W., and Mahaffey, J. W. (2000). Regulation by homeoproteins: a comparison of deformed-responsive elements. *Genetics* **156,** 677–686.

Penkov, D., Tanaka, S., Di Rocco, G., Berthelsen, J., Blasi, F., and Ramirez, F. (2000). Cooperative interactions between PBX, PREP, and HOX proteins modulate the activity of the alpha 2(V) collagen (COL5A2) promoter. *J. Biol. Chem.* **275,** 16681–16689.

Phelan, M. L., Rambaldi, I., and Featherstone, M. S. (1995). Cooperative interactions between HOX and PBX proteins mediated by a conserved peptide motif. *Mol. Cell Biol.* **15,** 3989–3997.

Pinsonneault, J., Florence, B., Vaessin, H., and McGinnis, W. (1997). A model for extradenticle function as a switch that changes HOX proteins from repressors to activators. *EMBO J.* **16,** 2032–2042.

Piper, D. E., Batchelor, A. H., Chang, C. P., Cleary, M. L., and Wolberger, C. (1999). Structure of a HoxB1-Pbx1 heterodimer bound to DNA: role of the hexapeptide and a fourth homeodomain helix in complex formation. *Cell* **96,** 587–597.

Plaza, S., Prince, F., Adachi, Y., Punzo, C., Cribbs, D. L., and Gehring, W. J. (2008). Cross-regulatory protein-protein interactions between Hox and Pax transcription factors. *Proc. Natl. Acad. Sci. USA* **105,** 13439–13444.

Pöpperl, H., Bienz, M., Studer, M., Chan, S. K., Aparicio, S., Brenner, S., Mann, R. S., and Krumlauf, R. (1995). Segmental expression of Hoxb-1 is controlled by a highly conserved autoregulatory loop dependent upon exd/pbx. *Cell* **81,** 1031–1042.

Potter, C. S., Peterson, R. L., Barth, J. L., Pruett, N. D., Jacobs, D. F., Kern, M. J., Argraves, W. S., Sundberg, J. P., and Awgulewitsch, A. (2006). Evidence that the satin hair mutant gene Foxq1 is among multiple and functionally diverse regulatory targets for Hoxc13 during hair follicle differentiation. *J. Biol. Chem.* **281,** 29245–29255.

Prévôt, D., Voeltzel, T., Birot, A. M., Morel, A. P., Rostan, M. C., Magaud, J. P., and Corbo, L. (2000). The leukemia-associated protein Btg1 and the p53-regulated protein Btg2 interact with the homeoprotein Hoxb9 and enhance its transcriptional activation. *J. Biol. Chem.* **275,** 147–153.

Prince, F., Katsuyama, T., Oshima, Y., Plaza, S., Resendez-Perez, D., Berry, M., Kurata, S., and Gehring, W. J. (2008). The YPWM motif links Antennapedia to the basal transcriptional machinery. *Development* **135,** 1669–1679.

Raman, V., Martensen, S. A., Reisman, D., Evron, E., Odenwald, W. F., Jaffee, E., Marks, J., and Sukumar, S. (2000a). Compromised HOXA5 function can limit p53 expression in human breast tumours. *Nature* **405,** 974–978.

Raman, V., Tamori, A., Vali, M., Zeller, K., Korz, D., and Sukumar, S. (2000b). HOXA5 regulates expression of the progesterone receptor. *J. Biol. Chem.* **275,** 26551–26555.

Rambaldi, I., Kovàcs, E. N., and Featherstone, M. S. (1994). A proline-rich transcriptional activation domain in murine HOXD-4 (HOX-4.2). *Nucleic Acids Res.* **22,** 376–382.

Randazzo, F. M., Cribbs, D. L., and Kaufman, T. C. (1991). Rescue and regulation of proboscipedia: a homeotic gene of the Antennapedia Complex. *Development* **113,** 257–271.

Rao, M. V., Donoghue, M. J., Merlie, J. P., and Sanes, J. R. (1996). Distinct regulatory elements control muscle-specific, fiber-type-selective, and axially graded expression of a myosin light-chain gene in transgenic mice. *Mol. Cell Biol.* **16,** 3909–3922.

Regulski, M., Harding, K., Kostriken, R., Karch, F., Levine, M., and McGinnis, W. (1985). Homeo box genes of the Antennapedia and bithorax complexes of Drosophila. *Cell* **43,** 71–80.

Robertson, L. K., Bowling, D. B., Mahaffey, J. P., Imiolczyk, B., and Mahaffey, J. W. (2004). An interactive network of zinc-finger proteins contributes to regionalization of the Drosophila embryo and establishes the domains of HOM-C protein function. *Development* **131,** 2781–2789.

Ronshaugen, M., McGinnis, N., and McGinnis, W. (2002). Hox protein mutation and macroevolution of the insect body plan. *Nature* **415,** 914–917.

Rousso, D. L., Gaber, Z. B., Wellik, D., Morrisey, E. E., and Novitch, B. G. (2008). Coordinated actions of the forkhead protein Foxp1 and Hox proteins in the columnar organization of spinal motor neurons. *Neuron* **59,** 226–240

Ryoo, H. D., and Mann, R. S. (1999). The control of trunk Hox specificity and activity by Extradenticle. *Genes Dev.* **13,** 1704–1716.

Ryoo, H. D., Marty, T., Casares, F., Affolter, M., and Mann, R. S. (1999). Regulation of Hox target genes by a DNA bound Homothorax/Hox/Extradenticle complex. *Development* **126,** 5137–5148.

Saleh, M., Rambaldi, I., Yang, X. J., and Featherstone, M. S. (2000). Cell signaling switches HOX-PBX complexes from repressors to activators of transcription mediated by histone deacetylases and histone acetyltransferases. *Mol. Cell Biol.* **20,** 8623–8633.

Salsi, V., and Zappavigna, V. (2006). Hoxd13 and Hoxa13 directly control the expression of the EphA7 Ephrin tyrosine kinase receptor in developing limbs. *J. Biol. Chem.* **281,** 1992–1999.

Samad, O. A., Geisen, M. J., Caronia, G., Varlet, I., Zappavigna, V., Ericson, J., Goridis, C., and Rijli, F. M. (2004). Integration of anteroposterior and dorsoventral regulation of Phox2b transcription in cranial motoneuron progenitors by homeodomain proteins. *Development* **131,** 4071–4083.

Serpente, P., Tümpel, S., Ghyselinck, N. B., Niederreither, K., Wiedemann, L. M., Dollé, P., Chambon, P., Krumlauf, R., and Gould, A. P. (2005). Direct crossregulation between retinoic acid receptor {beta} and Hox genes during hindbrain segmentation. *Development* **132,** 503–513.

Shen, W., Chrobak, D., Krishnan, K., Lawrence, H. J., and Largman, C. (2004). HOXB6 protein is bound to CREB-binding protein and represses globin expression in a DNA binding-dependent, PBX interaction-independent process. *J. Biol. Chem.* **279,** 39895–39904.

Shen, W. F., Rozenfeld, S., Lawrence, H. J., and Largman, C. (1997). The Abd-B-like Hox homeodomain proteins can be subdivided by the ability to form complexes with Pbx1a on a novel DNA target. *J. Biol. Chem.* **272,** 8198–81206.

Shi, X., Bai, S., Li, L., and Cao, X. (2001). Hoxa-9 represses transforming growth factor-beta-induced osteopontin gene transcription. *J. Biol. Chem.* **276,** 850–855.

Shi, X., Yang, X., Chen, D., Chang, Z., and Cao, X. (1999). Smad1 interacts with homeobox DNA-binding proteins in bone morphogenetic protein signaling. *J. Biol. Chem.* **274,** 13711–13717.

Stevens, K. E., and Mann, R. S. (2007). A balance between two nuclear localization sequences and a nuclear export sequence governs extradenticle subcellular localization. *Genetics* **175,** 1625–1636.

Streit, A., Kohler, R., Marty, T., Belfiore, M., Takacs-Vellai, K., Vigano, M. A., Schnabel, R., Affolter, M., and Müller, F. (2002). Conserved regulation of the Caenorhabditis elegans labial/Hox1 gene ceh-13. *Dev. Biol.* **242,** 96–108.

Stultz, B. G., Jackson, D. G., Mortin, M. A., Yang, X., Beachy, P. A., and Hursh, D. A. (2006). Transcriptional activation by extradenticle in the Drosophila visceral mesoderm. *Dev. Biol.* **290,** 482–494.

Sun, B., Hursh, D. A., Jackson, D., and Beachy, P. A. (1995). Ultrabithorax protein is necessary but not sufficient for full activation of decapentaplegic expression in the visceral mesoderm. *EMBO J.* **14,** 520–535.

Taghli-Lamallem, O., Gallet, A., Leroy, F., Malapert, P., Vola, C., Kerridge, S., and Fasano, L. (2007). Direct interaction between Teashirt and Sex combs reduced proteins, via Tsh's acidic domain, is essential for specifying the identity of the prothorax in Drosophila. *Dev. Biol.* **307,** 142–151.

Taghli-Lamallem, O., Hsia, C., Ronshaugen, M., and McGinnis, W. (2008). Context-dependent regulation of Hox protein functions by CK2 phosphorylation sites. *Dev. Genes Evol.* **218,** 321–332.

Tour, E. (2005). Evolutionarily conserved domains required for activation and repression functions of the Drosophila Hox protein Ultrabithorax. *Development* **132,** 5271–5281.

Troy, P. J., Daftary, G. S., Bagot, C. N., and Taylor, H. S. (2003). Transcriptional repression of peri-implantation EMX2 expression in mammalian reproduction by HOXA10. *Mol. Cell Biol.* **23,** 1–13.

Tümpel, S., Cambronero, F., Ferretti, E., Blasi, F., Wiedemann, L. M., and Krumlauf, R. (2007). Expression of Hoxa2 in rhombomere 4 is regulated by a conserved cross-regulatory mechanism dependent upon Hoxb1. *Dev. Biol.* **302,** 646–660.

Vachon, G., Cohen, B., Pfeifle, C., McGuffin, M. E., Botas, J., and Cohen, S. M. (1992). Homeotic genes of the Bithorax complex repress limb development in the abdomen of the Drosophila embryo through the target gene Distal-less. *Cell* **71,** 437–450.

Viganò, M. A., Di Rocco, G., Zappavigna, V., and Mavilio, F. (1998). Definition of the transcriptional activation domains of three human HOX proteins depends on the DNA-binding context. *Mol. Cell Biol.* **18,** 6201–6212.

Vincent, J. P., Kassis, J. A., and O'Farrell, P. H. (1990). A synthetic homeodomain binding site acts as a cell type specific, promoter specific enhancer in Drosophila embryos. *EMBO J.* **9,** 2573–2578.

Wagmaister, J. A., Miley, G. R., Morris, C. A., Gleason, J. E., Miller, L. M., Kornfeld, K., and Eisenmann, D. M. (2006). Identification of cis-regulatory elements from the *C. elegans* Hox gene lin-39 required for embryonic expression and for regulation by the transcription factors LIN-1, LIN-31 and LIN-39. *Dev. Biol.* **297,** 550–565.

Wagner, K., Mincheva, A., Korn, B., Lichter, P., and Popperl, H. (2001). Pbx4, a new Pbx family member on mouse chromosome 8, is expressed during spermatogenesis. *Mech. Dev.* **103,** 127–131.

Walsh, C. M., and Carroll, S. B. (2007). Collaboration between Smads and a Hox protein in target gene repression. *Development* **134,** 3585–3592.

Wang, N., Kim, H. G., Cotta, C. V., Wan, M., Tang, Y., Klug, C. A., and Cao, X. (2006). TGFbeta/BMP inhibits the bone marrow transformation capability of Hoxa9 by repressing its DNA-binding ability. *EMBO J.* **25,** 1469–1480.

Weatherbee, S. D., and Carroll, S. B. (1999). Selector genes and limb identity in arthropods and vertebrates. *Cell* **97,** 283–286.

Williams, T. M., Selegue, J. E., Werner, T., Gompel, N., Kopp, A., and Carroll, S. B. (2008). The regulation and evolution of a genetic switch controlling sexually dimorphic traits in Drosophila. *Cell* **134,** 610–623.

Williams, T. M., Williams, M. E., Heaton, J. H., Gelehrter, T. D., and Innis, J. W. (2005). Group 13 HOX proteins interact with the MH2 domain of R-Smads and modulate Smad transcriptional activation functions independent of HOX DNA-binding capability. *Nucleic Acids Res.* **33,** 4475–4484.

Wu, Y., Moser, M., Bautch, V. L., and Patterson, C. (2003). HoxB5 is an upstream transcriptional switch for differentiation of the vascular endothelium from precursor cells. *Mol. Cell Biol.* **23,** 5680–5691.

Yao, L. C., Liaw, G. J., Pai, C. Y., and Sun, Y. H. (1999). A common mechanism for antenna-to-Leg transformation in Drosophila: suppression of homothorax transcription by four HOM-C genes. *Dev. Biol.* **211,** 268–276.

Zeng, C., Pinsonneault, J., Gellon, G., McGinnis, N., and McGinnis, W. (1994). Deformed protein binding sites and cofactor binding sites are required for the function of a small segment-specific regulatory element in Drosophila embryos. *EMBO J.* **13,** 2362–2377.

Zhao, J. J., Lazzarini, R. A., and Pick, L. (1996). Functional dissection of the mouse Hox-a5 gene. *EMBO J.* **15,** 1313–1322.

Zhao, Y., and Potter, S. S. (2001). Functional specificity of the Hoxa13 homeobox. *Development* **128,** 3197–3207.

Zhao, Y., and Potter, S. S. (2002). Functional comparison of the Hoxa 4, Hoxa 10, and Hoxa 11 homeoboxes. *Dev. Biol.* **244,** 21–36.

Zhou, B., Chen, L., Wu, X., Wang, J., Yin, Y., and Zhu, G. (2008). MH1 domain of SMAD4 binds N-terminal residues of the homeodomain of Hoxc9. *Biochim. Biophys. Acta* **1784,** 747–752.

CHAPTER FOUR

Hox Genes and Segmentation of the Vertebrate Hindbrain

Stefan Tümpel,*[,1] Leanne M. Wiedemann,*[,†] *and* Robb Krumlauf*[,‡]

Contents

Abstract

In the vertebrate central nervous system, the hindbrain is an important center for coordinating motor activity, posture, equilibrium, sleep patterns, and essential unconscious functions, such as breathing rhythms and blood circulation. During development, the vertebrate hindbrain depends upon the process of segmentation or compartmentalization to create and organize regional properties essential for orchestrating its highly conserved functional roles. The process of segmentation in the hindbrain differs from that which functions in the paraxial mesoderm to generate somites and the axial skeleton. In the prospective hindbrain, cells in the neural epithelia transiently alter their ability to interact with their neighbors, resulting in the formation of seven lineage-restricted cellular compartments. These different segments or rhombomeres each go on to adopt unique characters in response to environmental signals. The *Hox* family of transcription factors is coupled to this process. Overlapping or

* Stowers Institute for Medical Research, Kansas City, Missouri, USA

† Department of Pathology and Laboratory Medicine, Kansas University Medical School, Kansas City, Kansas, USA

‡ Department of Anatomy and Cell Biology, Kansas University Medical School, Kansas City, Kansas, USA

1 Present address: Institut für Molekulare Medizin und Max-Planck-Forschungsgruppe Stammzellalterung, Universität Ulm, Ulm, Germany

Current Topics in Developmental Biology, Volume 88
ISSN 0070-2153, DOI: 10.1016/S0070-2153(09)88004-6

nested patterns of *Hox* gene expression correlate with segmental domains and provide a combinatorial code and molecular framework for specifying the unique identities of hindbrain segments. The segmental organization and patterns of *Hox* expression and function are highly conserved among vertebrates and, as a consequence, comparative studies between different species have greatly enhanced our ability to build a picture of the regulatory cascades that control early hindbrain development. The purpose of this chapter is to review what is known about the regulatory mechanisms which establish and maintain *Hox* gene expression and function in hindbrain development.

1. Introduction

During development, compartmentalization or segmentation is a frequently used cellular strategy for generating segregated groups of cells with unique properties that ultimately leads to the formation of distinct structures (Garcia Martinez *et al.*, 1993; Lawrence and Struhl, 1996). Segmentation in animal development is believed to have independently evolved many times, suggesting that there is not a common set of mechanisms, pathways, or processes which govern the formation of cellular compartments. Cellular and molecular events in segmentation of the vertebrate hindbrain have been extensively studied in teleost, avian, and murine model systems. By exploiting the experimental advantages of these systems, a wealth of information has been generated which strongly supports a fundamental and conserved role for hindbrain segmentation in specifying neuronal diversity, organizing the functional architecture of neural components, such as branchial motor nerves, and in impacting craniofacial development through formation of neural crest (reviewed in Chatonnet *et al.*, 2003; Clarke and Lumsden, 1993; Glover *et al.*, 2006; Keynes and Lumsden, 1990; Kiecker and Lumsden, 2005; Lumsden, 2004; Lumsden and Krumlauf, 1996; Moens and Prince, 2002; Trainor and Krumlauf, 2000b, 2001).

The *Hox* genes encode a highly conserved family of transcription factors with important roles in the regulation of axial patterning in many animal species and tissues (Carroll, 1995; Krumlauf, 1994; McGinnis and Krumlauf, 1992). An intriguing feature of *Hox* genes is the relationship between their organization within gene clusters and their highly ordered domains of expression and function along the anterior-posterior (A-P) axis during development (Duboule and Dollé, 1989; Graham *et al.*, 1989; Lewis, 1978). This property is termed colinearity, and the mechanisms which underlie this feature result in the generation of overlapping or nested patterns of *Hox* gene expression which provide a combinatorial code for specifying unique regional identities. The ordered expression of *Hox* genes has been coupled to A-P patterning in many developmental contexts, suggesting that it provides a robust mechanism for differential regulation of regional identity in many tissues and organs (Dollé *et al.*, 1991; Kessel and Gruss, 1991; Lumsden and Krumlauf, 1996; Maconochie *et al.*, 1996; Zacchetti *et al.*, 2007; Zakany and Duboule, 2007). In the vertebrate hindbrain, during the process of segmentation, nested domains of *Hox* gene expression are linked to the formation of rhombomeres and provide the basis for regulating segmental identity (Hunt *et al.*, 1991a; Keynes and Krumlauf, 1994; Moens and Prince, 2002; Prince *et al.*, 1998b; Wilkinson *et al.*, 1989b).

The segmental organization and patterns of *Hox* expression are highly conserved among vertebrates. This has facilitated analyses aimed at understanding the gene regulatory cascades that control segmental identity in early hindbrain development (Maconochie *et al.*, 1996; Nolte and Krumlauf, 2006; Parrish *et al.*, 2009). A detailed characterization of the *Hox* regulatory network involved in hindbrain segmentation will be important for comparing and contrasting how *Hox* genes regulate cellular properties in many different tissue contexts. Toward this end, a series of studies have begun to identify many of the *cis*-regulatory modules of *Hox* genes and their cognate upstream factors which govern segmental expression and function of the *Hox* family in hindbrain development. This chapter will review vertebrate hindbrain segmentation and *Hox* gene expression in hindbrain development and build upon existing regulatory information to construct a gene regulatory network to provide an initial framework for understanding the regulation of *Hox* genes in these events.

2. Hindbrain Segmentation

Following neural induction, in the prospective hindbrain territory cells are able to move freely within the epithelium along the A-P axis during the growth of the neural plate. Cell lineage analyses in chicken embryos indicate that clonal descendants of cells labeled in early somite stages are able to mix with their neighbors and cross future rhombomere boundaries, but this ability is lost when cells are labeled at slightly later stages (Birgbauer and Fraser, 1994; Birgbauer *et al.*, 1995; Fraser *et al.*, 1990; Kulesa and Fraser, 1998; Lumsden *et al.*, 1994). These experiments reveal that there is a temporally regulated restriction to the movement of cells, which results in the establishment of lineage-restricted cellular compartments within the hindbrain. Within a rhombomeric compartment cells retain the ability to mix with their neighbors, but cells do not move between compartments. These separate compartments are rapidly set up over the entire length of the hindbrain, subdividing the territory into seven rhombomeric (r) segments r1–r7 (Fig. 4.1A). It is interesting to note that this

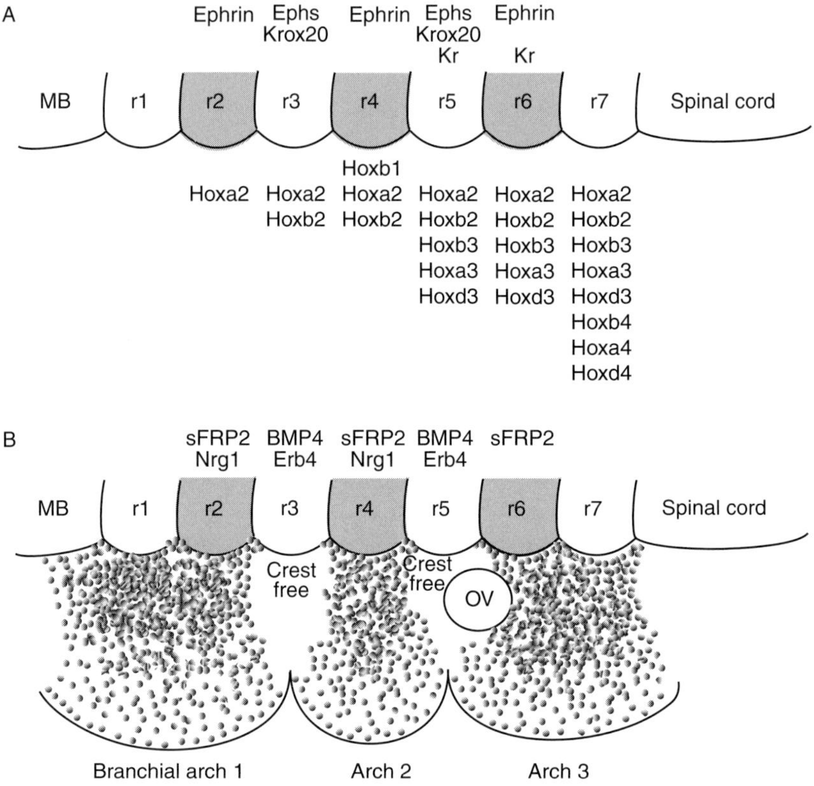

Figure 4.1 Diagram illustrating basic organization of the hindbrain and patterns of gene expression. The hindbrain is divided into seven rhombomeric segments (r1-r7) flanked anteriorly by the midbrain (MB) and posteriorly by the spinal cord. (A) Segmental patterns of *ephrin*, *Eph*, *Krox20*, and *kreisler (Kr)* expression are listed above the rhombomeres. The patterns and combinations of *Hox* gene expression in each segment is listed below the segments. (B) Segmental patterns of *sFRP2*, *BMP4*, *ErbB4*, and *Nrg1* expression, which are involved in inter-rhombomeric signaling are listed above the rhombomeres. The correlation between rhombomeres and neural crest formation and migration is shown at the bottom.

process of segmentation differs from that observed during somitogenesis, where there is a progressive formation of somitic segments from the presomitic mesoderm over an extended period of time in association with the elongation and anterior to posterior development of the embryonic axis (see Chapter 7 by Pourquié).

The lack of mixing between cells of adjacent rhombomeres is not the result of the formation of a mechanical or physical barrier to cell movement. Experimental removal of cells at a rhombomere boundary has no influence on the ability of cells to mix with each other at the new interface between adjacent rhombomeres (Guthrie and Lumsden, 1991). This implies that cells in adjacent rhombomeres have distinct properties (e.g., as cell adhesion) that prevent them from intermingling. This idea is directly supported by *in vivo* grafting experiments in avian embryos, which juxtapose rhombomeres from different A-P levels. When groups of cells from r3 are placed next to cells from r3 or r5, they readily mix with each other. However, if r3 cells are placed next to cells from r2 or r4 they are unable to intermingle (Guthrie *et al.*, 1993). In an analogous manner, cells from r4 are freely able to mix with cells from r2 or r4, but not with those derived from r3 or r5 (Guthrie *et al.*, 1993). Based on experiments using *in vitro* cell aggregation cultures in combination with lineage tracers, a mixture of dissociated cells from r3 and r5 or from r2 and r4 are remain intermingled and form a mixed aggregate. However, if dissociated cells from r3 or r5 are mixed with cells from r2 or r4, they sort out from each other to form separate aggregates (Wizenmann and Lumsden, 1997). Analyses with inhibitors reveal that these patterns of rhombomere-specific segregation require Ca^{2+}-dependent adhesion molecules, but not Ca^{2+}-independent adhesion molecules (Wizenmann and Lumsden, 1997). Together, these studies demonstrate that cells from adjacent rhombomeres have distinct differences in cell adhesion or affinity that play a role in segregating cells into discrete compartments. Furthermore, cells from odd-numbered rhombomeres (r3 and r5) share common cell adhesive properties with each other, but these are different from the cell affinities shared between even-numbered rhombomeres (r2 and r4). This sets up a repeating pattern of cellular properties that alternates in a two-segment periodicity. The restriction to cell movement and formation of rhombomere compartments is therefore a progressive process that involves cell sorting of intermingled populations with different properties from broadly defined regions.

Insight into the molecular mechanisms that underlie the differences in cellular interaction between rhombomeres has come through analyses of the Eph receptor tyrosine kinase family and their membrane-bound ligands, the ephrins (Gale *et al.*, 1996). The Ephs and ephrins form a bidirectional signaling pathway, whereby interaction between cells expressing a ligand (ephrin) with those expressing an Eph receptor can lead to the transduction of functionally relevant signals in both cells. The Eph/ephrin pathway can serve to potentiate both attraction and repulsion between cell populations and plays diverse roles in neural development (Flanagan and Vanderhaeghen, 1998; Holder and Klein, 1999; Mellitzer *et al.*, 2000; Sela-Donenfeld and Wilkinson, 2005; Wilkinson, 2001, 2003). In the vertebrate hindbrain, several members of the Eph receptors (e.g., *EphA4*) are highly expressed in r3 and r5, while several ephrin ligands (e.g., *ephrin-B1*) are strongly expressed in the adjacent r2, r4, and r6 segments (Xu *et al.*, 2000) (Fig. 4.1A). In zebrafish, these complementary patterns of *Eph/ephrin* expression have been shown to be functionally important in regulating cellular interactions in the hindbrain. Blocking Eph receptor activity and misexpression of Eph receptors or ephrin ligands in the developing hindbrain result in inappropriate sorting of cells between rhombomeric segments (Xu *et al.*, 1995, 1999). *In vitro* cell-sorting assays in zebrafish explant cultures have also demonstrated that bidirectional Eph/ephrin signaling restricts the intermingling of adjacent cell populations, while unidirectional activation of either pathway alters gap junctional cell communication (Mellitzer *et al.*, 1999). There is evidence that the activity of Eph/ephrin signaling is required for establishing the patterns of Ca^{2+}-dependent adhesion molecules, which also contribute to rhombomere-specific segregation of cell populations (Mellitzer *et al.*, 2000). These studies demonstrate that the Eph/ephrin pathway has multiple inputs into the regulation of cellular behaviors and affinities in the developing hindbrain.

Segmental compartments in the hindbrain serve to segregate cell populations with similar potential, enabling them to respond in a differential manner to environmental signals. This allows each rhombomere to adopt a unique character distinct from its neighboring segment, which ultimately leads to the generation of diverse neuronal components essential for organization and function of the hindbrain. However, in several developmental contexts (e.g., the *Drosophila* wing imaginal disc) compartments not only represent segregated groups of cells, but compartmental boundaries are local signaling centers that participate in controlling aspects of the segmental process. Analyses in zebrafish have shown that the distinct population of cells which form at rhombomere boundaries, function as signaling centers to regulate cell differentiation and spreading of boundary cell character (Amoyel *et al.*, 2005; Cheng *et al.*, 2004). The ability of boundary cells to exert these functions is mediated through coordinate modulation of the Notch and Wnt pathways. Furthermore, there is a remarkable degree of similarly in the regulatory hierarchy governing these signaling events in both the vertebrate hindbrain and the anterior compartment of the *Drosophila* wing disc, which raises interesting questions with respect to the evolution of these segmental processes (Amoyel *et al.*, 2005).

Rhombomeres not only receive signals from surrounding tissues, but there is increasing evidence that they can serve as an important source of signals in hindbrain patterning.

Grafting and misexpression experiments in the zebrafish hindbrain have revealed that r4 functions as an early signaling center that influences the development of the adjacent r5 and r6 territories (Maves *et al.*, 2002). Fgf signaling, potentiated by expression of Fgf ligands in r4, appears to be responsible for this effect. In avian embryos, there is also evidence that Fgf signaling within the hindbrain plays multiple roles in regulating segmentation (Marin and Charnay, 2000). These observations support the idea that the Fgf pathway participates in inter-rhombomeric signaling and illustrates a general role for local organizing centers in early hindbrain patterning. Transposition experiments in mouse and zebrafish embryos aimed at examining the degree of autonomy or plasticity of segmental identity in rhombomeric cells have uncovered an important role for cell-cell signaling within a rhombomere. When individual or very small groups of cells from one rhombomere are heterotopically transposed to different rhombomere, they change their identity to adopt the fate at the new location (Schilling, 2001; Trainor and Krumlauf, 2000a,b). However, when a larger cluster of cells is transposed, the grafted cells retain the identity characteristic of their point of origin. This shows that cell-cell signaling and community effects within a rhombomere play an important role in maintaining the specific identity and properties of cells in each segment and these signaling pathways vary between rhombomeres. Hence, signaling between and within rhombomeres acts to refine and elaborate an initial pattern set in motion by the segmentation process.

Hindbrain segmentation plays an important role in the regulation of patterning processes in other tissues during craniofacial development. In head development, most of the bone and connective tissue is derived from cranial neural crest cells, which delaminate from midbrain and hindbrain regions to migrate into the branchial arches (Le Douarin and Kalcheim, 1999). The migration of neural crest cells has been extensively studied in a number of species through a combination of grafting, lineage tracing, and time-lapse imaging approaches to follow the formation, movement, and fate of cells derived from the hindbrain segments (Kontges and Lumsden, 1996; Kulesa and Fraser, 2000; Kulesa *et al.*, 2000, 2008; Lumsden *et al.*, 1991; Schilling and Kimmel, 1994; Sechrist *et al.*, 1993; Teddy and Kulesa, 2004; Trainor *et al.*, 2002). The pattern of cranial neural crest cell formation and migration is tightly correlated with organization and segmental properties rhombomeres (Knecht and Bronner-Fraser, 2002). The neural crest cells migrate from hindbrain into the branchial arches 1-3 in three major streams (Fig. 4.1B). The stream moving into arch 1 arises from the posterior midbrain and anterior hindbrain segments (r1-r2). The stream which fills the second arch emigrates primarily from r4, but some cells are derived from r3 and r5. The stream entering arch 3 originates primarily from r6 and r7, but some cells from r5 contribute to this population. This stream also fills the more caudal branchial arches. Hence, cranial neural crest cells entering a branchial arch are derived from multiple rhombomeres, but not all rhombomeres generate an equivalent amount of neural crest.

Studies investigating the formation and migratory behaviors of cranial neural crest cells have uncovered important roles for inter-rhombomeric signaling and for signaling between rhombomeres and their environment. Large populations of neural crest cells delaminate from r1, r2, r4, r6, and r7, while considerably fewer cells migrate from r3 and r5, forming a "crest-free zone" around these segments (Fig. 4.1B). The difference in the ability of r3 and r5 to generate neural crest cells is associated with increased cell death (Ellies *et al.*, 2000; Graham *et al.*, 1993) and inhibition by environmental cues (Farlie *et al.*, 1999; Sechrist *et al.*, 1993; Trainor *et al.*, 2002). If r3 or r5 cells are grafted into r4 or are cultured *in vitro*, they are capable of producing neural crest cells (Farlie *et al.*, 1999; Graham *et al.*, 1993, 1994; Trainor *et al.*, 2002). However, when they are cultured next to even-numbered rhombomeres neural crest cells fail to form. Key regulators in this inhibitory phenomenon are the BMP and Wnt signaling pathways, which are involved in a feedback loop. The signaling molecule BMP4 and its target homeobox transcription factor Msx2 are implicated in modulating cell death in r3 and r5 in avian embryos and their expression is dependent upon signals from the adjacent rhombomeres (Graham *et al.*, 1994). The Wnt antagonist sFRP2 is expressed in r2, r4, and r6 and this factor inhibits BMP signaling and cell death in these segments (Ellies *et al.*, 2000) (Fig. 4.1B). In parallel, cues outside the hindbrain in the environment surrounding r3 and r5 inhibit the ability of neural crest cells to enter these territories (Farlie *et al.*, 1999; Sechrist *et al.*, 1993; Trainor *et al.*, 2002). This forces neural crest cells derived r3 and r5 to take anterior and posterior migratory routes, where they join cells emerging from adjacent even-numbered segments. These interactions create the characteristic "crest-free zone" in the branchial arch environment next to r3 and r5 (Fig. 4.1B). Interestingly, r3 and r5 appear to play a role in generating some of the inhibitory cues in their adjacent environment. The receptor tyrosine kinase receptor *ErbB4* is highly expressed in r3 and r5, while one of its ligands, *neuregulin-1 (Nrg1)*, is expressed in r2 and r4 (Fig. 4.1B). In the mouse, loss of *ErbB4* leads to the diversion and infiltration of cells from streams of neural crest cells migrating from r2 and r4 into the branchial environment next to r3, resulting in the formation of fused cranial sensory ganglia (Golding *et al.*, 2002, 2004). Since *ErbB4* is not expressed in the neural crest cells themselves or the branchial arch environment, these findings imply that r3 and r5 provide *ErbB4*-dependent cues to their surrounding tissues, which in turn modulate migratory behaviors of neural crest cells.

To date, most studies have focused on investigating the relationship between hindbrain segmentation and pattern formation in establishing the early neuronal architecture of the region (Clarke and Lumsden, 1993; Clarke *et al.*, 1998; Lumsden *et al.*, 1994; Simon and Lumsden, 1993). However, refined fate-mapping studies in avian embryos and the application of methods for genetic fate mapping in the mouse are beginning to unravel the lineage relationships between early segmental organization of the hindbrain and the generation of structures in the postsegmental and adult brain (Kontges and Lumsden, 1996; Pasqualetti *et al.*, 2007; Wintgate and Lumsden, 1996). In addition, phenotypic studies in mice with targeted mutations in genes implicated in processes associated with hindbrain segmentation are providing functional insight into how segmentation impacts neurogenesis, sequential generation of distinct neuronal subtypes, the selection of neuronal circuits, migration of pontine neurons, development of the somatosensory map, and regulation of rhythmic respiratory activity (Briscoe and Wilkinson, 2004; Chatonnet *et al.*, 2003, 2007; Davenne *et al.*, 1999; del Toro *et al.*, 2001; Geisen *et al.*, 2008; Matis *et al.*, 2007; Oury *et al.*, 2006; Pattyn *et al.*, 2003).

In summary, the extensive body of work described above illustrates how the carefully orchestrated series of events in segmentation of the vertebrate hindbrain play fundamentally important roles in a wide variety of processes essential for elaborating the basic body plan of head development. The challenge now is to understand the mechanisms and pathways, which form the components of the regulatory networks that govern the segregation of cells into segmental compartments and the formation and function of signaling centers that specify their unique fates. A common theme in many of the early events emerging from experimental findings is the generation of different cellular behaviors and properties that alter with a two-segment periodicity (Fig. 4.1). In general, odd-numbered rhombomeres share many common features while even-numbered segments share a different set of properties. This is illustrated by differences in cell adhesion, cell-sorting affinities, generation of neural crest, timing of neurogenesis, birth of motor neurons, and patterns of gene expression. Therefore, understanding how this two-segment periodicity is established and maintained during hindbrain segmentation is fundamentally important and the *Hox* family of transcription factors plays a key role in this process.

3. Expression of *Hox* Genes in the Hindbrain

In many vertebrate species, there are 39 *Hox* genes organized into four separate chromosomal clusters (*HoxA-HoxD*) (Krumlauf, 1992, 1994). The vertebrate *Hox* clusters have evolved from a single ancestral homeobox gene complex through two successive genome-wide duplication events (2R) approximately 500 Myr ago, followed by subsequent divergence (Duboule and Dollé, 1989; Graham *et al.*, 1989; Kappen *et al.*, 1989; Panopoulou and Poustka, 2005; Popovici *et al.*, 2001; Wolfe, 2001). In fish, there has been a further expansion and divergence in the number of *Hox* clusters, compared to other vertebrates, through a putative "fish-specific" additional round of genome duplication (3R), approximately 320 Myr ago (Amores *et al.*, 1998, 2004; Aparicio *et al.*, 1997; Hoegg and Meyer, 2005; Vandepoele *et al.*, 2004). As a result of these events, the *Hox* clusters share several common or conserved features. Genes within each cluster have the same 5′-3′ orientation with respect to transcription. Sequence alignments reveal that it is possible to identify 13 paralogous groups based upon the relative position of genes within each cluster and sequence similarity of their encoded proteins.

A hallmark of the family of clustered *Hox* genes is the property of spatial colinearity (Duboule and Dollé, 1989; Graham *et al.*, 1989; Lewis, 1978). In many tissues, genes from *Hox* clusters display ordered and nested patterns of expression along the A-P axis, which directly correlate with their relative positions in a cluster (Dollé *et al.*, 1989, 1991; Hunt and Krumlauf, 1991; Hunt *et al.*, 1991a; Kessel and Gruss, 1991; Wilkinson *et al.*, 1989b). This creates a combinatorial *Hox* code for the transcriptional specification of regional diversity along the A-P axis during development (Carroll, 1995; Krumlauf, 1994; McGinnis and Krumlauf, 1992). In the central nervous system (CNS), the colinear expression of *Hox* genes is tightly coupled to hindbrain segmentation (Lumsden and Krumlauf, 1996; Nolte and Krumlauf, 2006; Parrish *et al.*, 2009). Of the 39 clustered *Hox* genes, members from paralogous groups 1-4 have domains of expression in the CNS which map to the hindbrain and spinal cord, while genes from groups 4-13 only display expression in the spinal cord. There are 12 genes contained in paralogous groups 1-4, but *Hoxd1* and *Hoxc4* are not expressed in the hindbrain. Therefore, 10 *Hox* genes have temporally dynamic and spatially restricted patterns of expression that are associated with segmental processes in hindbrain development.

In general, the neural expression of each *Hox* gene is initiated in a posterior region and expands in an anterior direction to form a sharp and distinct anterior boundary (Deschamps and Wijgerde, 1993; Murphy and Hill, 1991; Murphy *et al.*,1989; Wilkinson *et al.*, 1989b). The sharp anterior boundaries of gene expression in the hindbrain were first characterized in detail in the mouse and found to precisely correspond to boundaries between rhombomeres (Frohman *et al.*, 1990; Hunt *et al.*, 1991a,b; Murphy and Hill, 1991; Murphy *et al.*, 1989; Wilkinson *et al.*, 1989b). Figure 4.1A summarizes the correlation between different anterior boundaries of *Hox* expression and specific rhombomeres. No *Hox* genes are expressed in r1 and *Hoxa2* is the only *Hox* gene expressed in r2. With the exceptions of *Hoxb1*, *Hoxa1*, and *Hoxa2*, the anterior boundary of expression for adjacent genes within a cluster are offset in a two-segment periodicity (Fig. 4.1A). For example, the anterior expression boundaries of *Hoxb2*, *Hoxb3*, and *Hoxb4* map to the junction between r2/r3, r4/r5, and r6/r7, respectively (Wilkinson *et al.*, 1989b). Furthermore, the anterior domains of expression of members of the same paralogous group tend to be in register with each other, as *Hoxa3*, *Hoxb3*, and *Hoxd3* have boundaries that map to r4/r5, while *Hoxa4*, *Hoxb4*, and *Hoxd4* have boundaries at the r6/r7 interface (Fig. 4.1A).

Following initial gene activation, there are often regional changes in *Hox* expression, which result in domains of gene expression with clearly delineated anterior and posterior boundaries or a complete loss of expression. Such dynamic changes are illustrated by the patterns of expression of *Hoxb1* and *Hoxa1*. Expression of both genes is initiated at 7.75 days postcoitum (dpc) and by 8.25 dpc reaches a sharp anterior boundary at the junction between r3 and r4. In the case of *Hoxb1*, expression then becomes progressively restricted to r4 (Fig. 4.1A) through a loss of expression in posterior regions, which leaves a sharp posterior border of expression at 9.5 dpc at the boundary between r4 and r5 (Frohman *et al.*, 1990; Murphy and Hill, 1991; Murphy *et al.*, 1989; Wilkinson *et al.*, 1989b). In contrast, there is a complete loss of *Hoxa1* expression in the hindbrain and spinal cord by 9.5 dpc. In addition to the formation of distinct anterior and posterior boundaries of expression, there can be a wide variation in levels of transcripts within the domain of neural expression for a single gene. Many of these differences in level of expression have also been found to correspond to specific rhombomeres or groups of rhombomeres (Hunt *et al.*, 1991a; Wilkinson *et al.*, 1989b). For example, *Hoxb2* is expressed at high levels in r3, r4, and r5 but at much lower levels in more posterior rhombomeres and the spinal cord (Hunt *et al.*, 1991a,b; Wilkinson *et al.*, 1989b). Together, such studies reveal that segment-specific modulation of *Hox* expression can be imposed upon on a gene once it is activated in the hindbrain. Hence, the nested boundaries of *Hox* expression in conjunction with segment-specific variation in levels of expression create a combinatorial *Hox* code which can be used to regulate rhombomere identity. Figure 4.1A lists the combinations of the nine mouse *Hox* genes which are expressed in specific rhombomeres at 9.5 dpc of development.

Analyses of *Hox* gene expression during hindbrain segmentation in other vertebrates have demonstrated that there is a high degree of conservation in rhombomeric expression patterns. In comparison to the mouse, where examined *Hox* orthologs in the chick hindbrain have identical anterior rhombomere boundaries of expression, but there are differences in the timing and relative levels of expression between rhombomeres (Grapin-Botton *et al.*, 1995; Manzanares *et al.*, 2001; Prince and Lumsden, 1994; Sundin and Eichele, 1990). Since these expression studies detect steady-state mRNA levels by in situ hybridization, it is possible that differential stability of transcripts could mask more similarities between the species. Ultimately, it will be essential to quantify the distribution and relative levels of Hox proteins in rhombomeres to fully document the degree of conservation. In zebrafish and other fish model systems, the further amplification and divergence of the *Hox* clusters has complicated analyses because expression domains of orthologous mouse genes may be differentially divided among several fish paralogs. Despite this problem, there is an excellent correlation between mouse, chick, and fish, in rhombomeric expression for clear orthologs and paralogous genes (Amores *et al.*, 2004; Davis *et al.*, 2008; Le Pabic *et al.*, 2007; Prince *et al.*, 1998a,b; Scemama *et al.*, 2002, 2006).

During the early phases of hindbrain segmentation, the expression of a *Hox* gene in a segment is observed in all cells within that rhombomere. However, in association with onset of neuronal development the patterns of *Hox* gene expression within segmental domains are progressively altered along the dorsal-ventral (D-V) axis. All members of the *Hoxb* cluster become progressively restricted to dorsal (sensory) regions, while members of the *Hoxc* and *Hoxd* complex display restrictions to specific ventral (motor) domains (Gaunt, 1991; Graham *et al.*, 1991). These D-V-restricted domains of expression in rhombomeres provide a combinatorial code for regulating axial differences in neuronal specification in the hindbrain and spinal cord (Dasen *et al.*, 2003, 2005, 2008; Davenne *et al.*, 1999; Liu *et al.*, 2001; Pattyn *et al.*, 2003). These studies illustrate that in addition to the early roles for *Hox* genes in controlling segmental identity, the segmentation process sets up domains of *Hox* expression that can modulated and coupled to the control of later aspects of neural architecture and function functions (Briscoe and Wilkinson, 2004; Chatonnet *et al.*, 2003, 2007; Davenne *et al.*, 1999; del Toro *et al.*, 2001; Geisen *et al.*, 2008; Matis *et al.*, 2007; Oury *et al.*, 2006; Pasqualetti *et al.*, 2007; Pattyn *et al.*, 2003; Samad *et al.*, 2004).

4. *Hox* Gene Regulatory Networks in Hindbrain Segmentation

Many transcription factors (e.g., Kreisler and Krox20), signaling pathways (e.g., Fgf, retinoid, Wnt, and BMP), and other molecular components are important for establishing and maintaining the restricted patterns of *Hox* expression essential for regulating regional identity of the hindbrain segments. Furthermore, these regulatory components frequently have input into multiple pathways and processes which generates the complex gene regulatory network that coordinates the tightly coupled events and cellular processes of hindbrain segmentation. Hindbrain segmentation is a dynamic and progressive process, which presents a challenge to understanding when a key regulatory influence is initially activated and when or how it exerts its function. In this section, *cis*-regulation of *Hox* expression will be used as basis for describing the temporal establishment of a hindbrain regulatory network. The regulatory model is based largely on analyses of gene expression and regulation in mouse development, although in light of the high degree of conservation between species, results from chicken, zebrafish, and other model systems are included.

Investigations into the mechanistic basis for regulation of *Hox* expression in rhombomeres have identified a wide range of *cis*-regulatory modules, upstream transcription factors and signaling pathways that underlie the generation of discrete domains of segmental expression (Maconochie *et al.*, 1996; Nolte and Krumlauf, 2006; Parrish *et al.*, 2009). Figure 4.2 summarizes what is known about the *cis*-elements and regulatory modules involved in controlling neural expression for 9 of the 10 *Hox* genes implicated in hindbrain segmentation. There is no information available on

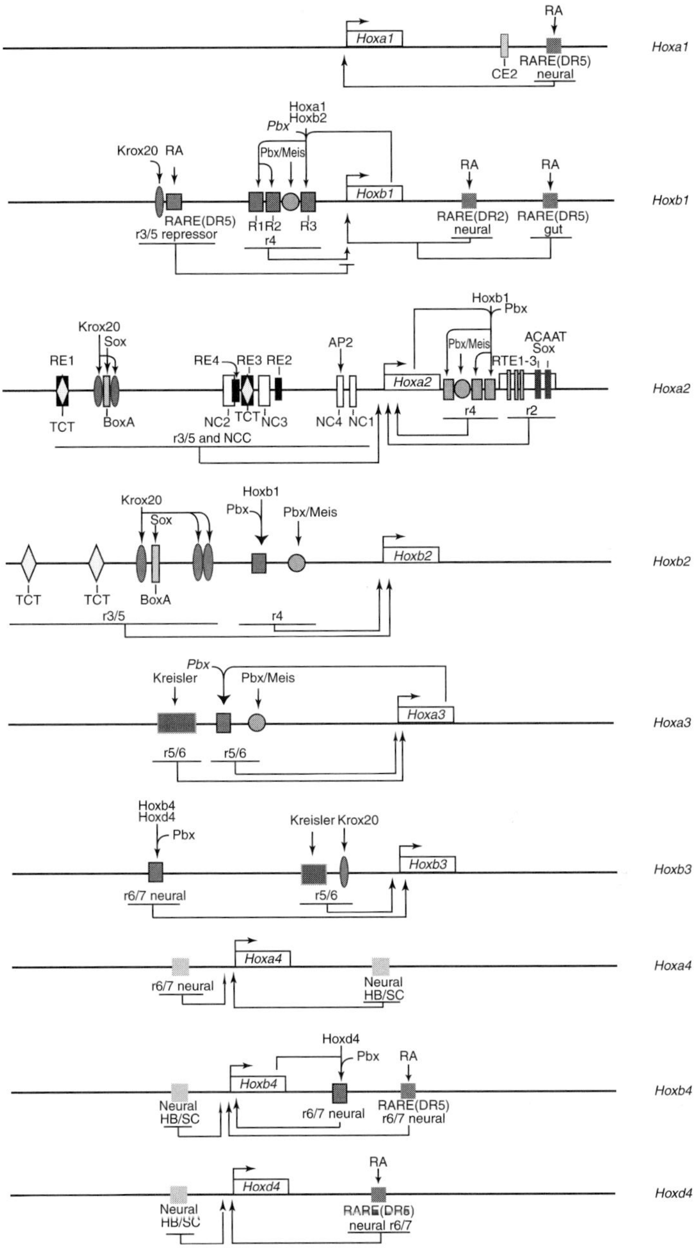

Figure 4.2 Schematic diagrams illustrating the organization of *cis*-regulatory modules of *Hox* genes expressed in hindbrain segmentation. The location and identity of *cis*-elements characterized from regulatory studies are presented on the line for each gene. *Trans*-acting factors and regulatory inputs for specific regulatory elements and modules are listed above each gene, when defined. Regulatory modules and their cognate domains of expression are noted below each gene. References for these modules are provided in the text.

regulation of *Hoxd3* in the hindbrain. The ordered and nested domains of *Hox* expression are not generated by a single control module for each gene. It is evident that the overall pattern of expression for any given gene is the result of the combined activity of many independent modules that control specific subsets of the expression pattern. For example, *Hoxa2* is expressed up to the r1/r2 boundary in the hindbrain, but distinct modules regulate expression in r2, r3 and r5, r4, and neural crest (Fig. 4.2). This modular organization of regulatory elements helps to explain variations in the levels of expression within specific segmental domains and the temporal dynamics of establishing *Hox* expression. Each module is subject to its own regulatory inputs and timing. Hence, distinct gene regulatory networks are formed in each segment and they confer the differential regulation of *Hox* genes. The organization and properties of this *cis*-circuitry can be used to build a picture of these regulatory networks and current progress toward this goal is described below. Analysis has been divided into several stages due to the progressive nature of hindbrain segmentation.

4.1. The early network (7.5–8.0 dpc)

At 7.5 dpc, the mouse hindbrain has not yet developed distinct morphological segments; however, domains of *Hox* expression have begun to be initiated. *Hoxa1* and *Hoxb1* are the earliest *Hox* genes expressed in the hindbrain and they both show an anterior limit of expression at the presumptive (pre-) r3/r4 border (Barrow *et al.*, 2000; Frohman *et al.*, 1990; Murphy and Hill, 1991; Wilkinson *et al.*, 1989b). The expression of both genes is initiated by retinoid signaling (RA), mediated through highly conserved retinoid response elements (RAREs) located in their 3′ regulatory regions (Frasch *et al.*, 1995; Langston and Gudas, 1992; Marshall *et al.*, 1994). The 3′ RARE of *Hoxb1* regulates expression in neural, mesodermal, and primitive streak tissues through RA-mediated binding of RAR/RXR heterodimers to this response element (Marshall *et al.*, 1994; Studer *et al.*, 1998). The RARE 3′ of *Hoxa1* activates expression in the notochord, floor plate, caudal neural tube, and gut epithelium (Frasch *et al.*, 1995). Adjacent to this RARE, a highly conserved region (CE2) has been identified, which is sufficient to regulate early somitic and mesenchymal *Hoxa1* expression (Thompson *et al.*, 1998). The expression of *Cyp26C* genes is induced in the future r1-r3 territory of early anterior hindbrain. These genes encode enzymes that degrade retinoic acid, which sets up opposing activities that restrict induction of *Hoxa1* and *Hoxb1* by RA to the future r3/r4 border (Hernandez *et al.*, 2007; Sirbu *et al.*, 2005).

During early development (8.0 dpc) expression of the zinc-finger transcription factor *Krox20* is activated in the pre-r3 region and expands to reach an anterior limit at the position of the preotic sulcus, the future r2/r3 boundary (Irving *et al.*, 1996; Wilkinson *et al.*, 1989a). Genetic studies have shown that *Krox20* is essential for the establishment and growth of r3 (Giudicelli *et al.*, 2001; Helmbacher *et al.*, 1998; Schneider-Maunoury *et al.*, 1993, 1997). This early expression of *Krox20* may be initiated by Fgf signaling, as it has been shown that Fgfs can ectopically induce *Krox20* expression and reducing levels of Fgfs can inhibit *Krox20* expression in the hindbrain (Marin and Charnay, 2000; Walshe *et al.*, 2002). The early expression domain of *Krox20* is shifted more posteriorly in compound mutants for *Hoxa1* and *Hoxb1*, suggesting that these *Hox* genes have a role in repressing *Krox20* expression in the pre-r4-r7 domain (Barrow *et al.*, 2000). At 8.0 dpc, transcripts of the tyrosine receptor kinase *EphA4* are detectable in the pre-r3 region and this early expression of *EphA4* is directly regulated by *Krox20* (Nieto *et al.*, 1992; Theil *et al.*, 1998). *Nab* genes begin to be expressed in the pre-r3 region. Nab1 and Nab2 repress transcriptional activation mediated by Krox20 and another zinc-finger transcription factor (NGFI-A) (Russo *et al.*, 1995; Svaren *et al.*, 1996). Genetic studies reveal that their expression is also dependent on *Krox20* (Mechta-Grigoriou *et al.*, 2000). Another early hindbrain marker is *follistatin*, which modulates BMP signaling. *Follistatin* is expressed at 8.0 dpc from the pre-r1/r2 boundary through pre-r4, leaving a gap in pre-r3 (Albano *et al.*, 1994). The factors that positively regulate *follistatin* expression are unknown, but it has been shown that *Krox20* downregulates the expression domain in r3 (Seitanidou *et al.*, 1997).

In posterior regions of the pre-r4-r7 region of the hindbrain, the Maf/BZip transcription factor *kreisler* is expressed in the presumptive region of r5 and r6 (Manzanares *et al.*, 1997, 1999a; McKay *et al.*, 1994). *kreisler* expression is initiated at 7.5 dpc in the pre-r5 region and is later expressed in r5 and r6 until 9.0 dpc, when it is quickly downregulated in both rhombomeres. There is evidence that *kreisler* expression also depends on Fgf signaling (Marin and Charnay, 2000). Genetic studies in the mouse and zebrafish have shown that *kreisler/valentino* plays a critical role in regulating the formation and properties of r5 and r6 (Cordes and Barsh, 1994; Manzanares *et al.*, 1999b; Moens *et al.*, 1996; Prince *et al.*, 1998b).

This information can be combined into a gene regulatory network for the hindbrain at 8.0 dpc (Fig. 4.3). It is apparent that key steps in this early phase relate to the activation of *Krox20*, *kreisler*, and *Hoxa1* and *Hoxb1* which set in motion differences that will lead to the formation of r3, r4, r5, and r6. In light of the important roles for *Krox20* and *kreisler* in activating multiple *Hox* genes and controlling other aspects of early segmentation, it will be important to understand the basis of their regulation. Toward this goal, studies in zebrafish and mouse are beginning to unravel some of these early inputs and have found that the *vhnf1*, *pou2*, and *irx* genes are involved in establishing domains of *Krox20* and *kreisler/valentino* expression in the early hindbrain (Choe and Sagerstrom, 2004; Hauptmann *et al.*, 2002;

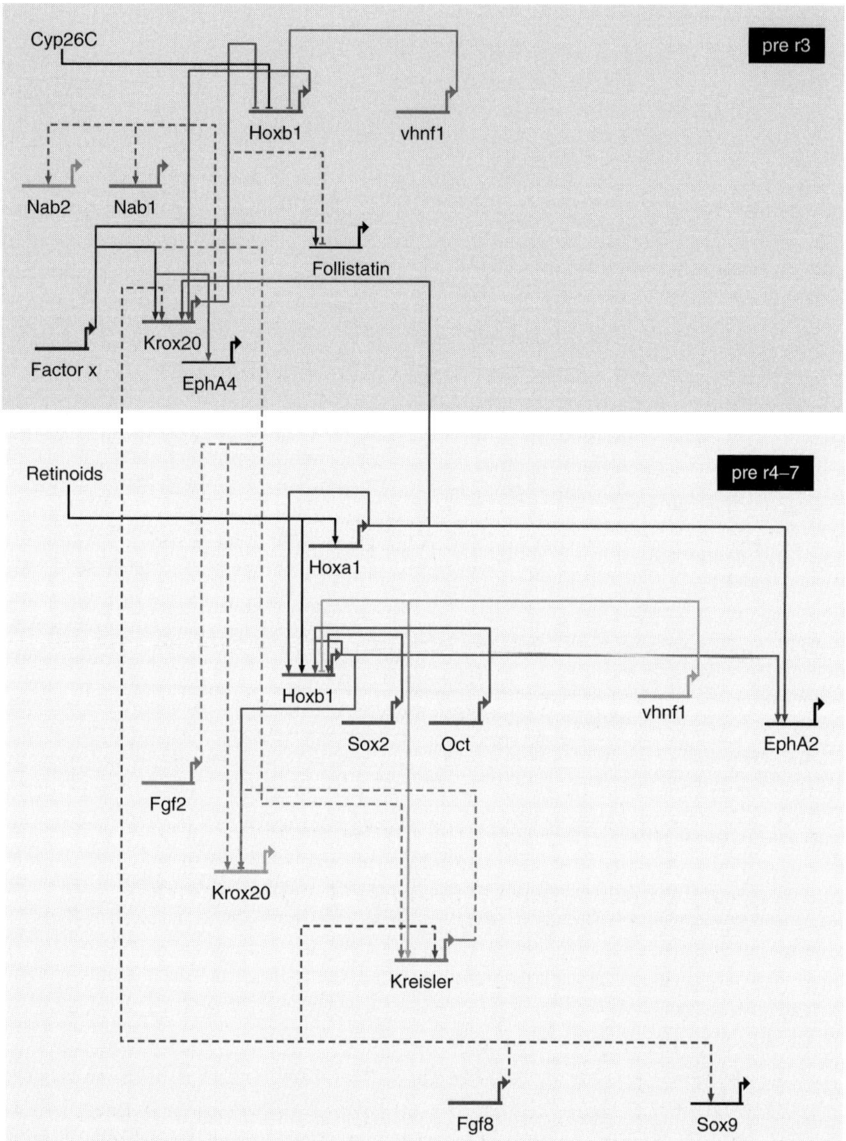

Figure 4.3 Genetic regulatory network in the hindbrain region at 8.0 dpc. At this stage, the hindbrain can be subdivided into two regions, the presumptive rhombomere 3 (pre-r3) regions and presumptive rhombomere 4-7 (pre-r4-r7) region. Solid lines indicate direct and dotted lines indirect interactions between the genes. Arrowheads at the end of a line represent activation, whereas a perpendicular bar at the end of the line illustrates repression between the connected genes. This and subsequent networks were drawn using the Biotapestry utility (Longabaugh *et al.*, 2005).

Hernandez *et al.*, 2004; Kim *et al.*, 2005; Maves and Kimmel, 2005; Sirbu *et al.*, 2005; Stedman *et al.*, 2009; Wiellette and Sive, 2003).

4.2. Intermediate genetic network (8.0–8.5 dpc)

During this time frame, *Hoxa1* expression is lost from the hindbrain in association with a reduction in the levels of retinoid signaling. However, the expression of *Hoxb1* is maintained in the region of r4 and becomes progressively restricted to r4 through the activity of an independent *cis*-element. *Hoxb1* has an r4 enhancer which contains multiple binding sites for Hox proteins and their Pbx, Meis, and Prep cofactors (Ferretti *et al.*, 2005; Pöpperl *et al.*, 1995). The vertebrate *Pbx* genes (*Pbx1-3*) are homologous to the *Drosophila extradenticle* (*exd*) and represent members of the three amino acid extension (TALE) class of homeodomain proteins. Pbx/Exd proteins cooperatively interact with a subset of Hox proteins to bind to bipartite Hox/Pbx sites (Chan *et al.*, 1994; Mann, 1995; Mann and Chan, 1996). This interaction modulates the affinity of Hox proteins for potential binding sites and activates promoters containing bipartite *Hox/Pbx*-responsive elements (Chang *et al.*, 1996). Variations in the central core of consensus binding sequence can modulate affinities for the respective Hox partner (Manzanares *et al.*, 1999b). Meis and Prep also belong to the TALE subfamily of homeodomain proteins (Burglin, 1997). Interaction of Pbx/Exd with these TALE family members modifies both the transcriptional activity and subcellular localization of Hox proteins (Mann and Abu-Shaar, 1996). Heterodimeric complexes of Hox/Pbx and Pbx/Meis bind to separate motifs, allowing the formation of Meis/Pbx/Hox ternary complexes important for gene activation (Ferretti *et al.*, 2000, 2005).

Analysis of the r4 enhancer from *Hoxb1* has revealed that it can be activated by cooperative binding of Hoxb1, Hoxa1, and Hoxb2 with members of the Pbx and Meis family of *Hox* cofactors (Ferretti *et al.*, 2005; Gavalas *et al.*, 2003; Pöpperl *et al.*, 1995; Studer *et al.*, 1998). This highly conserved enhancer also contains a binding site for a Sox2/Oct heterodimers that modulate regulatory activity (Di Rocco *et al.*, 2001). Genetic studies have shown that both *Hoxa1* and *Hoxb1* participate in the initial activation of expression mediated by this *Hoxb1* r4 enhancer, but in later stages, when expression of *Hoxa1* is lost, *Hoxb1* is required to maintain its own expression with input from *Hoxb2* (Gavalas *et al.*, 1998, 2001, 2003; Pattyn *et al.*, 2003; Pöpperl *et al.*, 1995; Studer *et al.*, 1996, 1998). Therefore, this r4 enhancer serves as a *Hox* response element which receives input by auto- (*Hoxb1*), para- (*Hoxa1*), and crossregulatory (*Hoxb2*) interactions.

An RARE element, located 5′ of the coding region of *Hoxb1* (Fig. 4.2), functions as part of an r3/r5 repressor module to help restrict expression of *Hoxb1* to r4 (Studer *et al.*, 1994). Mutation of the RARE motif in this repressor module results in the expansion of *Hoxb1* expression into r3 and r5

(Studer *et al.*, 1994). Krox20 also binds to the repressor module and is important for inhibition of *Hoxb1*. *Hoxa1* lacks this combination of an autoregulatory and repressor modules; hence, it is not maintained in r4 following its early activation by retinoids. However, placing a copy of the r4 module from *Hoxb1* upstream of *Hoxa1* in the endogenous mouse locus is sufficient to generate *Hoxa1* expression that can substitute for *Hoxb1* function in hindbrain development (Tvrdik and Capecchi, 2006).

The expression of *Hox* genes from paralogous group 2 and group 3 is activated between 8.0 and 8.5 dpc. Regulatory studies on *Hoxa2* and *Hoxb2* have found that expression of these genes in r4 is mediated by enhancers that also function as *Hox* response elements (Maconochie *et al.*, 1997; Tümpel *et al.*, 2006, 2007). The r4 regulatory module of *Hoxb2* contains conserved bipartite binding sites for Hox/Pbx and Pbx/Meis heterodimers necessary for its expression and transactivation experiments have shown that these sites mediate a response to expression *Hoxb1* (Ferretti *et al.*, 2000; Jacobs *et al.*, 1999; Maconochie *et al.*, 1997). Therefore, in r4 *Hoxb2* is a direct target of *Hoxb1*, and genetic studies have revealed in turn, *Hoxb2* itself feeds back to help maintain expression of *Hoxb1* in r4 (Davenne *et al.*, 1999; Gavalas *et al.*, 2003; Pattyn *et al.*, 2003). Similarly, *Hoxa2* contains an r4 enhancer in its intron (Fig. 4.2), which is comprised of multiple Hox/Pbx and Pbx/Meis binding motifs necessary for regulatory activity (Tümpel *et al.*, 2006, 2007). This module mediates a response to expression of *Hoxb1* and *Hoxa2* itself, suggesting that *Hoxa2* is initially activated in r4 as a direct target of *Hoxb1* and then maintained in an autoregulatory manner by its own expression (Tümpel *et al.*, 2007).

Additional downstream targets of *Hoxb1* at this stage have also been identified. *Gata3* is a zinc-finger transcription factor whose expression is downregulated in *Hoxb1* mutants (Pata *et al.*, 1999). The transcription factor *Phox2b* and the receptor tyrosine kinase *EphA2* are directly regulated by *Hoxb1* (Chen and Ruley, 1998; Samad *et al.*, 2004) and a direct interaction between *Hoxa1* and the regulatory region of *Epha2* has also been described (Chen and Ruley, 1998). In zebrafish, the *prickle* gene which encodes a modulator of the noncanonical Wnt pathway is dependent upon *Hoxb1a* activity (Rohrschneider *et al.*, 2007).

These experiments illustrate the pivotal role *Hoxb1* plays in regulating segmental identity in r4. Genetic analyses in mouse have demonstrated that *Hoxa1* and *Hoxb1* synergize in establishing *Hoxb1* expression in r4 and then *Hoxb1* is required to maintain the identity of r4 (Goddard *et al.*, 1996; Studer *et al.*, 1996, 1998). In the absence of *Hoxb1*, r4 adopts an r2 character. Furthermore, consistent with the regulatory relationships *Hoxa1*, *Hoxb1*, *Hoxa2*, and *Hoxb2*, all participate in regulating aspects of patterning in r4 and its derivatives (Barrow *et al.*, 2000; Davenne *et al.*, 1999; Gavalas *et al.*, 1998, 2001, 2003; Rossel and Capecchi, 1999). Figure 4.4 presents a gene regulatory network focused on the activation

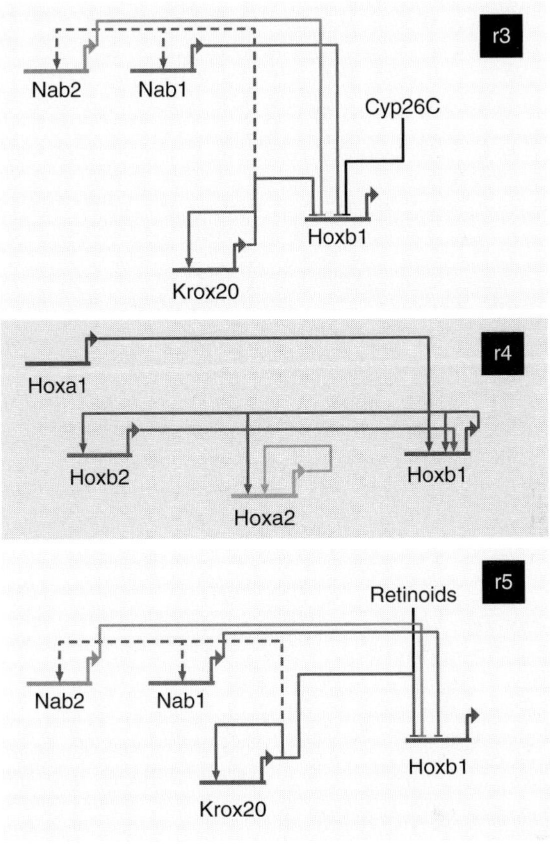

Figure 4.4 A genetic regulatory network for *Hoxb1* in r3, r4, and r5 at 8.75 dpc. At this stage, rhombomeres are becoming morphologically distinct territories with their own regulatory interactions. This regulatory network is focused on *Hoxb1* to illustrate inputs that activate and repress its expression in r3, r4, and r5 and show its input into regulation of *Hoxa2* and *Hoxb2* in r4. Solid lines indicate direct and dotted lines indirect interactions between the genes. Arrowheads at the end of a line represent activation, whereas a perpendicular bar at the end of the line illustrates repression between the connected genes.

of *Hoxb1* and its crossregulatory interactions with group 2 *Hox* genes in r3, r4, and r5 segments at 8.75 dpc. This network highlights the important and diverse roles played by auto-, para-, and crossregulatory interactions between *Hox* genes in the developing hindbrain.

At approximately 8.25 dpc, *Krox20* expression is observed in pre-r5. The domain of *Krox20* expression in pre-r3 and pre-r5 has irregular borders, with some of the cells expressing *Krox20* observed in the presumptive even-numbered rhombomeres. The expression patterns sharpen by 8.5 dpc and become restricted to r3 and r5 (Irving *et al.*, 1996). The expression of

Krox20 in r5 is dependent on the basic domain-leucine zipper transcription factor *kreisler*, since *Krox20* expression is seen in r3, but not in r5 of *kreisler* mutants (Manzanares *et al.*, 1999b). Regulatory studies have demonstrated that the expression of *Hoxa2* and *Hoxb2* in r3 and r5 is directly controlled by *Krox20* (Maconochie *et al.*, 1999, 2001; Nonchev *et al.*, 1996a,b; Sham *et al.*, 1993; Vesque *et al.*, 1996). In addition to binding sites for *Krox20* in the r3/r5 enhancers, there is evidence that many other conserved elements work in concert with *Krox20* sites to potentiate activity of the r3/r5 module. For example, in *Hoxa2*, four conserved elements (RE1-RE4) and a BoxA motif (Sox site) are important for expression in r3 and r5 (Fig. 4.2). Several of these associated motifs, particularly the BoxA (Sox) site, have been identified in the *Krox20*-dependent control modules of the *Hoxb2* and *EphA4* genes (Theil *et al.*, 1998; Vesque *et al.*, 1996). This suggests that these elements may form part of the core architecture of *Krox20* regulatory modules on other target genes and may aid in the identification of more *in vivo* targets (Tümpel *et al.*, 2002). There is evidence that *Hoxb1* and possibly *Hoxa2* and *Hoxb2* can bind to an r3 regulatory module in the *Krox20* gene and form feedback loops between these genes that maintains their expression (Wassef *et al.*, 2008).

Hoxa2 is the only *Hox* gene expressed in r2 and genetic studies have shown that it plays an important role regulating the identity of this segment (Gavalas *et al.*, 1997; Oury *et al.*, 2006; Ren *et al.*, 2002). The r2 domain of *Hoxa2* expression is activated between 8.5 and 9.5 dpc, and transgenic analyses have identified an r2 enhancer that mediates this expression (Frasch *et al.*, 1995; Tümpel *et al.*, 2006, 2008). Intriguingly, this r2 regulatory module is embedded in the second coding exon of *Hoxa2*, and contains two binding sites for Sox proteins which have been shown to be essential for enhancer activity (Tümpel *et al.*, 2006, 2008). Three other motifs (Fig. 4.2, RE1-3) also cooperate with the Sox sites to direct the r2 expression of *Hoxa2*. While the factors interacting with these sites are unknown, Hoxa2 itself is not required for the activity of this enhancer.

With respect to expression of members of the third paralogous group, *Hoxa3*, *Hoxb3*, and *Hoxd3*, there is evidence that segmental expression of *Hoxa3* and *Hoxb3* in r5 and r6 is directly initiated by *kreisler* (Manzanares *et al.*, 1997, 2001, 2002). Nothing is yet known about the regulation of *Hoxd3* expression in r5 and r6. The r5/r6 enhancer in the *Hoxb3* locus contains conserved binding sites for Kreisler and there is evidence that Krox20 also cooperates with Kreisler to potentiate regulatory activity of this module (Manzanares *et al.*, 1997, 2002). The *Hoxa3* also contains Kreisler binding sites and is directly upregulated in r5 and r6 in a *Kreisler*-dependent manner (Manzanares *et al.*, 1999a, 2001). *Kreisler* expression in r5 and r6 is very transient (Cordes and Barsh, 1994; Manzanares *et al.*, 1997, 2001; McKay *et al.*, 1994; Moens *et al.*, 1996) so there must be a means of maintaining expression of its targets in the hindbrain. Analysis of *Hoxb3* and

Hoxa3 in mouse and chick embryos has shown that expression of *Hoxb3* regresses into more posterior regions following initial activation by *Kreisler*, but that *Hoxa3* expression is maintained in r5 and r6 (Manzanares *et al.*, 2001). The maintenance of Hoxa3 in r5/r6 is mediated by a conserved auto- and crossregulatory region, dependent on Hoxa3 and Hoxb3 and Pbx and Prep/Meis cofactors (Manzanares *et al.*, 2001). This represents another example of a key role for auto- and crossregulation between *Hox* genes in regulating segmental identity.

4.3. Late genetic network (8.5–9.5 dpc)

During this period the group 4 genes, *Hoxa4*, *Hoxb4*, and *Hoxd4*, are activated and have an anterior limit of expression at the r6/r7 border (Fig. 4.1). The boundary of *Hoxc4* expression has a more posterior limit (Geada *et al.*, 1992). All of these genes respond to RA, and a direct response has been demonstrated for *Hoxb4* and *Hoxd4* (Gould *et al.*, 1998; Moroni *et al.*, 1993; Nolte *et al.*, 2003; Pöpperl and Featherstone, 1993; Whiting *et al.*, 1991). In *Hoxb4*, a 3′ RARE is required for regulating early neural expression up to the r6/r7 boundary (Gould *et al.*, 1998). This enhancer is activated retinoids initially generated by somites that later diffuse to the neural tube (Gould *et al.*, 1998; Linville *et al.*, 2004; Sirbu *et al.*, 2005). A conserved RARE has also been described 5′ of Hoxd4 and in important for potentiating neural regulation of *Hoxd4* (Moroni *et al.*, 1993; Nolte *et al.*, 2003; Pöpperl and Featherstone, 1993). Neural enhancers have also been found flanking the *Hoxa4* genes but although they respond to RA the basis of their activity has not been established (Morrison *et al.*, 1997; Packer *et al.*, 1998).

Following initial activation of the group 4 *Hox* genes, expression begins to regress in the most anterior regions of the hindbrain due to the transient nature of the inducing signal or factor. However, expression of *Hoxb4* is maintained at the r6/r7 boundary (Gould *et al.*, 1997, 1998). In a manner analogous to *Hoxb1* and *Hoxa3*, the maintenance of this segmental pattern of *Hoxb4* expression has been found to be mediated by auto- and cross-regulatory input from Hox proteins and their Pbx and Meis/Prep cofactors (Gould *et al.*, 1997, 1998). In the *Hoxb4* locus, there is separate neural enhancer (Fig. 4.2) that functions as a Hox response element to maintain expression at the r6/r7 boundary in later stages (9.5 dpc). Hoxb4 itself is able to bind to this element, and genetically it has been shown that activity depends upon both *Hoxb4* and *Hoxd4* (Serpente *et al.*, 2005). There is also evidence for a positive feedback loop between *Hoxb4* and retinoid signaling. The *RARβ* gene contains a *Hox* response element that binds Hoxb4, Hoxd4, and their Pbx and Meis/Prep cofactors (Serpente *et al.*, 2005). Therefore, RA signaling triggers neural expression of *Hoxb4* and *Hoxd4* which in turn upregulates expression of the RARβ receptor.

At 9.5 dpc, there are clear morphological boundaries and the seven rhombomeric segments have formed distinct domains. Many of the early regulatory interactions and inputs that establish the sharp domains of segmental expression are lost or have been stabilized by components that maintain expression domains. For example, *Krox20* expression is downregulated in r3 and then in r5 so that it is no longer expressed in the hindbrain (Irving *et al.*, 1996). In association with this change, expression of *EphA4* and the *Nab* corepressors, direct targets of Krox20, also begins to be downregulated (Kanzler *et al.*, 1998; Mechta-Grigoriou *et al.*, 2000; Theil *et al.*, 1998). Therefore, 9.5 dpc is a relevant stage for building a gene regulatory network that integrates the culminations of regulatory inputs that govern establishment of segmental *Hox* expression that forms the basis of their control of regional identities. Figure 4.5 provides a working model for this *Hox* regulatory network.

It is clear that there are many gaps to fill with respect to identifying missing components and clarifying direct and indirect interactions. Assumptions have also been made in extrapolating results from one species to another, and there are likely to be species-specific aspects to the core network. Nonetheless, several interesting features arise from this model. Feedforward and feedback loops are frequently utilized to modulate and enhance pathways of transcriptional activation. This is illustrated by the *cis*-circuitry that mediates direct crosstalk between *Hoxa1*, *Hoxb1*, *Hoxb2*, and *Hoxa2*. Such interactions complicate understanding phenotypes generated by a mutation of one *Hox* gene, because it can indirectly influence the expression of many other *Hox* genes. In *Drosophila*, transient early inputs into *Hox* genes from the segmentation pathway are maintained by regulation of chromatin, via Polycomb and trithorax. There is evidence for auto- and crossregulation but this is frequently negative. In vertebrates, auto-, para-, and crossregulatory interactions appear to play a frequent role in serving to maintain segment-restricted domains of *Hox* expression in the hindbrain.

Many genes have multiple inputs and functions. *Krox20* represses *Hoxb1* expression in r4 but activates expression of *Hoxa2* and *Hoxb2* in r3/r5 and *Hoxa3* in r5. Such differences are in part due to modulators like the Nab or PIASxβ factors that can switch Krox20 from an activator to a repressor. *Krox20* activates *Hox* genes to control regional identity but it also regulates *EphA4* as part of regulation of cell sorting. Similarly, Hox proteins can be involved in the early activation of factors like Kreisler and Krox20 and have later roles in regulating segmental identity. They also can directly regulate members of the Eph and ephrin family, to participate in cell sorting. Different Hox proteins can activate or repress activity from the same Hox response element. To distinguish between context-dependent functions of the transcriptional regulators, it will be important to build a more complete understanding of the cofactors and components that gate their distinct regulatory activities during development. The network shows that there is

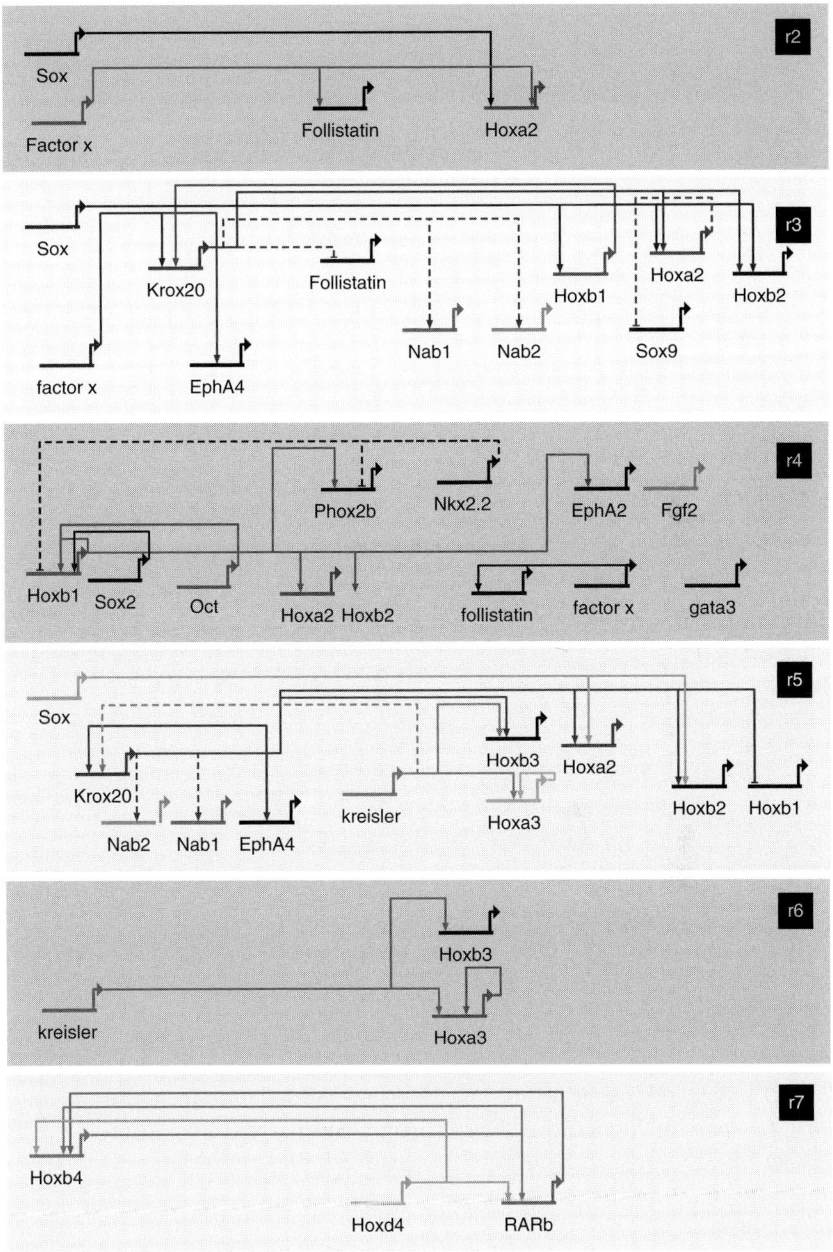

Figure 4.5 Genetic regulatory network in the hindbrain at 9.5 dpc. At this stage, the hindbrain can be subdivided into seven rhombomeres and segment-restricted expression of *Hox* genes is fully established. Solid lines indicate direct and dotted lines indirect interactions between the genes. Arrowheads at the end of a line represent activation, whereas a perpendicular bar at the end of the line illustrates repression between the connected genes.

not a strict regulatory hierarchy or segregation in the roles of *Hox* genes and their regulators during hindbrain segmentation. The complex overlaps and multiple inputs may serve to generate a robust system that reproducibly generates segmental compartments and specifies regional identity of this key region of the CNS. With the improvement of technologies for measuring global changes in gene expression, it should be possible to quantify how changes in specific components alter the network. This will lead to a more accurate model of the functional interactions within the network.

Finally, the regulation of *Hox* genes in other tissues, such as mesoderm and neural crest cells, are mediated by separate regulatory modules, some of which cooperate with hindbrain modules. For example, *Hoxa2* and *Hoxb2* are expressed in r4-derived neural crest cells migrating into the second branchial arch (Fig. 4.1B), but their expression is governed by different mechanisms. *Hoxb2* expression in neural crest cells is regulated by cross-regulation of *Hoxb1* and Pbx/Prep cofactors. *Hoxa2* expression in neural crest is dependent on a number of different regulatory elements (Fig. 4.2, NC1-NC4), one of which has been shown to be bound by the transcription factor AP2 (Maconochie *et al.*, 1997, 1999). In addition, rhombomeres signal to their adjacent environments and influence how cells derived from rhombomeres respond to signals in the branchial environment. Some of these signals and responses may be under the regulation of *Hox* genes in the hindbrain (Gavalas *et al.*, 2001). Therefore, it will be valuable to build an understanding of gene regulatory networks and their interactions for the all of the tissues that play a role in head development.

ACKNOWLEDGMENTS

REK, LMW, and SWT were supported by the Stowers Institute for Medical Research and SWT was also supported by a fellowship from Boehringer Ingelheim Funds. SWT was registered for the Degree of Doctor of Philosophy with the Open University (UK).

REFERENCES

Albano, R. M., Arkell, R., Beddington, R. S., and Smith, J. C. (1994). Expression of inhibin subunits and follistatin during postimplantation mouse development: Decidual expression of activin and expression of follistatin in primitive streak, somites and hindbrain. *Development* **120,** 803–813.

Amores, A., Force, A., Yan, Y. L., Joly, L., Amemiya, C., Fritz, A., Ho, R., Langeland, J., Prince, V., Wang, Y. L., Westerfield, M., Ekker, M., *et al.* (1998). Zebrafish *hox* clusters and vertebrate genome evolution. *Science* **282,** 1711–1714.

Amores, A., Suzuki, T., Yan, Y. L., Pomeroy, J., Singer, A., Amemiya, C., and Postlethwait, J. H. (2004). Developmental roles of pufferfish Hox clusters and genome evolution in ray-fin fish. *Genome Res.* **14,** 1–10.

Amoyel, M., Cheng, Y. C., Jiang, Y. J., and Wilkinson, D. G. (2005). Wnt1 regulates neurogenesis and mediates lateral inhibition of boundary cell specification in the zebrafish hindbrain. *Development* **132,** 775–785.

Aparicio, S., Hawker, K., Cottage, A., Mikawa, Y., Zuo, L., Chen, E., Krumlauf, R., and Brenner, S. (1997). Organisation of the *Fugu rubripes Hox* clusters, evidence for continuing evolution of vertebrate *Hox* complexes. *Nat. Genet.* **16,** 79–84.

Barrow, J. R., Stadler, H. S., and Capecchi, M. R. (2000). Roles of Hoxa1 and Hoxa2 in patterning the early hindbrain of the mouse. *Development* **127,** 933–944.

Birgbauer, E., and Fraser, S. E. (1994). Violation of cell lineage restriction compartments in the chick hindbrain. *Development* **120,** 1347–1356.

Birgbauer, E., Sechrist, J., Bronner-Fraser, M., and Fraser, S. (1995). Rhombomeric origin and rostrocaudal reassortment of neural crest cells revealed by intravital microscopy. *Development* **121,** 935–945.

Briscoe, J., and Wilkinson, D. G. (2004). Establishing neuronal circuitry: Hox genes make the connection. *Genes Dev.* **18,** 1643–1648.

Burglin, T. R. (1997). Analysis of TALE superclass homeobox genes (MEIS, PBC, KNOX, Iroquois, TGIF) reveals a novel domain conserved between plants and animals. *Nucleic Acids Res.* **25,** 4173–4180.

Carroll, S. B. (1995). Homeotic genes and the evolution of arthropods and chordates. *Nature* **376,** 479–485.

Chan, S. K., Jaffe, L., Capovilla, M., Botas, J., and Mann, R. S. (1994). The DNA binding specificity of Ultrabithorax is modulated by cooperative interactions with extradenticle, another homeoprotein. *Cell* **78,** 603–615.

Chang, C. P., Brocchieri, L., Shen, W. F., Largman, C., and Cleary, M. L. (1996). Pbx modulation of Hox homeodomain amino-terminal arms establishes different DNA-binding specificities across the *Hox* locus. *Mol. Cell. Biol.* **16,** 1734–1745.

Chatonnet, F., Dominguez del Toro, E., Thoby-Brisson, M., Champagnat, J., Fortin, G., Rijli, F. M., and Thaeron-Antono, C. (2003). From hindbrain segmentation to breathing after birth: Developmental patterning in rhombomeres 3 and 4. *Mol. Neurobiol.* **28,** 277–294.

Chatonnet, F., Wrobel, L. J., Mezieres, V., Pasqualetti, M., Ducret, S., Taillebourg, E., Charnay, P., Rijli, F. M., and Champagnat, J. (2007). Distinct roles of Hoxa2 and Krox20 in the development of rhythmic neural networks controlling inspiratory depth, respiratory frequency, and jaw opening. *Neural Dev.* **2,** 19.

Chen, J., and Ruley, H. (1998). An enhancer element in the *EphA2 (Eck)* gene sufficient for rhombomere-specific expression is activated by *Hoxa1* and *Hoxb1* homeobox proteins. *J. Biol. Chem.* **273,** 24670–24675.

Cheng, Y. C., Amoyel, M., Qiu, X., Jiang, Y. J., Xu, Q., and Wilkinson, D. G. (2004). Notch activation regulates the segregation and differentiation of rhombomere boundary cells in the zebrafish hindbrain. *Dev. Cell* **6,** 539–550.

Choe, S. K., and Sagerstrom, C. G. (2004). Paralog group 1 hox genes regulate rhombomere 5/6 expression of vhnf1, a repressor of rostral hindbrain fates, in a meis-dependent manner. *Dev. Biol.* **271,** 350–361.

Clarke, J. D., and Lumsden, A. (1993). Segmental repetition of neuronal phenotype sets in the chick embryo hindbrain. *Development* **118,** 151–162.

Clarke, J. D. W., Erskine, L., and Lumsden, A. (1998). Differential progenitor dispersal and the spatial origin of early neurons can explain the predominance of single-phenotype clones in the chick hindbrain. *Dev. Dyn.* **212,** 14–26.

Cordes, S. P., and Barsh, G. S. (1994). The mouse segmentation gene *kr* encodes a novel basic domain-leucine zipper transcription factor. *Cell* **79,** 1025–1034.

Dasen, J. S., Liu, J. P., and Jessell, T. M. (2003). Motor neuron columnar fate imposed by sequential phases of Hox-c activity. *Nature* **425,** 926–933.

Dasen, J. S., Tice, B. C., Brenner-Morton, S., and Jessell, T. M. (2005). A Hox regulatory network establishes motor neuron pool identity and target-muscle connectivity. *Cell* **123,** 477–491.

Dasen, J. S., De Camilli, A., Wang, B., Tucker, P. W., and Jessell, T. M. (2008). Hox repertoires for motor neuron diversity and connectivity gated by a single accessory factor, FoxP1. *Cell* **134,** 304–316.

Davenne, M., Maconochie, M. K., Neun, R., Pattyn, A., Chambon, P., Krumlauf, R., and Rijli, F. M. (1999). *Hoxa2* and *Hoxb2* control dorsoventral patterns of neuronal development in the rostral hindbrain. *Neuron* **22,** 677–691.

Davis, A., Scemama, J. L., and Stellwag, E. J. (2008). Japanese medaka Hox paralog group 2: Insights into the evolution of Hox PG2 gene composition and expression in the Osteichthyes. *J. Exp. Zool. B* **310,** 623–641.

del Toro, E. D., Borday, V., Davenne, M., Neun, R., Rijli, F. M., and Champagnat, J. (2001). Generation of a novel functional neuronal circuit in Hoxa1 mutant mice. *J. Neurosci.* **21,** 5637–5642.

Deschamps, J., and Wijgerde, M. (1993). Two phases in the establishment of *Hox* expression domains. *Dev. Biol.* **156,** 473–480.

Di Rocco, G., Gavalas, A., Popperl, H., Krumlauf, R., Mavilio, F., and Zappavigna, V. (2001). The recruitment of SOX/OCT complexes and the differential activity of HOXA1 and HOXB1 modulate the Hoxb1 auto-regulatory enhancer function. *J. Biol. Chem.* **1,** 1.

Dollé, P., Izpisùa-Belmonte, J. C., Falkenstein, H., Renucci, A., and Duboule, D. (1989). Co-ordinate expression of the murine *Hox-5* complex homeobox-containing genes during limb pattern formation. *Nature* **342,** 767–772.

Dollé, P., Izpisùa-Belmonte, J. C., Brown, J. M., Tickle, C., and Duboule, D. (1991). *Hox-4* genes and the morphogenesis of mammalian genitalia. *Genes Dev.* **5,** 1767–1776.

Duboule, D., and Dollé, P. (1989). The structural and functional organization of the murine *HOX* gene family resembles that of *Drosophila* homeotic genes. *EMBO J.* **8,** 1497–1505.

Ellies, D. L., Church, V., Francis-West, P., and Lumsden, A. (2000). The WNT antagonist cSFRP2 modulates programmed cell death in the developing hindbrain. *Development* **127,** 5285–5295.

Farlie, P. G., Kerr, R., Thomas, P., Symes, T., Minichiello, J., Hearn, C. J., and Newgreen, D. (1999). A paraxial exclusion zone creates patterned cranial neural crest cell outgrowth adjacent to rhombomeres 3 and 5. *Dev. Biol.* **213,** 70–84.

Ferretti, E., Marshall, H., Pöpperl, H., Maconochie, M., Krumlauf, R., and Blasi, F. (2000). Segmental expression of Hoxb2 in r4 requires two separate sites that integrate cooperative interactions between Prep1, Pbx and Hox proteins. *Development* **127,** 155–166.

Ferretti, E., Cambronero, F., Tümpel, S., Longobardi, E., Wiedemann, L. M., Blasi, F., and Krumlauf, R. (2005). The *Hoxb1* enhancer and control of rhombomere 4 expression: Complex interplay between PREP1-PBX1-HOXB1 binding sites. *Mol. Cell. Biol.* **25,** 8541–8552.

Flanagan, J. G., and Vanderhaeghen, P. (1998). The ephrins and Eph receptors in neural development. *Annu. Rev. Neurobiol.* **21,** 309–345.

Frasch, M., Chen, X., and Lufkin, T. (1995). Evolutionary-conserved enhancers direct region-specific expression of the murine *Hoxa-1* and *Hoxa-2* loci in both mice and *Drosophila. Development* **121,** 957–974.

Fraser, S., Keynes, R., and Lumsden, A. (1990). Segmentation in the chick embryo hindbrain is defined by cell lineage restrictions. *Nature* **344,** 431–435.

Frohman, M. A., Boyle, M., and Martin, G. R. (1990). Isolation of the mouse *Hox-2.9* gene: Analysis of embryonic expression suggests that positional information along the anterior-posterior axis is specified by mesoderm. *Development* **110,** 589–607.

Gale, N. W., Holland, S. J., Valenzuela, D. M., Flenniken, A., Pan, L., Ryan, T. E., Henkemeyer, M., Strebhardt, K., Hirai, H., Wilkinson, D. G., Pawson, T., Davis, S., *et al.* (1996). Eph receptors and ligands comprise two major specificity subclasses and are reciprocally compartmentalized during embryogenesis. *Neuron* **17,** 9–19.

Garcia Martinez, V., Macias, D., Ganan, Y., Garcia Lobo, J. M., Francia, M. V., Fernandez Teran, M. A., and Hurle, J. M. (1993). Internucleosomal DNA fragmentation and programmed cell death (apoptosis) in the interdigital tissue of the embryonic chick leg bud. *J. Cell Sci.* **106,** 201–208.

Gaunt, S. J. (1991). Expression patterns of mouse Hox genes: Clues to an understanding of developmental and evolutionary strategies. *Bioessays* **13,** 505–513.

Gavalas, A., Davenne, M., Lumsden, A., Chambon, P., and Rijli, F. M. (1997). Role of Hoxa-2 in axon pathfinding and rostral hindbrain patterning. *Development* **124,** 3693–3702.

Gavalas, A., Studer, M., Lumsden, A., Rijli, F. M., Krumlauf, R., and Chambon, P. (1998). *Hoxa1* and *Hoxb1* synergize in patterning the hindbrain, cranial nerves and second pharyngeal arch. *Development* **125,** 1123–1136.

Gavalas, A., Trainor, P., Ariza-McNaughton, L., and Krumlauf, R. (2001). Synergy between Hoxa1 and Hoxb1: The relationship between arch patterning and the generation of cranial neural crest. *Development* **128,** 3017–3027.

Gavalas, A., Ruhrberg, C., Livet, J., Henderson, C. E., and Krumlauf, R. (2003). Neuronal defects in the hindbrain of Hoxa1, Hoxb1 and Hoxb2 mutants reflect regulatory interactions among these Hox genes. *Development* **130,** 5663–5679.

Geada, A. M. C., Gaunt, S. J., Azzawi, M., Shimeld, S. M., Pearce, J., and Sharpe, P. T. (1992). Sequence and embryonic expression of the murine Hox-3.5 gene. *Development* **116,** 497–506.

Geisen, M. J., Di Meglio, T., Pasqualetti, M., Ducret, S., Brunet, J. F., Chedotal, A., and Rijli, F. M. (2008). Hox paralog group 2 genes control the migration of mouse pontine neurons through slit-robo signaling. *PLoS Biol.* **6,** e142.

Giudicelli, F., Taillebourg, E., Charnay, P., and Gilardi-Hebenstreit, P. (2001). Krox-20 patterns the hindbrain through both cell-autonomous and non cell-autonomous mechanisms. *Genes Dev.* **15,** 567–580.

Glover, J. C., Renaud, J. S., and Rijli, F. M. (2006). Retinoic acid and hindbrain patterning. *J. Neurobiol.* **66,** 705–725.

Goddard, J., Rossel, M., Manley, N., and Capecchi, M. (1996). Mice with targeted disruption of *Hoxb1* fail to form the motor nucleus of the VIIth nerve. *Development* **122,** 3217–3228.

Golding, J. P., Dixon, M., and Gassmann, M. (2002). Cues from neuroepithelium and surface ectoderm maintain neural crest-free regions within cranial mesenchyme of the developing chick. *Development* **129,** 1095–1105.

Golding, J. P., Sobieszczuk, D., Dixon, M., Coles, E., Christiansen, J., Wilkinson, D., and Gassmann, M. (2004). Roles of erbB4, rhombomere-specific, and rhombomere-independent cues in maintaining neural crest-free zones in the embryonic head. *Dev. Biol.* **266,** 361–372.

Gould, A., Morrison, A., Sproat, G., White, R. A., and Krumlauf, R. (1997). Positive cross-regulation and enhancer sharing: Two mechanisms for specifying overlapping *Hox* expression patterns. *Genes Dev.* **11,** 900–913.

Gould, A., Itasaki, N., and Krumlauf, R. (1998). Initiation of rhombomeric *Hoxb4* expression requires induction by somites and a retinoid pathway. *Neuron* **21,** 39–51.

Graham, A., Papalopulu, N., and Krumlauf, R. (1989). The murine and *Drosophila* homeobox gene complexes have common features of organization and expression. *Cell* **57,** 367–378.

Graham, A., Maden, M., and Krumlauf, R. (1991). The murine Hox-2 genes display dynamic dorsoventral patterns of expression during central nervous system development. *Development* **112,** 255–264.

Graham, A., Heyman, I., and Lumsden, A. (1993). Even-numbered rhombomeres control the apoptotic elimination of neural crest cells from odd-numbered rhombomeres in the chick hindbrain. *Development* **119,** 233–245.

Graham, A., Francis-West, P., Brickell, P., and Lumsden, A. (1994). The signalling molecule BMP4 mediates apoptosis in the rhombencephalic neural crest. *Nature* **372,** 684–686.

Grapin-Botton, A., Bonnin, M. A., McNaughton, L. A., Krumlauf, R., and Le Douarin, N. M. (1995). Plasticity of transposed rhombomeres: Hox gene induction is correlated with phenotypic modifications. *Development* **121,** 2707–2721.

Guthrie, S., and Lumsden, A. (1991). Formation and regeneration of rhombomere boundaries in the developing chick hindbrain. *Development* **112,** 221–229.

Guthrie, S., Prince, V., and Lumsden, A. (1993). Selective dispersal of avian rhombomere cells in orthotopic and heterotopic grafts. *Development* **118,** 527–538.

Hauptmann, G., Belting, H. G., Wolke, U., Lunde, K., Soll, I., Abdelilah-Seyfried, S., Prince, V., and Driever, W. (2002). spiel ohne grenzen/pou2 is required for zebrafish hindbrain segmentation. *Development* **129,** 1645–1655.

Helmbacher, F., Pujades, C., Desmarquet, C., Frain, M., Rijli, F. M., Chambon, P., and Charnay, P. (1998). *Hoxa1* and *Krox20* synergize to control the development of rhombomere 3. *Development* **125,** 4739–4748.

Hernandez, R. E., Rikhof, H. A., Bachmann, R., and Moens, C. B. (2004). vhnf1 integrates global RA patterning and local FGF signals to direct posterior hindbrain development in zebrafish. *Development* **131,** 4511–4520.

Hernandez, R. E., Putzke, A. P., Myers, J. P., Margaretha, L., and Moens, C. B. (2007). Cyp26 enzymes generate the retinoic acid response pattern necessary for hindbrain development. *Development* **134,** 177–187.

Hoegg, S., and Meyer, A. (2005). Hox clusters as models for vertebrate genome evolution. *Trends Genet.* **21,** 421–424.

Holder, N., and Klein, R. (1999). Eph receptors and ephrins: Effectors of morphogenesis. *Development* **126,** 2033–2044.

Hunt, P., and Krumlauf, R. (1991). Deciphering the Hox code: Clues to patterning the branchial region of the head. *Cell* **66,** 1075–1078.

Hunt, P., Gulisano, M., Cook, M., Sham, M. H., Faiella, A., Wilkinson, D., Boncinelli, E., and Krumlauf, R. (1991a). A distinct Hox code for the branchial region of the vertebrate head. *Nature* **353,** 861–864.

Hunt, P., Wilkinson, D., and Krumlauf, R. (1991b). Patterning the vertebrate head: Murine Hox 2 genes mark distinct subpopulations of premigratory and migrating neural crest. *Development* **112,** 43–51.

Irving, C., Nieto, M., Das Gupta, R., Charnay, P., and Wilkinson, D. (1996). Progressive spatial restriction of Sek1 and Krox20 gene expression during hindbrain segmentation. *Dev. Biol.* **173,** 26–38.

Jacobs, Y., Schnabel, C. A., and Cleary, M. L. (1999). Trimeric association of Hox and TALE homeodomain proteins mediates *Hoxb2* hindbrain enhancer activity. *Mol. Cell. Biol.* **19,** 5134–5142.

Kanzler, B., Kuschert, S. J., Liu, Y. H., and Mallo, M. (1998). *Hoxa2* restricts the chondrogenic domain and inhibits bone formation during development of the branchial area. *Development* **125,** 2587–2597.

Kappen, C., Schughart, K., and Ruddle, F. H. (1989). Two steps in the evolution of antennapedia-class vertebrate homeobox genes. *Proc. Natl. Acad. Sci. USA* **86,** 5459–5463.

Kessel, M., and Gruss, P. (1991). Homeotic transformations of murine prevertebrae and concommitant alteration of Hox codes induced by retinoic acid. *Cell* **67,** 89–104.

Keynes, R., and Krumlauf, R. (1994). Hox genes and regionalization of the nervous system. *Annu. Rev. Neurosci.* **17,** 109–132.

Keynes, R., and Lumsden, A. (1990). Segmentation and the origins of regional diversity in the vertebrate central nervous system. *Neuron* **4,** 1–9.

Kiecker, C., and Lumsden, A. (2005). Compartments and their boundaries in vertebrate brain development. *Nat. Rev.* **6,** 553–564.

Kim, F. A., Sing, I. A., Kaneko, T., Bieman, M., Stallwood, N., Sadl, V. S., and Cordes, S. P. (2005). The vHNF1 homeodomain protein establishes early rhombomere identity by direct regulation of Kreisler expression. *Mech. Dev.* **122,** 1300–1309.

Knecht, A. K., and Bronner-Fraser, M. (2002). Induction of the neural crest: A multigene process. *Nat. Rev. Genet.* **3,** 453–461.

Kontges, G., and Lumsden, A. (1996). Rhombencephalic neural crest segmentation is preserved throughout craniofacial ontogeny. *Development* **122,** 3229–3242.

Krumlauf, R. (1992). Evolution of the vertebrate Hox homeobox genes. *Bioessays* **14,** 245–252.

Krumlauf, R. (1994). *Hox* genes in vertebrate development. *Cell* **78,** 191–201.

Kulesa, P. M., and Fraser, S. E. (1998). Segmentation of the vertebrate hindbrain: A time-lapse analysis. *Int. J. Dev. Biol.* **42,** 385–392.

Kulesa, P. M., and Fraser, S. E. (2000). In ovo time-lapse analysis of chick hindbrain neural crest cell migration shows cell interactions during migration to the branchial arches. *Development* **127,** 1161–1172.

Kulesa, P., Bronner-Fraser, M., and Fraser, S. (2000). In ovo time-lapse analysis after dorsal neural tube ablation shows rerouting of chick hindbrain neural crest. *Development* **127,** 2843–2852.

Kulesa, P. M., Teddy, J. M., Stark, D. A., Smith, S. E., and McLennan, R. (2008). Neural crest invasion is a spatially-ordered progression into the head with higher cell proliferation at the migratory front as revealed by the photoactivatable protein, KikGR. *Dev. Biol.* **316,** 275–287.

Langston, A. W., and Gudas, L. J. (1992). Identification of a retinoic acid responsive enhancer 3′ of the murine homeobox gene *Hox-1.6. Mech. Dev.* **38,** 217–228.

Lawrence, P. A., and Struhl, G. (1996). Morphogens, compartments, and pattern: Lessons from Drosophila? *Cell* **85,** 951–961.

Le Douarin, N. M., and Kalcheim, C. (1999). "The Neural Crest." Developmental and Cell Biology Series 36. (J.B.L. Bard, P.W. Barlow, and D.L. Kirk, Eds.), 2nd ed., Cambridge University Press, Cambridge, UK; New York, USA.

Le Pabic, P., Stellwag, E. J., Brothers, S. N., and Scemama, J. L. (2007). Comparative analysis of Hox paralog group 2 gene expression during Nile tilapia (*Oreochromis niloticus*) embryonic development. *Dev. Genes Evol.* **217,** 749–758.

Lewis, E. B. (1978). A gene complex controlling segmentation in Drosophila. *Nature* **276,** 565–570.

Linville, A., Gumusaneli, E., Chandraratna, R. A., and Schilling, T. F. (2004). Independent roles for retinoic acid in segmentation and neuronal differentiation in the zebrafish hindbrain. *Dev. Biol.* **270,** 186–199.

Liu, J. P., Laufer, E., and Jessell, T. M. (2001). Assigning the positional identity of spinal motor neurons. Rostrocaudal patterning of Hox-c expression by FGFs, Gdf11, and retinoids. *Neuron* **32,** 997–1012.

Longabaugh, W. J., Davidson, E. H., and Bolouri, H. (2005). Computational representation of developmental genetic regulatory networks. *Dev. Biol.* **283,** 1–16.

Lumsden, A. (2004). Segmentation and compartition in the early avian hindbrain. *Mech. Dev.* **121,** 1081–1088.

Lumsden, A., and Krumlauf, R. (1996). Patterning the vertebrate neuraxis. *Science* **274,** 1109–1115.

Lumsden, A., Sprawson, N., and Graham, A. (1991). Segmental origin and migration of neural crest cells in the hindbrain region of the chick embryo. *Development* **113,** 1281–1291.

Lumsden, A., Clarke, J., Keynes, R., and Fraser, S. E. (1994). Early phenotypic choices by neuronal precursors, revealed by clonal analysis of the chick embryo hindbrain. *Development* **120,** 1581–1589.

Maconochie, M., Nonchev, S., Morrison, A., and Krumlauf, R. (1996). Paralogous *Hox* genes: Function and regulation. *Annu. Rev. Genet.* **30,** 529–556.

Maconochie, M. K., Nonchev, S., Studer, M., Chan, S. K., Popperl, H., Sham, M. H., Mann, R. S., and Krumlauf, R. (1997). Cross-regulation in the mouse *HoxB* complex: The expression of *Hoxb2* in rhombomere 4 is regulated by *Hoxb1*. *Genes Dev.* **11,** 1885–1896.

Maconochie, M., Krishnamurthy, R., Nonchev, S., Meier, P., Manzanares, M., Mitchell, P. J., and Krumlauf, R. (1999). Regulation of *Hoxa2* in cranial neural crest cells involves members of the *AP-2* family. *Development* **126,** 1483–1494.

Maconochie, M. K., Nonchev, S., Manzanares, M., Marshall, H., and Krumlauf, R. (2001). Differences in Krox20-dependent regulation of *Hoxa2* and *Hoxb2* during hindbrain development. *Dev. Biol.* **233,** 468–481.

Mann, R. S. (1995). The specificity of homeotic gene function. *Bioessays* **17,** 855–863.

Mann, R., and Abu-Shaar, M. (1996). Nuclear import of the homeodomain protein extradenticle in response to Wg and Dpp signalling. *Nature* **383,** 630–633.

Mann, R., and Chan, S. K. (1996). Extra specificity from extradenticle: The partnership between HOX and PBX/EXD homeodomain proteins. *Trends Genet.* **12,** 258–262.

Manzanares, M., Cordes, S., Kwan, C. T., Sham, M. H., Barsh, G., and Krumlauf, R. (1997). Segmental regulation of *Hoxb3* by kreisler. *Nature* **387,** 191–195.

Manzanares, M., Cordes, S., Ariza-McNaughton, L., Sadl, V., Maruthainar, K., Barsh, G., and Krumlauf, R. (1999a). Conserved and distinct roles of kreisler in regulation of the paralogous Hoxa3 and Hoxb3 genes. *Development* **126,** 759–769.

Manzanares, M., Trainor, P. A., Nonchev, S., Ariza-McNaughton, L., Brodie, J., Gould, A., Marshall, H., Morrison, A., Kwan, C. T., Sham, M. H., Wilkinson, D. G., and Krumlauf, R. (1999b). The role of *kreisler* in segmentation during hindbrain development. *Dev. Biol.* **211,** 220–237.

Manzanares, M., Bel-Vialer, S., Ariza-McNaughton, L., Ferretti, E., Marshall, H., Maconochie, M. K., Blasi, F., and Krumlauf, R. (2001). Independent regulation of initiation and maintenance phases of *Hoxa3* expression in the vertebrate hindbrain involves auto and cross-regulatory mechanisms. *Development* **128,** 3595–3607.

Manzanares, M., Nardelli, J., Gilardi-Hebenstreit, P., Marshall, H., Giudicelli, F., Martinez-Pastor, M. T., Krumlauf, R., and Charnay, P. (2002). Krox20 and kreisler co-operate in the transcriptional control of segmental expression of Hoxb3 in the developing hindbrain. *EMBO J.* **21,** 365–376.

Marin, F., and Charnay, P. (2000). Hindbrain patterning: FGFs regulate Krox20 and mafB/kr expression in the otic/preotic region. *Development* **127,** 4925–4935.

Marshall, H., Studer, M., Pöpperl, H., Aparicio, S., Kuroiwa, A., Brenner, S., and Krumlauf, R. (1994). A conserved retinoic acid response element required for early expression of the homeobox gene *Hoxb-1*. *Nature* **370,** 567–571.

Matis, C., Oury, F., Remacle, S., Lampe, X., Gofflot, F., Picard, J. J., Rijli, F. M., and Rezsohazy, R. (2007). Identification of Lmo1 as part of a Hox-dependent regulatory network for hindbrain patterning. *Dev. Dyn.* **236,** 2675–2684.

Maves, L., and Kimmel, C. B. (2005). Dynamic and sequential patterning of the zebrafish posterior hindbrain by retinoic acid. *Dev. Biol.* **285,** 593–605.

Maves, L., Jackman, W., and Kimmel, C. B. (2002). FGF3 and FGF8 mediate a rhombomere 4 signaling activity in the zebrafish hindbrain. *Development* **129,** 3825–3837.

McGinnis, W., and Krumlauf, R. (1992). Homeobox genes and axial patterning. *Cell* **68,** 283–302.

McKay, I. J., Muchamore, I., Krumlauf, R., Maden, M., Lumsden, A., and Lewis, J. (1994). The kreisler mouse: A hindbrain segmentation mutant that lacks two rhombomeres. *Development* **120,** 2199–2211.

Mechta-Grigoriou, F., Garel, S., and Charnay, P. (2000). Nab proteins mediate a negative feedback loop controlling Krox-20 activity in the developing hindbrain. *Development* **127,** 119–128.

Mellitzer, G., Xu, Q., and Wilkinson, D. (1999). Eph receptors and ephrins restrict cell intermingling and communication. *Nature* **400,** 77–81.

Mellitzer, G., Xu, Q., and Wilkinson, D. G. (2000). Control of cell behaviour by signalling through Eph receptors and ephrins. *Curr. Opin. Neurobiol.* **10,** 400–408.

Moens, C. B., and Prince, V. E. (2002). Constructing the hindbrain: Insights from the zebrafish. *Dev. Dyn.* **224,** 1–17.

Moens, C. B., Yan, Y. L., Appel, B., Force, A. G., and Kimmel, C. B. (1996). Valentino: A zebrafish gene required for normal hindbrain segmentation. *Development* **122,** 3981–3990.

Moroni, M., Vigano, M., and Mavilio, F. (1993). Regulation of the human *HOXD4* gene by retinoids. *Mech. Dev.* **44,** 139–154.

Morrison, A., Ariza-McNaughton, L., Gould, A., Featherstone, M., and Krumlauf, R. (1997). *HOXD4* and regulation of the group 4 paralog genes. *Development* **124,** 3135–3146.

Murphy, P., and Hill, R. E. (1991). Expression of the mouse labial-like homeobox-containing genes, Hox 2.9 and Hox 1.6, during segmentation of the hindbrain. *Development* **111,** 61–74.

Murphy, P., Davidson, D. R., and Hill, R. E. (1989). Segment-specific expression of a homeobox-containing gene in the mouse hindbrain. *Nature* **341,** 156–159.

Nieto, M. A., Gilardi-Hebenstreit, P., Charnay, P., and Wilkinson, D. G. (1992). A receptor protein tyrosine kinase implicated in the segmental patterning of the hindbrain and mesoderm. *Development* **116,** 1137–1150.

Nolte, C., and Krumlauf, R. (2006). Expression of Hox genes in the nervous system of vertebrates. *In* "HOX Gene Expression" (S. Papageorgiou, Ed.), pp. 14–41. Landes Bioscience and Springer, Austin, TX.

Nolte, C., Amores, A., Nagy Kovacs, E., Postlethwait, J., and Featherstone, M. (2003). The role of a retinoic acid response element in establishing the anterior neural expression border of Hoxd4 transgenes. *Mech. Dev.* **120,** 325–335.

Nonchev, S., Maconochie, M., Vesque, C., Aparicio, S., Ariza-McNaughton, L., Manzanares, M., Maruthainar, K., Kuroiwa, A., Brenner, S., Charnay, P., and Krumlauf, R. (1996a). The conserved role of *Krox-20* in directing *Hox* gene expression during vertebrate hindbrain segmentation. *Proc. Natl. Acad. Sci. USA* **93,** 9339–9345.

Nonchev, S., Vesque, C., Maconochie, M., Seitanidou, T., Ariza-McNaughton, L., Frain, M., Marshall, H., Sham, M. H., Krumlauf, R., and Charnay, P. (1996b). Segmental expression of Hoxa-2 in the hindbrain is directly regulated by Krox-20. *Development* **122,** 543–554.

Oury, F., Murakami, Y., Renaud, J. S., Pasqualetti, M., Charnay, P., Ren, S. Y., and Rijli, F. M. (2006). Hoxa2- and rhombomere-dependent development of the mouse facial somatosensory map. *Science* **313,** 1408–1413.

Packer, A. I., Crotty, D. A., Elwell, V. A., and Wolgemuth, D. J. (1998). Expression of the murine Hoxa4 gene requires both autoregulation and a conserved retinoic acid response element. *Development* **125,** 1991–1998.

Panopoulou, G., and Poustka, A. J. (2005). Timing and mechanism of ancient vertebrate genome duplications—The adventure of a hypothesis. *Trends Genet.* **21,** 559–567.

Parrish, M., Nolte, C., and Krumlauf, R. (2009). Hox genes expression. *In* "Encyclopedia of Neuroscience" (L.R. Squire, Ed.), Vol. 4, pp. 1221–1231. Academic Press, Oxford, UK.

Pasqualetti, M., Diaz, C., Renaud, J. S., Rijli, F. M., and Glover, J. C. (2007). Fate-mapping the mammalian hindbrain: Segmental origins of vestibular projection neurons assessed using rhombomere-specific Hoxa2 enhancer elements in the mouse embryo. *J. Neurosci.* **27,** 9670–9681.

Pata, I., Studer, M., van Doorninck, J. H., Briscoe, J., Kuuse, S., Engel, J. D., Grosveld, F., and Karis, A. (1999). The transcription factor GATA3 is a downstream effector of Hoxb1 specification in rhombomere 4. *Development* **126,** 5523–5531.

Pattyn, A., Vallstedt, A., Dias, J. M., Samad, O. A., Krumlauf, R., Rijli, F. M., Brunet, J. F., and Ericson, J. (2003). Coordinated temporal and spatial control of motor neuron and serotonergic neuron generation from a common pool of CNS progenitors. *Genes Dev.* **17,** 729–737.

Popovici, C., Leveugle, M., Birnbaum, D., and Coulier, F. (2001). Coparalogy: Physical and functional clusterings in the human genome. *Biochem. Biophys. Res. Commun.* **288,** 362–370.

Pöpperl, H., and Featherstone, M. (1993). Identification of a retinoic acid response element upstream of the murine *Hox-4.2* gene. *Mol. Cell. Biol.* **13,** 257–265.

Pöpperl, H., Bienz, M., Studer, M., Chan, S., Aparicio, S., Brenner, S., Mann, R., and Krumlauf, R. (1995). Segmental expression of *Hoxb1* is controlled by a highly conserved autoregulatory loop dependent upon *exd/Pbx*. *Cell* **81,** 1031–1042.

Prince, V., and Lumsden, A. (1994). Hox-a2 expression in normal and transposed rhombomeres: Independent regulation in the neural tube and neural crest. *Development* **120,** 911–923.

Prince, V. E., Joly, L., Ekker, M., and Ho, R. K. (1998a). Zebrafish hox genes: Genomic organization and modified colinear expression patterns in the trunk. *Development* **125,** 407–420.

Prince, V. E., Moens, C. B., Kimmel, C. B., and Ho, R. K. (1998b). Zebrafish *hox* genes: Expression in the hindbrain region of wild-type and mutants of the segmentation gene, *valentino*. *Development* **125,** 393–406.

Ren, S. Y., Angrand, P. O., and Rijli, F. M. (2002). Targeted insertion results in a Rhombomere 2-specific Hoxa2 knockdown and ectopic activation of Hoxa1 expression. *Dev. Dyn.* **225,** 305–315.

Rohrschneider, M. R., Elsen, G. E., and Prince, V. E. (2007). Zebrafish Hoxb1a regulates multiple downstream genes including prickle1b. *Dev. Biol.* **309,** 358–372.

Rossel, M., and Capecchi, M. R. (1999). Mice mutant for both *Hoxa1* and *Hoxb1* show extensive remodeling of the hindbrain and defects in craniofacial development. *Development* **126,** 5027–5040.

Russo, M. W., Sevetson, B. R., and Milbrandt, J. (1995). Identification of NAB1, a repressor of NGFI-A- and Krox20-mediated transcription. *Proc. Natl. Acad. Sci. USA* **92,** 6873–6877.

Samad, O. A., Geisen, M. J., Caronia, G., Varlet, I., Zappavigna, V., Ericson, J., Goridis, C., and Rijli, F. M. (2004). Integration of anteroposterior and dorsoventral regulation of Phox2b transcription in cranial motoneuron progenitors by homeodomain proteins. *Development* **131,** 4071–4083.

Scemama, J. L., Hunter, M., McCallum, J., Prince, V., and Stellwag, E. (2002). Evolutionary divergence of vertebrate Hoxb2 expression patterns and transcriptional regulatory loci. *J. Exp. Zool.* **294,** 285–299.

Scemama, J. L., Vernon, J. L., and Stellwag, E. J. (2006). Differential expression of hoxa2a and hoxa2b genes during striped bass embryonic development. *Gene Expr. Patterns* **6,** 843–848.

Schilling, T. (2001). Plasticity of zebrafish Hox expression in the hindbrain and cranial neural crest hindbrain. *Dev. Biol.* **231,** 201–216.

Schilling, T. F., and Kimmel, C. B. (1994). Segment and cell type lineage restrictions during pharyngeal arch development in the zebrafish embryo. *Development* **120,** 483–494.

Schneider-Maunoury, S., Topilko, P., Seitandou, T., Levi, G., Cohen-Tannoudji, M., Pournin, S., Babinet, C., and Charnay, P. (1993). Disruption of *Krox-20* results in alteration of rhombomeres 3 and 5 in the developing hindbrain. *Cell* **75,** 1199–1214.

Schneider-Maunoury, S., Seitanidou, T., Charnay, P., and Lumsden, A. (1997). Segmental and neuronal architecture of the hindbrain of *Krox-20* mouse mutants. *Development* **124,** 1215–1226.

Sechrist, J., Serbedzija, G. N., Scherson, T., Fraser, S. E., and Bronner-Fraser, M. (1993). Segmental migration of the hindbrain neural crest does not arise from its segmental generation. *Development* **118**(3), 691–703.

Seitanidou, T., Schneider-Maunoury, S., Desmarquet, C., Wilkinson, D., and Charnay, P. (1997). *Krox20* is a key regulator of rhombomere-specific gene expression in the developing hindbrain. *Mech. Dev.* **65,** 31–42.

Sela-Donenfeld, D., and Wilkinson, D. G. (2005). Eph receptors: Two ways to sharpen boundaries. *Curr. Biol.* **15,** R210-R212.

Serpente, P., Tümpel, S., Ghyselinck, N. B., Niederreither, K., Wiedemann, L. M., Dollé, P., Chambon, P., Krumlauf, R., and Gould, A. P. (2005). Direct crossregulation between retinoic acid receptor *β* and Hox genes during hindbrain segmentation. *Development* **132,** 503–513.

Sham, M. H., Vesque, C., Nonchev, S., Marshall, H., Frain, M., Das Gupta, R. D., Whiting, J., Wilkinson, D., Charnay, P., and Krumlauf, R. (1993). The zinc finger gene *Krox-20* regulates *Hoxb-2 (Hox2.8)* during hindbrain segmentation. *Cell* **72,** 183–196.

Simon, H., and Lumsden, A. (1993). Rhombomere-specific origin of the contralateral vestibulo-acoustic efferent neurons and their migration across the embryonic midline. *Neuron* **11,** 209–220.

Sirbu, I. O., Gresh, L., Barra, J., and Duester, G. (2005). Shifting boundaries of retinoic acid activity control hindbrain segmental gene expression. *Development* **132,** 2611–2622.

Stedman, A., Lecaudey, V., Havis, E., Anselme, I., Wassef, M., Gilardi-Hebenstreit, P., and Schneider-Maunoury, S. (2009). A functional interaction between Irx and Meis patterns the anterior hindbrain and activates krox20 expression in rhombomere 3. *Dev. Biol.* **327,** 566–577.

Studer, M., Pöpperl, H., Marshall, H., Kuroiwa, A., and Krumlauf, R. (1994). Role of a conserved retinoic acid response element in rhombomere restriction of *Hoxb-1. Science* **265,** 1728–1732.

Studer, M., Lumsden, A., Ariza-McNaughton, L., Bradley, A., and Krumlauf, R. (1996). Altered segmental identity and abnormal migration of motor neurons in mice lacking *Hoxb-1. Nature* **384,** 630–635.

Studer, M., Gavalas, A., Marshall, H., Ariza-McNaughton, L., Rijli, F., Chambon, P., and Krumlauf, R. (1998). Genetic interaction between *Hoxa1* and *Hoxb1* reveal new roles in regulation of early hindbrain patterning. *Development* **125,** 1025–1036.

Sundin, O., and Eichele, G. (1990). A homeo domain protein reveals the metameric nature of the developing chick hindbrain. *Genes Dev.* **4,** 1267–1276.

Svaren, J., Sevetson, B. R., Apel, E. D., Zimonjic, D. B., Popescu, N. C., and Milbrandt, J. (1996). NAB2, a corepressor of NGFI-A (Egr-1) and Krox20, is induced by proliferative and differentiative stimuli. *Mol. Cell. Biol.* **16,** 3545–3553.

Teddy, J. M., and Kulesa, P. M. (2004). *In vivo* evidence for short- and long-range cell communication in cranial neural crest cells. *Development* **131,** 6141–6151.

Theil, T., Frain, M., Gilardi-Hebenstreit, P., Flenniken, A., Charnay, P., and Wilkinson, D. G. (1998). Segmental expression of the EphA4 (Sek-1) receptor tyrosine kinase in the hindbrain is under direct transcriptional control of Krox-20. *Development* **125,** 443–452.

Thompson, J. R., Chen, S. W., Ho, L., Langston, A. W., and Gudas, L. J. (1998). An evolutionary conserved element is essential for somite and adjacent mesenchymal expression of the Hoxa1 gene. *Dev. Dyn.* **211,** 97–108.

Trainor, P., and Krumlauf, R. (2000a). Plasticity in mouse neural crest cells reveals a new patterning role for cranial mesoderm. *Nat. Cell Biol.* **2,** 96–102.

Trainor, P. A., and Krumlauf, R. (2000b). Patterning the cranial neural crest: Hindbrain segmentation and *Hox* gene plasticity. *Nat. Rev. Neurosci.* **1,** 116–124.

Trainor, P. A., and Krumlauf, R. (2001). Hox genes, neural crest cells and branchial arch patterning. *Curr. Opin. Cell Biol.* **13,** 698–705.

Trainor, P. A., Sobieszczuk, D., Wilkinson, D., and Krumlauf, R. (2002). Signalling between the hindbrain and paraxial tissues dictates neural crest migration pathways. *Development* **129,** 433–442.

Tümpel, S., Maconochie, M., Wiedemann, L. M., and Krumlauf, R. (2002). Conservation and diversity in the *cis*-regulatory networks that integrate information controlling expression of Hoxa2 in hindbrain and cranial neural crest cells in vertebrates. *Dev. Biol.* **246,** 45–56.

Tümpel, S., Cambronero, F., Wiedemann, L. M., and Krumlauf, R. (2006). Evolution of *cis* elements in the differential expression of two *Hoxa2* coparalogous genes in pufferfish (*Takifugu rubripes*). *Proc. Natl. Acad. Sci. USA* **103,** 5419–5424.

Tümpel, S., Cambronero, F., Ferretti, E., Blasi, F., Wiedemann, L. M., and Krumlauf, R. (2007). Expression of Hoxa2 in rhombomere 4 is regulated by a conserved cross-regulatory mechanism dependent upon Hoxb1. *Dev. Biol.* **302,** 646–660.

Tümpel, S., Cambronero, F., Sims, C., Krumlauf, R., and Wiedemann, L. M. (2008). A regulatory module embedded in the coding region of Hoxa2 controls expression in rhombomere 2. *Proc. Natl. Acad. Sci. USA* **105,** 20077–20082.

Tvrdik, P., and Capecchi, M. R. (2006). Reversal of hox1 gene subfunctionalization in the mouse. *Dev. Cell* **11,** 239–250.

Vandepoele, K., De Vos, W., Taylor, J. S., Meyer, A., and Van de Peer, Y. (2004). Major events in the genome evolution of vertebrates: Paranome age and size differ considerably between ray-finned fishes and land vertebrates. *Proc. Natl. Acad. Sci. USA* **101,** 1638–1643.

Vesque, C., Maconochie, M., Nonchev, S., Ariza-McNaughton, L., Kuroiwa, A., Charnay, P., and Krumlauf, R. (1996). Hoxb-2 transcriptional activation in rhombomeres 3 and 5 requires an evolutionarily conserved *cis*-acting element in addition to the Krox-20 binding site. *EMBO J.* **15,** 5383–5896.

Walshe, J., Maroon, H., McGonnell, I. M., Dickson, C., and Mason, I. (2002). Establishment of hindbrain segmental identity requires signaling by FGF3 and FGF8. *Curr. Biol.* **12,** 1117–1123.

Wassef, M. A., Chomette, D., Pouilhe, M., Stedman, A., Havis, E., Desmarquet-Trin Dinh, C., Schneider-Maunoury, S., Gilardi-Hebenstreit, P., Charnay, P., and Ghislain, J. (2008). Rostral hindbrain patterning involves the direct activation of a Krox20 transcriptional enhancer by Hox/Pbx and Meis factors. *Development* **135,** 3369–3378.

Whiting, J., Marshall, H., Cook, M., Krumlauf, R., Rigby, P. W. J., Stott, D., and Allemann, R. K. (1991). Multiple spatially specific enhancers are required to reconstruct the pattern of *Hox-2.6* gene expression. *Genes Dev.* **5,** 2048–2059.

Wiellette, E. L., and Sive, H. (2003). vhnf1 and Fgf signals synergize to specify rhombomere identity in the zebrafish hindbrain. *Development* **130,** 3821–3829.

Wilkinson, D. G. (2001). Multiple roles of EPH receptors and ephrins in neural development. *Nat. Rev.* **2,** 155–164.

Wilkinson, D. G. (2003). How attraction turns to repulsion. *Nat. Cell Biol.* **5,** 851–853.

Wilkinson, D. G., Bhatt, S., Chavrier, P., Bravo, R., and Charnay, P. (1989a). Segment-specific expression of a zinc finger gene in the developing nervous system of the mouse. *Nature* **337,** 461–464.

Wilkinson, D. G., Bhatt, S., Cook, M., Boncinelli, E., and Krumlauf, R. (1989b). Segmental expression of Hox-2 homeobox-containing genes in the developing mouse hindbrain. *Nature* **341,** 405–409.

Wintgate, R., and Lumsden, A. (1996). Persistence of rhombomeric organisation in the postsegmental avian hindbrain. *Development* **122,** 2143–2152.

Wizenmann, A., and Lumsden, A. (1997). Segregation of rhombomeres by differential chemoaffinity. *Mol. Cell. Neurosci.* **9,** 448–459.

Wolfe, K. H. (2001). Yesterday's polyploids and the mystery of diploidization. *Nat. Rev. Genet.* **2,** 333–341.

Xu, Q., Alldus, G., Holder, N., and Wilkinson, D. G. (1995). Expression of truncated *Sek-1* receptor tyrosine kinase disrupts the segmental restriction of gene expression in the *Xenopus* and zebrafish hindbrain. *Development* **121,** 4005–4016.

Xu, Q., Mellitzer, G., Robinson, V., and Wilkinson, D. G. (1999). *In vivo* cell sorting in complementary segmental domains mediated by *Eph* receptors and *ephrins*. *Nature* **399,** 267–271.

Xu, Q., Mellitzer, G., and Wilkinson, D. G. (2000). Roles of Eph receptors and ephrins in segmental patterning. *Philos. Trans. R. Soc. Lond. B Biol. Sci.* **355,** 993–1002.

Zacchetti, G., Duboule, D., and Zakany, J. (2007). Hox gene function in vertebrate gut morphogenesis: The case of the caecum. *Development* **134,** 3967–3973.

Zakany, J., and Duboule, D. (2007). The role of Hox genes during vertebrate limb development. *Curr. Opin. Genet. Dev.* **17,** 359–366.

CHAPTER FIVE

Hox Genes in Neural Patterning and Circuit Formation in the Mouse Hindbrain

Yuichi Narita*,† *and* Filippo M. Rijli*,†

Contents

Abstract

The mammalian hindbrain is the seat of regulation of several vital functions that involve many of the organ systems of the body. Such functions are controlled through the activity of intricate arrays of neuronal circuits and connections. The establishment of ordered patterns of neuronal specification, migration, and axonal topographic connectivity during development is crucial to build such a complex network of circuits and functional connectivity in the mature hindbrain. The early development of the vertebrate hindbrain proceeds according to a fundamental metameric partitioning along the anteroposterior axis into cellular

* Friedrich Miescher Institute for Biomedical Research, Basel, Switzerland
† Institut de Génétique et de Biologie Moléculaire et Cellulaire, Illkirch Cedex, C.U. de Strasbourg, France

Current Topics in Developmental Biology, Volume 88
ISSN 0070-2153, DOI: 10.1016/S0070-2153(09)88005-8

compartments known as rhombomeres. Such an organization has been highly conserved in vertebrate evolution and has a fundamental impact on the hindbrain adult structure, nuclear organization, and connectivity. Here, we review the cellular and molecular mechanisms underlying hindbrain neuronal circuitry in the mouse, with a specific focus on the role of the homeodomain transcription factors of the *Hox* gene family. The *Hox* genes are crucial determinants of rhombomere segmental identity and anteroposterior patterning. However, recent findings suggest that, in addition to their well-known roles at early embryonic stages, the *Hox* genes may play important roles also in later aspect of neuronal circuit development, including stereotypic neuronal migration, axon pathfinding, and topographic mapping of connectivity.

1. Introduction

During the development of the central nervous system (CNS), a large variety of neurons are generated at appropriate times and locations along the main axes of the developing brain. An ordered pattern of neuronal specification, migration, and axonal topographic mapping is crucial to achieve the extraordinary complexity of functional connections in the mature brain. The possibility that at least part of the CNS is patterned from a reiterated set of repeated units has received much attention in the last two decades or so with the discovery that the development of the vertebrate hindbrain proceeds according to a fundamental metameric cellular organization which has been highly conserved in vertebrate evolution (Fraser *et al.*, 1990; Hanneman *et al.*, 1988; Lumsden, 1990; Lumsden and Keynes, 1989; Metcalfe *et al.*, 1986). This review focuses on the cellular and molecular mechanisms underlying hindbrain neuronal circuitry development, with a specific focus on the role of the homeodomain-containing transcription factors of the *Hox* gene family in the mouse.

2. Basic Anatomical Background and Cellular Mechanisms of Hindbrain Development

The mammalian hindbrain, or rhombencephalon, is composed of the cerebellum, pons, and medulla, and it is the site of regulation of several vital functions through a complex system of reflex arcs and neuronal networks. The list of hindbrain nuclei encompasses many important neuronal structures including, for instance, the cranial nerves that innervate the muscles of the head and neck and relay somatosensory inputs from the face and information on hearing, balance and taste, as well as control the cardiovascular and gastrointestinal systems. Through the hindbrain also navigate the major sensory tracts ascending from the spinal cord to thalamic and cortical areas, and the motor pathways that descend from the cortex to the hindbrain and spinal cord. In addition, the hindbrain contains relay centers that control respiration and blood pressure, as well as centers that mediate arousal and wakefulness, which are part of the reticular formation. Finally, the hindbrain is also the major source of noradrenergic and serotonergic neuron inputs to most parts of the brain.

How such a complex circuitry is generated? Although the anatomical organization of the adult vertebrate hindbrain is sufficiently well defined, relatively little is still known about the ontogeny of hindbrain nuclei and neuronal circuits. In the last years, significant progress has been made in identifying the cellular mechanisms of early hindbrain development. At early embryonic stages, the hindbrain is segmented along the antero-posterior (AP) axis into 7-8 compartments, depending on the species, known as rhombomeres (r) (Kiecker and Lumsden, 2005). They contain clonally related populations showing limited intermixing between neighboring segments (Birgbauer and Fraser, 1994), suggesting that restricting the position of a neural progenitor during the segmentation stage may be a mechanism to determine its future neuronal fate and connectivity pattern in the mature brain. Rhombomeres form in turn segmentally iterated, discrete, cell groups displaying heterochronic patterns of neurogenesis (e.g., Carpenter *et al.*, 1993; Clarke and Lumsden, 1993; Lumsden and Keynes, 1989), which will give rise to motor or sensory nuclei, depending on their position along the dorsoventral (DV) axis. Namely, sensory neurons are generated in the alar plate and form columns spanning several rhombomeres along the AP axis, which include the trigeminal, vestibular, and the viscerosensory, gustatory, and solitary nuclei (e.g., Gaufo *et al.*, 2004; Glover, 2003; Oury *et al.*, 2006; Pasqualetti *et al.*, 2007; Sieber *et al.*, 2007). In addition, the rhombic lip (Wingate and Hatten, 1999), a stripe of neuroepithelium in the dorsal alar plate running the entire AP length of the hindbrain, generates the cochlear complex from r2 to r5 (Farago *et al.*, 2006). In the posterior r6-r8 hindbrain, the rhombic lip gives rise instead to specific sets of long-distance migrating neurons contributing to the nuclei of the precerebellar system, including the pontine and inferior olivary nuclei that constitute the main fiber input to the cerebellum (Farago *et al.*, 2006; Geisen *et al.*, 2008, Rodriguez and Dymecki, 2000; Sotelo, 2004). In the basal plate, discrete complexes of branchio- (BM) and visceromotor (VM) neurons are generated with a two-rhombomere periodicity and contribute to the distinct nuclei of the cranial nerves (Lumsden, 1990). Trigeminal (V), facial (VII), and glossopharyngeal (IX) nerve motor nuclei are generated in r2-r3, r4-r5, and r6-r7, respectively. Their cell bodies display specific migratory behaviors and axon trajectories to peripheral muscle targets in the face and neck region. Serotonergic neurons of the raphe nuclei are also sequentially generated from the motoneuron progenitor domain in the ventral basal plate of the rhombomeres following extinction of BM and VM neuron generation, with the exception of r4 (Jacob *et al.*, 2007; Jensen *et al.*, 2008; Pattyn *et al.*, 2003). Hindbrain premotor neurons with specific axon projections and synaptic connections onto motoneuron targets (Auclair *et al.*, 1999; Diaz *et al.*, 1998; Glover, 2000a,b) are also grouped into segment-like clusters or columns that correlate with specific rhombomeres. Moreover, central pattern generators (CPGs), that is, interneuronal circuits that coordinate and synchronize rhythmic motor neuron activities, arise within defined rhombomeric pairs at early stages (e.g., Fortin *et al.*, 1999, 2000; Jacquin *et al.*, 1996). Finally, hindbrain partitioning underlies segmental specification of neural crest cells (NCCs) contributing to cranial sensory ganglia and pharyngeal arches (Kontges and Lumsden, 1996; Lumsden and Guthrie, 1991; Lumsden and Krumlauf, 1996; Santagati and Rijli, 2003; Sechrist *et al.*, 1993; Serbedzija *et al.*, 1992; Trainor and Krumlauf, 2001). Thus, the early cellular partitioning of the hindbrain has a fundamental impact on its adult structure and nuclear organization.

3. The Impact of Segmental Patterning on Sensory Nuclei Columnar Organization and Projection Patterns

Long-term fate-mapping studies of rhombomeres in both chick and mouse revealed that individual rhombomeres form transverse stripes of cell progenies running through the ventriculopial axis of the mature hindbrain (Farago *et al.*, 2006; Marin and Puelles, 1995; Oury *et al.*, 2006; Pasqualetti *et al.*, 2007; Wingate and Lumsden, 1996). As a result, longitudinal columns of conventionally identified nuclei in the adult hindbrain, particularly in the sensory system, appear as "modular" multisegmental structures in which the deployment of individual neuronal populations generally respects the AP order of origin during development. One possibility is that such regional subdivisions correspond to different projections patterns (Glover, 2000b). Two recent studies in the mouse appear to support this idea.

In the first study, Pasqualetti *et al.* (2007) investigated connectivity patterns of a specific complex of sensorimotor nuclei, the vestibular nuclei, in relation with the mouse hindbrain segmental organization. Retrograde axonal tracing was combined with rhombomere-specific reporter gene expression in transgenic mice to create developmental fate maps of functionally identified vestibular neuron groups. This permitted visualization of rhombomere-derived domains well after morphological signs of rhombomere segmentation have disappeared, during stages when specific vestibular projection neuron phenotypes can be labeled and identified. The vestibular nuclear complex spans all the rhombomeres and contains well-characterized neuron populations that can be identified by their axon trajectories and connectivity (Cambronero and Puelles, 2000; Diaz *et al.*, 1998; Glover, 2003; Marin and Puelles, 1995). It was found that each rhombomere contains a unique set of vestibular projection neuron phenotypes, suggesting that within functionally related longitudinal neuron columns such as the vestibular nuclei, differentially distributed AP determinants assign different axonal projection patterns.

Moreover, rhombomere-related AP diversity may underlie the establishment of specific patterns of topographic connectivity within distinct components of individual nuclei, as recently shown for the trigeminal principal sensory (PrV) nucleus, which is part of the somatosensory pathway (Oury *et al.*, 2006). Sensory inputs from the face are topographically mapped onto the somatosensory cortex, via relay stations in the thalamus and hindbrain. Somatotopic representations are generated at each level of the neuraxis in which distinct facial structures, such as whiskers or lower jaw, are mapped at different scales. In the developing hindbrain, the main relay station is the PrV nucleus that receives input from the trigeminal nerve. In the mature hindbrain, the PrV dorsal portion is innervated by the mandibular branch and contains the lower jaw-lip representation, whereas the ventral component is organized in neuronal modules or "barrelettes" innervated by the maxillary branch and replicating the array of whiskers on the rodent snout (Fig. 5.3E). Similar neuronal arrangements exist at thalamic and cortical levels both for lower jaw-lip and whisker-related representations, the latter being known as "barreloids" and "barrels," respectively (Fig. 5.3E). By permanently labeling r2 and r3 derivatives using *ad hoc* transgenic tools, Oury *et al.* (2006) found that the segregation of rhombomere progenies sets out an early underlying cellular framework within the developing PrV nucleus eventually resulting in the somatotopic segregation of lower jaw-lip (r2-derived) and whisker-related "barrelette" (r3-derived) maps. Such a spatial arrangement results in rhombomere-specific topographic axonal mapping of PrV neurons to specific areas of the ventral posterior medial (VPM) nucleus of the thalamus, the next relay station in the circuit. In addition, such AP variations correlated with the arborization patterns of mandibular (lower jaw-lip) or maxillary (whisker) incoming trigeminal nerve afferents. Thus, both the topography of pontothalamic efferent projections and the spatial patterns of arborization of peripheral afferents onto target neurons are related to, and might be imposed by, the AP rhombomeric origin of PrV neurons.

4. Rostrocaudal Profiles and Sequential Phases of *Hox* Gene Expression: From Progenitor Patterning to Postmitotic Neuron Connectivity

The conservation of the segmental organization of the hindbrain in all vertebrates ensures that a similar cellular "scaffold" would be available on which to build topographically ordered connectivity, prompting the question of what are the molecular mechanisms involved in establishing and maintaining AP variation in hindbrain nuclei. Neuronal progenitors acquire their phenotypes according to their position within an orthogonal grid of coordinates established by AP and DV inductive signals (Briscoe *et al.*, 2000; Dessaud *et al.*, 2008; Ericson *et al.*, 1997; Glover *et al.*, 2006; Grapin-Botton *et al.*, 1997; Itasaki *et al.*, 1996; Jessell, 2000; Lee and Jessell, 1999; Pierani *et al.*, 1999; Ye *et al.*, 1998). This is a key mechanism to ensure appropriate spatial patterns of neuronal specification and connectivity. The position of a progenitor cell along the DV axis determines the specific class of postmitotic neurons that particular progenitor cell will give rise to (e.g., sensory rather than motor). The mechanisms generating cell diversity and specification along the DV axis have been extensively studied in the spinal cord (Jessell, 2000). Recent work suggests that similar processes are likely to be at work in the hindbrain (Briscoe and Ericson, 1999; Ericson *et al.*, 1997; Landsberg *et al.*, 2005; Machold and Fishell, 2005; Sander *et al.*, 2000; Sieber *et al.*, 2007; Wang *et al.*, 2005). A detailed account of the molecular determinants of DV patterning goes beyond the scope of the present review and will not be extensively treated here (but see below). On the other hand, the position of a progenitor cell along the AP axis is related to the specification of neuron subtype identity (e.g., trigeminal vs facial motoneuron), its specific migratory behavior, and axon trajectory to central or peripheral targets. Positional information is in turn translated into specific neuronal fates and patterns of connectivity by the expression of specific sets of transcription factors. These are expressed by positionally restricted populations of embryonic progenitors and/or postmitotic neurons mediating the response to inductive signals (Briscoe and Ericson, 1999; Briscoe *et al.*, 2000; Eisen, 1999; Goridis and Brunet, 1999; Irving and Mason, 2000; Jacob *et al.*, 2007; Jessell, 2000; Kageyama and Nakanishi, 1997; Lee and Jessell, 1999; Lumsden and Krumlauf, 1996; McMahon, 2000; Pattyn *et al.*, 2003; Pfaff and Kintner, 1998; Ye *et al.*, 1998).

In the developing hindbrain, the *Hox* family of homeobox-containing genes has been shown to play a key role in conferring segmental identity and patterning information to each neuroepithelial rhombomere compartment. Recent work suggests that, in addition to their roles in early hindbrain patterning, they may be critically involved in the control of stereotyped migratory behavior of neurons and the establishment of spatially restricted patterns of axonal connectivity. The mammalian *Hox* genes belong to a large family of 39 members. They are organized in four clusters (*a-d*) located on different chromosomes. Each cluster, derived from successive duplications during evolution of an ancestral complex, contains similar series of paralogous genes (*Hox* paralogue group 1-13). In the hindbrain, *Hox* genes from paralogue group (PG) 1-4 display nested expression patterns with offset rostral expression boundaries, according to their spatial colinearity (Glover *et al.*, 2006; Krumlauf, 1993b; Lumsden and Krumlauf, 1996; Rijli *et al.*, 1998; Wilkinson, 1995). Each rhombomere expresses a specific *Hox* combinatorial code, except r1 that is devoid of *Hox* expression (Hunt *et al.*, 1991; Kessel and Gruss, 1991). Although most of the previous studies focused on the expression of *Hox* genes during early rhombomere patterning, it is becoming increasingly clear that expression of some of these genes can be followed through advanced stages in specific neuronal populations and nuclei. *Hox* gene expression is activated early before hindbrain segmentation and transcripts are initially present uniformly throughout the DV extent of rhombomere neuroepithelium (Fig. 5.1A and B). Following this early phase, some *Hox* genes display dynamic expression patterns along the hindbrain DV axis that correlate with events in neurogenesis (Davenne *et al.*, 1999; Gaufo *et al.*, 2000), as it was shown in the spinal cord (Graham *et al.*, 1991), while maintaining sharp anterior expression boundaries. For instance, at E10.5, the two paralogues *Hoxa2* and *Hoxb2* display alternating regions of higher and lower expression in broad longitudinal columns located at distinct DV levels and running the length of the hindbrain (Fig. 5.1C and D; Davenne *et al.*, 1999). *Hoxa2* and *Hoxb2* are expressed in similar dorsal columns, but in r3 *Hoxa2* display higher levels than *Hoxb2*. Conversely, high levels of *Hoxb2*, though not of *Hoxa2*, are present in two ventral columns adjacent to the floor plate that include progenitors and/or differentiating motor neurons. In r4, *Hoxa2*-expressing cells abut the ventral columns of *Hoxb2* expression. These patterns define a set of complementary, partially overlapping and quantitatively different distributions, indicating that *Hoxa2* and *Hoxb2* functions may impinge on DV patterning and neurogenesis. DV regionalization of *Hox* transcripts is due to differential down- or upregulation along this axis, indicating that during this phase *Hox* gene regulation is under the control of DV signaling pathways that may be distinct from those maintaining their AP borders. Additionally, our recent data show that *Hox* gene expression can be maintained through advanced prenatal and even postnatal stages and may differentially label distinct subdivisions of hindbrain relay nuclei contributing to defined sensory circuits. For instance, in the trigeminal circuit, *Hoxa2* is expressed at high levels in the r3-derived, though not the r2-derived, portion of the PrV nucleus (Fig. 5.1E; Oury *et al.*, 2006). *Hoxa2* is also strongly expressed in the ventral

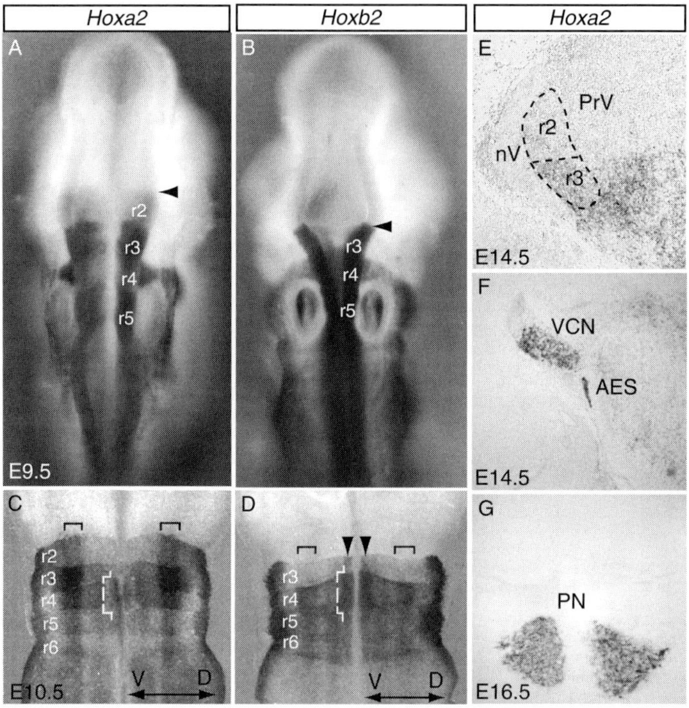

Figure 5.1 Dynamic expression patterns of *Hox* paralogue group 2 (PG2) genes during mouse hindbrain development. (A-G) Whole-mount and brain section in situ hybridizations with antisense probes on embryos and fetuses at various stages. (A, B) *Hoxa2* and *Hoxb2* expression patterns in the E9.5 hindbrain. Note that each gene has a specific anterior expression limit mapping at the r1/r2 (*Hoxa2*) and r2/r3 (*Hoxb2*) borders, respectively (arrowheads). (C, D) Expression domains of *Hoxa2* and *Hoxb2* on flat-mounted hindbrains during neurogenesis at E10.5. Note that at this stage, the *Hoxa2* and *Hoxb2* transcripts are differentially restricted along the mediolateral (i.e., dorso-ventral) axis in columns of higher and lower expression levels (black and white brackets), in addition to maintaining their normal anteroposterior patterns. (E) Late *Hoxa2* expression in the trigeminal principal sensory (PrV) nucleus at E14.5. High expression is selectively observed in the r3-derived domain whereas the r2-derived portion of the PrV nucleus displays low expression levels (see also Oury *et al.*, 2006). (F) *Hoxa2* is also strongly expressed in the ventral cochlear nucleus (VCN) and migrating pontine neurons (AES—anterior extramural stream) at E14.5. (G) Restricted expression of *Hoxa2* is observed in pontine nuclei (PN) at E16.5.

portion of the cochlear nucleus (VCN) up to postnatal stages (Fig. 5.1F; and data not shown). Similarly, *Hoxb2* transcripts are found in newborns in the main hindbrain relay nuclei of the auditory circuit such as the superior olivary complex (SOC), the ventral nucleus of lateral lemniscus (VLL), and the VCN (Fig. 5.2). There is also a remarkable expression of *Hox* genes

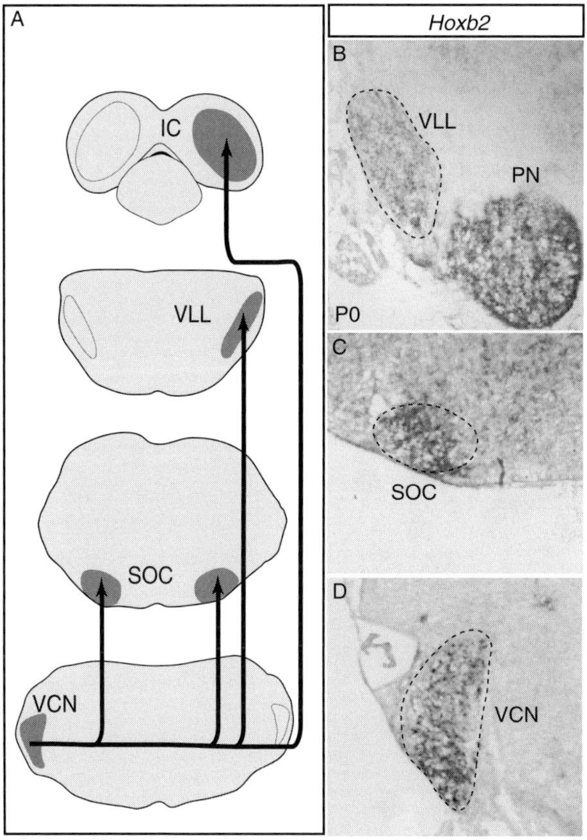

Figure 5.2 Expression of *Hoxb2* in the hindbrain projection nuclei of the auditory circuit at postnatal stages. (A) Major projection pathways from ventral cochlear nucleus (VCN). VCN neurons send parallel projections to both ipsilateral and contralateral superior olivary complex (SOC), contralateral ventral nucleus of lateral lemniscus (VLL), and inferior colliculus in the midbrain (Cant and Benson, 2003). (B-D) At P0, *Hoxb2* expression is maintained in all of the relay nuclei of auditory information in the hindbrain, namely VLL, SOC, and VCN.

during the migration of rhombic lip-derived neurons contributing to the pontine gray (PGN) and reticulotegmental nuclei (RTN), collectively referred to as pontine nucleus (PN). These neurons undergo a stereotyped long-distance tangential migration rostrally from the posterior rhombic lip in r6-r8 to a final location in ventral r4 (Farago *et al.*, 2006; Geisen *et al.*, 2008). We found that PN express paralogue group 2-5 *Hox* genes throughout their caudorostral migration and settling, thus expressing and maintaining a code characteristic of their axial origin posterior to r5 (Fig. 5.1F; Geisen *et al.*, 2008). Moreover, restricted *Hox* gene expression is observed within pontine nuclei at late prenatal and postnatal stages during the establishment of ordered connectivity (Figs. 5.1G and 5.2B; Geisen *et al.*, 2008; and data not shown).

Overall, the *Hox* gene expression patterns are highly suggestive of these factors controlling several steps in building functional neuronal circuitry within the developing hindbrain, from progenitors to neurons with specific patterns of topographic connectivity.

5. *Hox* Gene Function: Lessons from Mouse Knockouts

5.1. *Hox* genes in segmentation and specification of segmental identity: The two sides of the same coin

The advent of targeted deletion via homologous recombination in embryo stem (ES) cells (Capecchi, 2005) has made possible to investigate the involvement of *Hox* genes in mouse hindbrain development. Over the last 17 years, the generation of *Hox* mutants in the mouse has provided an invaluable model system for understanding the molecular basis of hindbrain segmentation and patterning. Namely, the analysis of single and compound knockout mice has undoubtedly shown that *Hox* factors carry out fundamental roles throughout the early steps of hindbrain development (Barrow and Capecchi, 1996; Barrow *et al.*, 2000; Carpenter *et al.*, 1993; Chisaka *et al.*, 1992; Davenne *et al.*, 1999; Dupe *et al.*, 1997; Gaufo *et al.*, 2000, 2003, 2004; Gavalas *et al.*, 1997, 1998, 2003; Goddard *et al.*, 1996; Helmbacher *et al.*, 1998; Lufkin *et al.*, 1991; Mark *et al.*, 1993; Rijli *et al.*, 1998; Rossel and Capecchi, 1999; Studer *et al.*, 1996), indicating that rhombomere segmentation and rhombomere-specific patterning may be interrelated aspects of the same process controlled by the *Hox* genes.

In this respect, grafting studies in the chick embryo have established that, after a period of plasticity, the definitive commitment to a rhombomere-specific fate is accompanied by the establishment of a unique code of *Hox* gene expression in each given compartment of neuroepithelial cells, which is maintained through subsequent cell divisions and directs the specific patterning program of each segment (Grapin-Botton *et al.*, 1995, 1997; Guthrie and Lumsden, 1992; Itasaki *et al.*, 1996; Kuratani and Eichele, 1993; Simon *et al.*, 1995). Definitive commitment to a rhombomere-specific fate may temporally coincide with cell-lineage restriction and inter-rhombomeric segregation of neuroepithelial cells. Therefore, segment formation and specification of segmental identity may be mutually linked and under the control of the same set of early expressed *Hox* selector genes.

Indeed, analysis of *Hoxa1*, *Hoxb1*, *Hoxa2*, and *Hoxb2* single and compound mutants revealed that PG1 and PG2 genes act by coupling the processes of segment formation, cell segregation, and specification of regional identity (Barrow and Capecchi, 1996; Carpenter *et al.*, 1993; Chisaka *et al.*, 1992; Davenne *et al.*, 1999; Dupe *et al.*, 1997; Gavalas *et al.*, 1997, 1998, 2003; Lufkin *et al.*, 1991; Mark *et al.*, 1993; Rijli *et al.*, 1998; Rossel and Capecchi, 1999; Studer *et al.*, 1998). *Hox* PG1 expression domains are established at early neural plate stages (E7.5–8.0 in the mouse), that is, about 1 day before the formation of definitive rhombomeres, with an anterior boundary at the r3/r4 border. In turn, *Hoxb1* expression is maintained in r4 by an autoregulatory loop (Popperl *et al.*, 1995), whereas *Hoxa1* is eventually downregulated (Murphy and Hill, 1991). *Hoxa2* expression instead occurs up to the r1/r2 boundary, whereas its paralogue *Hoxb2* is expressed only up to the r2/r3 border (Fig. 5.1A and B; Davenne *et al.*, 1999; Hunt *et al.*, 1991; Krumlauf, 1993a; Tan *et al.*, 1992). Thus, *Hoxa2* is the only *Hox* gene expressed in r2, *Hoxa2* and *Hoxb2* are coexpressed in r3, and r4 expresses both *Hox* PG1 and PG2 genes, though with distinct temporal and quantitative patterns.

Specific hindbrain defects were observed in mutants, including segmentation abnormalities and homeotic changes of identity with misspecifications of neurons. In particular, the *Hoxa1* knockout resulted in the failure to form specific rhombomeres. Histological and molecular analyses revealed that r4 and r5 are very reduced (Barrow *et al.*, 2000; Carpenter *et al.*, 1993; Chisaka *et al.*, 1992; Dolle *et al.*, 1993; Gavalas *et al.*, 1998; Lufkin *et al.*, 1991; Mark *et al.*, 1993; Rijli *et al.*, 1998). Similar function in rhombomere segmentation was reported for zebrafish *Hoxb1b* gene, whose expression is also similar to that of mammalian *Hoxa1* (McClintock *et al.*, 2002). On the other hand, mice lacking *Hoxb1* activity fail to upregulate r4-specific molecular markers while displaying ectopic expression of r2 markers, indicating that r4 undergoes a homeotic change of segmental identity (Goddard *et al.*, 1996; Studer *et al.*, 1996). Conversely, *Hoxb1*-targeted misexpression in r2 resulted in the reverse homeotic transformation of rhombomere identity (Bell *et al.*, 1999). Importantly, *Hoxb1* activation *in vivo* during normal development requires both retinoic acid signaling and *Hoxa1* itself (Studer *et al.*, 1998; reviewed in Glover *et al.*, 2006). Thus, in *Hoxa1* mutants, r4 cells failing to activate sufficient levels of *Hoxb1* may be respecified to a more rostral phenotype, linking segmentation and segmental identity specification processes (Gavalas *et al.*, 2003; Rijli *et al.*, 1998). This is supported by the appearance of ectopic patches of cells presenting a r2-like identity at r3–r4 levels in *Hoxa1* mutants (del Toro *et al.*, 2001; Helmbacher *et al.*, 1998). Notably, some of such ectopic cells are stably integrated in the mature hindbrain and contribute to a novel rhythmic respiratory network, which is functional during postnatal life (del Toro *et al.*, 2001). In *Hoxa2*-null mutants, the r1/r2 boundary is absent (Barrow *et al.*, 2000; Davenne *et al.*, 1999; Gavalas *et al.*, 1997). Moreover, r2–r3 size is reduced and r1 is enlarged resulting in posterior expansion of the r1 derived cerebellar territory (Gavalas *et al.*, 1997). This suggested a (partial) switch in cell fate of dorsal r2-r3 neuroepithelium toward a r1 identity, which may therefore represent a "ground patterning program" for hindbrain compartments (Gavalas *et al.*, 1997; Waskiewicz *et al.*, 2002). On the other hand, the *Hoxb2* mutation did not apparently affect normal hindbrain segmentation (Barrow and Capecchi, 1996; Davenne *et al.*, 1999). One possibility is that the *Hoxb2* inactivation is functionally compensated at early neural plate stages by the function of its paralogue *Hoxa2* in presumptive r3. Consistently, the r2/r3 and r3/r4 borders are absent or significantly affected, respectively, in compound *Hoxa2/Hoxb2* mutants (Davenne *et al.*, 1999). In addition, r3 patterning abnormalities observed in single *Hoxa2* mutants are aggravated in these mutants, showing synergistic genetic interactions between PG2 genes (Davenne *et al.*, 1999).

Morphological identification of homeotic changes in identity of rhombomeres and their derivatives in *Hox* PG1 and PG2 mutants has mainly focused on the analysis of pathfinding and migratory features of BM neurons of the r4-derived facial (FBM) and r2-r3-derived trigeminal (TBM) subtypes, respectively, innervating expression and mastication muscles in the pharyngeal arches (e.g., Auclair *et al.*, 1996; Carpenter *et al.*, 1993). In *Hoxa2* mutants, r2-r3-derived TBM neurons show abnormal axonal pathfinding and pharyngeal arch target selection, projecting into the second instead of first pharyngeal arch (Gavalas *et al.*, 1997). The axons of r3 TBM neurons normally project dorsally, turn anteriorly to enter r2, and exit the hindbrain through the r2 exit point together with the axons of r2 TBM neurons. TBM neurons eventually innervate the first, mandibular, pharyngeal arch. However, in *Hoxa2* mutant mouse, a significant proportion of TBM axons originated from both r2 and r3 turn posteriorly instead of anteriorly and ectopically exit the hindbrain from the r4 nerve exit point, thus incorrectly targeting the second, hyoid, arch. There are no molecular evidences of a TBM to FBM identity switch in such mutants (Gavalas *et al.*, 1997). Instead, this is likely a noncell autonomous effect, as *Hoxa2* does not appear to be expressed in TBM (data not shown). *Hoxa2* inactivation may result in the alteration of environmental cues that normally guide TBM axon pathfinding toward the r2 exit point and allow first arch innervation correctly (Gavalas *et al.*, 1997). It was shown that the nerve exit points, constituted by subpopulations of NCCs with rhombomere-specific origin (Niederlander and Lumsden, 1996), serve as attractive cues for motor axons (Chang *et al.*, 1992; Guthrie and Lumsden, 1992). One possibility is that *Hoxa2* provide appropriate segmental identity to the NCCs contributing the r4 exit point. Since NCCs emigrating from r2 are *Hox*-negative and those from r4 express *Hoxa2*, the inactivation of *Hoxa2* may result in r4-to-r2 homeotic transformation of exit point NCC fate, thus attracting some axons from TBM neurons. Indeed, the ectopic extension of the TBM nerve into the second pharyngeal arch of *Hoxa2* mutant fetuses is also consistent with the segmental fate change of r4-derived mesenchymal NCCs leading to second (hyoid) to first (mandibular) arch homeotic transformation of skeletal and muscle derivatives (Gendron-Maguire *et al.*, 1993; Rijli *et al.*, 1993; Santagati and Rijli, 2003).

Hoxb1- and *Hoxb2*-null mutants display changes of FBM toward TBM-like motoneuron progenitor identity (Gavalas *et al.*, 1998, 2003; Studer *et al.*, 1996). This interpretation is supported by the observation that the migratory behavior of r4-derived BM neurons, normally undergoing a caudal migration to reach their final location in ventrolateral r6, is dorsally directed, similar to that of r2-derived TBM neurons (Goddard *et al.*, 1996; Studer *et al.*, 1996). In addition, gain-of-function experiments demonstrated that ectopic expressions of *Hox* PG1 or PG2 in r1 are sufficient to confer specific BM neuron identity to progenitor cells (Jungbluth *et al.*, 1999; Samad *et al.*, 2004). Arenkiel *et al.* (2004) have reported an additional role of *Hoxb1* in the neurogenic NCCs that emigrate from r4 and contribute to the glial cells that eventually myelinate the facial motor axons. By conditional inactivation of *Hoxb1* in NCCs, the authors show that such expression of *Hoxb1* is required to provide environmental cues to control normal axon outgrowth and pathfinding of the FBM neurons. The defects observed in *Hoxb1* and *Hoxb2* knockout animals eventually result in impairment of FBM nucleus migration and formation causing a facial paralysis resembling the clinical signs associated with Bell's Palsy and Moebius syndrome in humans (Barrow and Capecchi, 1996; Davenne *et al.*, 1999; Gavalas *et al.*, 1998; Goddard *et al.*, 1996). This is due to the failure to form a normal somatic motor component of the facial (VII) nerve which controls the muscles of facial expression. In addition to FBM neurons, another r4-derived population, the contralateral vestibuloacoustic efferent (CVA) neurons, is affected in *Hoxb1* mutants due to the loss of GATA proteins, a class of zinc-finger transcription factors (Bruce and Fritzsch, 1997; Goddard *et al.*, 1996; Karis *et al.*, 2001; Pata *et al.*, 1999; Simon and Lumsden, 1993; Studer *et al.*, 1996).

Further support for a homeotic role for *Hox* genes in establishing rhombomere identities comes from the analysis of mice deficient for compound *Hox* PG3 genes, namely *Hoxa3*, *Hoxb3*, and *Hoxd3*, whose anterior expression boundary is at the interface between r4 and r5 (Gaufo *et al.*, 2003). In the absence of any combination of PG3 genes r6 undertakes an r4-like developmental program. In such mutants, ectopic expression of the r4-specific marker *Hoxb1* is associated to the ectopic generation of facial-like BM neurons, indicating that the activity of *Hox* PG3 genes normally suppress a r4-like program in r6. Finally, the apparent absence of hindbrain defects in PG4 *Hox* gene mutants (reviewed in Maconochie *et al.*, 1996) may reflect functional compensation by other paralogous and nonparalogous *Hox* genes.

The above targeted deletions have shown that the primary role of *Hox* genes during early hindbrain regionalization may be to restrict proliferating neural progenitors and their progeny in compartment-like blocks of cells segregated from each other (Wizenmann and Lumsden, 1997) that will eventually follow segment-specific developmental pathways appropriate for their AP axial levels. Neuroepithelial cells which are not correctly specified according to their AP axial level may not be restricted at their appropriate position therefore not forming a normal segment but becoming intermingled with cells from adjacent AP positions, eventually acquiring the same fate as their neighbors possibly through a segmental community effect (Gurdon, 1988; Wilkinson, 1995). Thus, mutating a given *Hox* gene may not only cause respecification to alternative AP developmental pathways of neuroepithelial cells, that is, homeosis, but also result in segmentation defects due to altered adhesive and/or signaling properties of group of cells (e.g., Barrow *et al.*, 2000; Goddard *et al.*, 1996; Rijli *et al.*, 1998; Studer *et al.*, 1996). In this respect, abnormal regulation of *cadherin-6* (Inoue *et al.*, 1997) or members of the *Eph* subfamily of receptor tyrosine kinase (RTK) class of recognition molecules, including *EphA2*, *EphA4*, and *EphA7* have been reported in *Hox* mutants (Chen and Ruley, 1998; Gavalas *et al.*, 1997; Taneja *et al.*, 1996).

5.2. *Hox*-dependent control of DV patterns of neuronal development

We mentioned above that, while generally maintaining their anterior expression boundaries, *Hox* genes display dynamic patterns of expression during neurogenesis that become progressively restricted along the DV axis and encompass specific populations of neuronal progenitors and early differentiating neurons (Fig. 5.1C and D; Davenne *et al.*, 1999; Gaufo *et al.*, 2000). Unveiling *Hox* gene function at such stages, that is, bypassing earlier requirements for *Hox* genes, will have to wait for the systematic analysis of spatiotemporally controlled knockout mouse models. However, such expression patterns underscore the idea that, in addition to their roles in conferring AP segmental identities to rhombomeres, *Hox*-encoded positional information and neurogenic programs are integrated in progenitor cells to specify distinct neuronal fates. Molecular analysis of single and compound *Hox* mutants provided strong support to such a proposal, suggesting that some molecular and morphological alterations observed in *Hox* mutants could not simply result from changes of the AP identities of rhombomeres (Davenne *et al.*, 1999; Pattyn *et al.*, 2003).

Unlike in other rhombomeres, the prolonged generation of FBM in ventral r4 prevents successive generation of serotonergic (S) neurons from the same progenitor domain (Pattyn *et al.*, 2003). This is achieved through *Hoxb1*- and *Hoxb2*-dependent maintenance of expression of the BM progenitor determinant *Phox2b* in ventral r4 beyond E10.5, through direct positive regulation (Samad *et al.*, 2004). At other axial levels, downregulation of *Phox2b* after E10.5 allows the BM/S switch, through a repression mechanism mediated by the forkhead transcription factor *Foxa2* (Jacob *et al.*, 2007; Pattyn *et al.*, 2003). Analysis of mutants provided evidence of late roles of *Hoxb1* and *Hoxb2* in repressing S and promoting BM fate in ventral r4 (Pattyn *et al.*, 2003). In compound mutants for the homeodomain transcription factors Nkx6.1 and Nkx6.2 that have a central role in ventral progenitor fate determination, *Hoxb1* expression is gradually lost in ventral r4 resulting in a BM to S neuron switch (Pattyn *et al.*, 2003). These data imply that *Hoxb1*, in addition to its role in establishing r4 identity at early stages, is continuously required to maintain progenitor BM identity and select the appropriate fate throughout development. Such a proposal is supported by the analysis of *Hoxb2* mutants (Pattyn *et al.*, 2003). In such mutants, *Hoxb1* expression is present in r4 at normal levels at early stages (Davenne *et al.*, 1999). In contrast, a significant ventral loss of *Hoxb1* expression is observed starting around E11.0 in *Hoxb2* mutants, resulting in a predicted BM to S ectopic switch of neuronal fate (Pattyn *et al.*, 2003). These data also suggest that *Hoxb2* may be involved in the maintenance of appropriate *Hoxb1* expression levels in late BM progenitors. However, alternative explanations are possible as the presence of the selectable marker cassette in the *Hoxb2* locus might interfere with normal *Hoxb1* regulation at late stages (e.g., Rijli *et al.*, 1994). Nonetheless, these data support the idea of the late involvement of *Hoxb1* and *Hoxb2* in the specification of neuronal fate at specific DV locations.

In *Hoxa2* mutants, the downregulation, though not absence, of *Pax6* in r2 and r3 indicated that *Hoxa2* is required, directly or indirectly, to maintain high levels of *Pax6* (Davenne *et al.*, 1999). The additional absence of *Hoxb2* resulted in the lack of *Pax6* expression in r3 of the double mutants, supporting a role for *Hox* PG2 genes in *Pax6* maintenance, which have been shown as an important determinant of ventral differentiation pathways of neural progenitors (e.g., Ericson *et al.*, 1997; Osumi *et al.*, 1997). Because low *Pax6* expression levels remained in the r1-r2 region of *Hoxa2/Hoxb2* compound mutants, such alterations may not simply result from changes in progenitor AP identities. Additionally, *Hoxa2* and *Hoxb2* differentially regulate the spatial distributions of transcripts encoding transcription factors such as *Mash1*, *Ngn2*, or *Phox2b* defining dorsoventrally restricted longitudinal columns of neuronal progenitors running the length of several rhombomeres (Davenne *et al.*, 1999). While *Hoxa2* appears to control the distribution of subsets of neuronal precursors in r2 and r3, *Hoxb2* exerts its function mainly in r4. Thus, distinct *Hox* genes could be involved in establishing neuronal fate of rostrocaudally restricted pools of progenitors contributing to multisegmental columnar nuclei. Furthermore, within a given rhombomere *Hoxa2* or *Hoxb2* function may be restricted along the DV axis. Namely, *Hoxa2* controls molecular regulation in the alar and "dorsal" basal plates of r2 and r3. This is suggested, for instance, by the selective lack of *Phox2b* expression in specific dorsal columns in r2-r3, while *Phox2b* expression in the ventral columns at the same axial level is unaffected in *Hoxa2* mutants (Davenne *et al.*, 1999). In contrast, *Hoxb2* is essential for motoneuron and serotonergic development in the "ventral" basal plate of r4. These observations are consistent with the highest relative expression levels along the DV axis of these genes during neurogenesis. In addition, *Hoxa2* and *Hoxb2* can functionally synergize in r3, as suggested by the lack of expression of *Evx1*, a marker for a ventral interneuron subtype (Burrill *et al.*, 1997), in double but not in single mutants. Finally, recently Matis *et al.* (2007) identified *Lmo1*, a member of the LIM-only protein, as another gene differentially regulated by *Hoxa2* and *Hoxb2*. *Lmo1* is expressed in both dorsal and ventral multisegmental progenitor columns. In keeping with what observed with the molecular markers described above, in *Hoxa2* mutant mice only the *Lmo1*-expressing dorsal column was lost at r2-r3 levels. In contrast, only the *Lmo1*-expressing ventral column was affected in r4 of *Hoxb2* mutants.

Further support for a role of *Hox* genes in controlling rhombomere-restricted DV patterns of neuronal specification came from recent studies of *Hox* PG1 and PG3 mutants (Gaufo *et al.*, 2000, 2003, 2004). For instance, as development proceeds *Hox* PG3 expression domains, whose anterior expression boundaries map to the r4/r5 junction, become progressively restricted along DV axis. At E11.5, the expression of *Hoxa3* and *Hoxb3* are only maintained in the ventral r5 domain giving rise to the somatic motor neurons (SMNs) of the abducens nucleus (Gaufo *et al.*, 2003). In *Hoxa3/Hoxb3* double mutants, the abducens nucleus is absent due to a DV transformation of SMN progenitors to a V2 interneuron identity. The expression of *Olig2*, a marker of SMN progenitors (Lu *et al.*, 2002; Novitch *et al.*, 2001), was lost and replaced by ectopic expression of *Chx10*, whose expression is normally restricted in progenitors of V2 interneurons. Furthermore, *Pax6*, which is weakly expressed in SMN progenitors and strongly expressed in V2 interneuron progenitors (Ericson *et al.*, 1997; Osumi *et al.*, 1997; Takahashi and Osumi, 2002) was strongly expressed in the SMN progenitor domain of the double-mutant mice. These results suggest that *Hoxa3* and *Hoxb3* are involved in the specification of SMNs as upstream regulators of *Olig2* and HB9 expression. This view is further supported by *Hoxa3* gain-of-function experiments in chick (Guidato *et al.*, 2003). Ectopic expression of *Hoxa3* in r1-r4, where SMN differentiation is normally not observed, resulted in repression of V2 interneuron markers such as *Pax6* and *Irx3* (Briscoe *et al.*, 2000; Ericson *et al.*, 1997; Osumi *et al.*, 1997; Takahashi and Osumi, 2002) and upregulation of *Hb9*, followed by ectopic differentiation of SMNs at the expense of V2 interneurons. More recently, Gaufo *et al.* (2004) also reported the loss of *Phox2b*-expressing visceral interneuron progenitors in r4 and r5 of *Hoxb1* and *Hoxa3/Hoxb3* double-mutant mice, respectively, and concomitant expansion of *Lim1/2*-positive neighboring somatic interneuron domain suggesting the involvement of PG1 and PG3 genes in the specification of progenitors of the visceral interneuron column.

Altogether, these findings suggest that rhombomere-specific, DV-restricted, pools of neurons contributing to distinct portions of multisegmental columnar nuclei may need prolonged *Hox* gene expression to continuously reassess their AP positional values in order to maintain correct DV specification.

5.3. *Hox* gene involvement in hindbrain neural circuit formation

In the spinal cord, recent elegant evidence has shown that *Hox* gene function is crucial to specify distinct AP motor pool identity within limb-innervating motor neuron columns that span several vertebral levels, as well as to control specific patterns of target muscle connectivity (Dasen *et al.*, 2003, 2005, 2008). Such a *Hox* regulatory network directs the downstream transcriptional identity of motor neuron pools regulating expression of dedicated transcription factors, guidance, and adhesion molecules. The late expression patterns and restricted distributions of *Hox* gene transcripts in selected nuclei of the developing hindbrain (Fig. 5.1E–G) strongly indicate their implication in late aspects of neuronal differentiation, including stereotypical neuronal migration and topographic axon pathfinding and mapping, during the establishment of sensory circuitry. Indeed, recent reports support such a view and unveil late roles for *Hox* genes in controlling precerebellar neuron migration (Geisen *et al.*, 2008) and the development of a facial somatosensory map (Oury *et al.*, 2006).

In the developing hindbrain, neurons migrate from their birthplace to their final destination, where they form specific nuclei. Since neuronal function depends on precise connection of neurons with their targets, the correct positioning of neurons is critical to build a topographically ordered neuronal connectivity between pre- and postsynaptic partners. Neuronal migration is thought to be controlled by the same set of attractive and repulsive guidance cues that regulate axonal pathfinding and topographical mapping (Wong *et al.*, 2002). However, little is still known about how exposure of migrating neurons to several simultaneous extrinsic inputs along the orthogonal axes of the brain may be integrated at the transcriptional level and in turn translated into directional migratory responses specific for each neuronal population. Recently, Geisen *et al.* (2008) provided novel insights into how *Hox* genes control long-distance migration of pontine neurons, which are part of the precerebellar system (see above). The migratory route of pontine neurons through rhombomere territories was precisely mapped by combining in situ hybridization with selected molecular markers and long-term fate mapping of r3 and r5 progenies.

After they migrated out from r6 to r8 rhombic lip (between E13.5 and E17.5), pontine neurons undertake a short ventral migration (Fig. 5.3A). Then, pontine neurons turn anteriorly and travel rostrally through the r5–r3 region, until they reach the r2-derived territory. However, they never enter the r2 domain, as they turn again ventrally, leave r3 and re-enter the r4-derived domain where they finally settle near the floor plate. Such a complex and stereotyped roadmap suggests that, in addition to responding to DV positional cues, pontine neurons may express and/or respond to molecular determinants of positional information along the AP axis. The *Hox* genes are prime candidates as AP positional determinants; moreover, the expression of *Hox* PG2-PG5 genes is maintained in pontine neurons throughout their migration and settling in pontine nuclei (Geisen *et al.*, 2008; Fig. 5.1F and G). Such observations indicated that pontine neurons might be endowed with molecular information as to their origin and relative position along the AP axis throughout their migration, suggesting a functional involvement of *Hox* genes. In both *Hoxa2* and *Hoxb2* mutant mice, subsets of pontine neurons turn prematurely toward the ventral midline during their caudorostral phase of migration (Fig. 5.3B; Geisen *et al.*, 2008). Thus, these data identify *Hox* PG2 genes as important players in regulating caudorostral migration of pontine neurons. On the other hand, the fraction of neurons displaying migration errors varied both in spatial distribution and number among individuals, both in *Hoxa2* and *Hoxb2* mutants, while the bulk of neurons followed a normal migration pathway. These observations indicate that such a molecular guidance system is quite robust and buffered against a limited degree of variation, such as loss of function of one or two *Hox* genes. In fact, even in double PG2 *Hox* mutants many pontine neurons still migrated normally, although these mutant brains appeared to display more ectopically migrating neurons as compared to single mutants. Thus, these results indicate redundant functional gene effects among PG2 genes, and that their loss may be stochastically compensated by other members of the *Hox* family during pontine neuron migration through rhombomeric domains.

How *Hox* PG2 genes control neuronal migration in this system? They may be required cell autonomously in pontine neurons throughout their migration. Alternatively, or in addition, *Hox* genes may be required noncell autonomously to pattern the environment through which neurons migrate and to which they respond in order to maintain their rostral migration. Analysis of conditional mutants in which Cre-mediated *Hoxa2* deletion was achieved in rhombic lip derivatives also results in ectopic streams of prematurely ventrally migrating pontine neurons, supporting a cell autonomous role of *Hox* PG2 genes. Nonetheless, a less pronounced phenotype as compared to the null mutants suggests that *Hoxa2* is required both in a cell autonomous and noncell autonomous fashion to regulate the response of pontine neurons to guidance cues during their rostral migration. Further molecular and functional analysis in single and compound knockout mice reveals that repulsive signaling mediated through the Robo2 guidance receptor expressed in migrating pontine neurons and its ligand Slit2/3 secreted from the FBM nucleus are key components of the *Hox*-dependent molecular guidance system that maintains caudorostral migration (Fig. 5.3A and B). Notably, *Robo2* is a direct transcriptional target gene of *Hoxa2* (Geisen *et al.*, 2008). These authors propose a model in which *Netrin1/ Dcc*-mediated attraction along the DV axis is antagonized through *Hox*-mediated regulation of sustained levels of Slit-Robo signaling during the caudorostral phase of migration, thereby preventing premature ventral migration of pontine neurons until they reach r3 and undertake their final ventral migration to ventral r4. These results highlight novel late roles for *Hox* genes in controlling complex neuronal migration processes and provide insights on how transcriptional regulation controls coherent neuronal migratory responses to multiple and simultaneous guidance cues in the developing mammalian hindbrain.

Another fundamental step in the assembly of functional neural circuits is the control of precise guidance of axons toward their target cells. Neurons extend their axons sometimes over long distances before approaching the correct target. It is believed that the environment provides some extracellular guidance cues, including diffusible and cell surface adhesion molecules (Dickson, 2002; Guan and Rao, 2003), to which axons respond by expressing specific receptors on their surface in order to detect and interpret such guidance cues. To add further complexity, neuronal circuits are characterized by topographic maps of coordinated axonal connections in which the positional coordinates of spatially ordered input neurons are precisely copied onto spatially ordered sets of target neurons. Little is still known about the transcriptional control of topographic axon pathfinding and mapping.

Recent work has started to address the involvement of *Hox* genes in regulating the molecular mechanisms of topographic map formation in the mouse whisker-to-barrel circuit (Oury *et al.*, 2006). In particular, *Hoxa2* has multiple spatiotemporal roles in the building of a topographic trigeminal circuit. At early embryonic stages, the absence of *Hoxa2* in the r2 neuroepithelium results in axonal pathfinding defects of incoming trigeminal afferents. The trigeminal nerve gives ascending and descending branches after entering the hindbrain at the level of r2. The ascending branch normally stops at the r1/r2 boundary. In *Hoxa2* mutant mice, however, the ascending branch of the trigeminal nerve extends through r1 and ectopically project into cerebellum. Thus, a *Hoxa2* dependent molecular barrier may exist in r2 that prevents entering of peripheral afferents in r1. It is noteworthy that inner ear vestibular afferents, running just lateral to the trigeminal axons, are not stopped at the r1/r2 border and normally project to the cerebellum (Maklad and Fritzsch, 2003). *Hoxa2* therefore controls environmental cues specifically required for stopping trigeminal, but not vestibular, sensory afferents at the r1/r2 boundary. Future studies will be

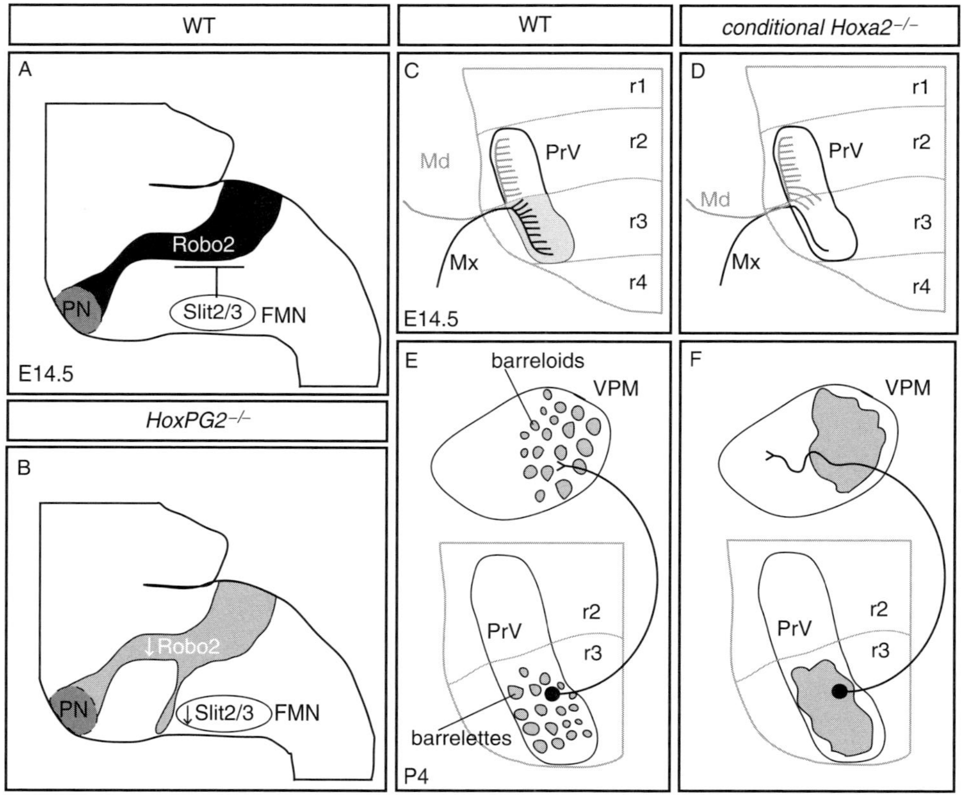

Figure 5.3 Summary diagrams of neural circuit developmental defects in *Hox* PG2 mutants. (A) Tangential migration pathway of pontine neurons toward pontine nucleus (PN) in wild-type mice. Migrating pontine neurons express the Robo2 receptor as well as Robo1 and Robo3. The facial motor nucleus (FMN) secretes their ligands, Slit2 and Slit3, preventing the premature ventral migration of the pontine

required to identify the molecular mechanisms involved. Furthermore, as mentioned above, domains of high or low *Hoxa2* expression in the PrV nucleus correlate with the arborization patterns of whisker-related or mandibular afferents onto their targets, respectively (Figs. 5.1E and 5.3C). Indeed, in addition to the early requirement, spatially and temporally induced Cre-mediated inactivations have established a specific late role for *Hoxa2* in regulating collateral formation from whisker-related afferents onto PrV target neurons (Fig. 5.3D). Finally, Oury *et al.* (2006) demonstrate a direct involvement of *Hoxa2* in the topographic wiring of PrV axons to the VPM thalamus. While pathfinding of PrV axons to the thalamus is not affected in *Hoxa2* mutants, selective changes occur in the topographic specificity of axonal mapping within the VPM nucleus (Fig. 5.3E and F). Altogether, these results indicate that *Hoxa2* could regulate the distribution of molecules providing positional mapping labels simultaneously regulating the wiring between brainstem (PrV) and thalamus (VPM), as well as between the periphery (incoming whisker-related afferents) and PrV during development of the trigeminal circuit. In this respect, *Hoxa2* positively regulates *EphA4* and *EphA7* expressions in the PrV nucleus, indicating that the *Hoxa2*-dependent control of connectivity could be mediated through Eph receptor/ephrin ligand signaling.

Finally, in some case, *Hox* gene expression is maintained in selected nuclei up to postnatal stages (Fig. 5.2B-D; see also Geisen *et al.*, 2008). Interestingly, final downregulation of *Hox* expression may temporally coincide with circuit maturation (Y. Narita and F. M. Rijli, unpublished

---

neurons through repulsive signaling (Geisen *et al.*, 2008). (B) In *Hox* PG2 mutant mice, selective downregulation of Robo2 is observed in migrating pontine neurons as well as of Slit2/3 in FMN. By the combination of cell autonomous and noncell autonomous alterations, the lack of *Hox* PG2 function affects the migration trajectory of the pontine neurons in mutant mice resulting in premature ectopic ventral attraction of such neurons toward the midline (Geisen *et al.*, 2008). (C) Trigeminal nerve afferent projections onto the trigeminal principal sensory (PrV) nucleus at E14.5. After entering the hindbrain, the mandibular (Md) branch of the trigeminal nerve sends collaterals specifically to the r2-derived PrV, whereas the r3-derived portion of PrV receives selective input from the maxillary (Mx) branch, correlating with high *Hoxa2* expression levels (gray shading) (Oury *et al.*, 2006). (D) In spatiotemporally induced conditional *Hoxa2* knockout mice, the arborization of Mx to the r3-derived region is specifically inhibited. The r3-derived domain is partly ectopically targeted by Mx (Oury *et al.*, 2006). (E) In wild-type mice, the r3-derived PrV component and the dorsolateral region of the ventral posterior medial (VPM) nucleus in the thalamus are organized in neuronal modules, barrelettes and barreloids, respectively, which replicates the spatial array of whiskers on the snout. Barrelette neurons send their projections to the barreloids in VPM thalamus with a one-to-one topography. (F) In the conditional knockout mice of *Hoxa2*, both barrelette and barreloid organization is profoundly affected, while axon projections from the r3-derived PrV mistargeted to the ventromedial VPM (Oury *et al.*, 2006).

observations). This suggests the potential involvement of *Hox* genes during the refinement of topographic connectivity, thus acting during processes that are thought to be regulated by experience-dependent neural activity during early postnatal life, such as synaptic pruning or maturation. Moreover, recently nonconventional roles for homeoproteins as soluble intercellular signaling molecules have been described, suggesting additional mechanisms through which these factors can regulate axon connectivity or the postnatal refinement of circuits (Brunet *et al.*, 2005; Sugiyama *et al.*, 2008). Future analyses using sophisticated spatial and temporal genetic tools and conditional mutations in mice will allow to test whether such predictions will find experimental support. These are definitely exciting times for research on *Hox* gene function in the developing mammalian brain.

ACKNOWLEDGMENTS

We thank C. Laumonnerie and Y. Murakami for kindly providing unpublished images and advice. YN was supported by the Association pour la Recherche sur le Cancer (ARC). Work in FMR laboratory is supported by the Agence Nationale pour la Recherche (ANR), Fondation pour la Recherche Médicale (Equipe labelisée FRM), ARC, Federation pour la Recherche sur le Cerveau (FRC), Association Française contre les Myopathies (AFM), Ministère de la Recherche, CNRS and INSERM, and the Novartis Research Foundation.

REFERENCES

Arenkiel, B. R., Tvrdik, P., Gaufo, G. O., and Capecchi, M. R. (2004). Hoxb1 functions in both motoneurons and in tissues of the periphery to establish and maintain the proper neuronal circuitry. *Genes Dev.* **18,** 1539–1552.

Auclair, F., Valdes, N., and Marchand, R. (1996). Rhombomere-specific origin of branchial and visceral motoneurons of the facial nerve in the rat embryo. *J. Comp. Neurol.* **369,** 451–461.

Auclair, F., Marchand, R., and Glover, J. C. (1999). Regional patterning of reticulospinal and vestibulospinal neurons in the hindbrain of mouse and rat embryos. *J. Comp. Neurol.* **411,** 288–300.

Barrow, J. R., and Capecchi, M. R. (1996). Targeted disruption of the Hoxb-2 locus in mice interferes with expression of Hoxb-1 and Hoxb-4. *Development* **122,** 3817–3828.

Barrow, J. R., Stadler, H. S., and Capecchi, M. R. (2000). Roles of Hoxa1 and Hoxa2 in patterning the early hindbrain of the mouse. *Development* **127,** 933–944.

Bell, E., Wingate, R. J., and Lumsden, A. (1999). Homeotic transformation of rhombomere identity after localized Hoxb1 misexpression. *Science* **284,** 2168–2171.

Birgbauer, E., and Fraser, S. E. (1994). Violation of cell lineage restriction compartments in the chick hindbrain. *Development* **120,** 1347–1356.

Briscoe, J., and Ericson, J. (1999). The specification of neuronal identity by graded Sonic Hedgehog signalling. *Semin. Cell Dev. Biol.* **10,** 353–362.

Briscoe, J., Pierani, A., Jessell, T. M., and Ericson, J. (2000). A homeodomain protein code specifies progenitor cell identity and neuronal fate in the ventral neural tube. *Cell* **101,** 435–445.

Bruce, L. L., and Fritzsch, B. (1997). The development of vestibular connections in rat embryos in microgravity. *J. Gravit. Physiol.* **4,** P59–P62.

Brunet, I., Weinl, C., Piper, M., Trembleau, A., Volovitch, M., Harris, W., Prochiantz, A., and Holt, C. (2005). The transcription factor Engrailed-2 guides retinal axons. *Nature* **438,** 94–98.

Burrill, J. D., Moran, L., Goulding, M. D., and Saueressig, H. (1997). PAX2 is expressed in multiple spinal cord interneurons, including a population of EN1+ interneurons that require PAX6 for their development. *Development* **124,** 4493–4503.

Cambronero, F., and Puelles, L. (2000). Rostrocaudal nuclear relationships in the avian medulla oblongata: A fate map with quail chick chimeras. *J. Comp. Neurol.* **427,** 522–545.

Cant, N. B., and Benson, C. G. (2003). Parallel auditory pathways: Projection patterns of the different neuronal populations in the dorsal and ventral cochlear nuclei. *Brain Res. Bull.* **60,** 457–474.

Capecchi, M. R. (2005). Gene targeting in mice: Functional analysis of the mammalian genome for the twenty-first century. *Nat. Rev. Genet.* **6,** 507–512.

Carpenter, E. M., Goddard, J. M., Chisaka, O., Manley, N. R., and Capecchi, M. R. (1993). Loss of Hox-A1 (Hox-1.6) function results in the reorganization of the murine hindbrain. *Development* **118,** 1063–1075.

Chang, S., Fan, J., and Nayak, J. (1992). Pathfinding by cranial nerve VII (facial) motor-neurons in the chick hindbrain. *Development* **114,** 815–823.

Chen, J., and Ruley, H. E. (1998). An enhancer element in the epha2 (Eck) gene sufficient for rhombomere-specific expression is activated by HOXA1 and HOXB1 homeobox proteins. *J. Biol. Chem.* **273,** 24670–24675.

Chisaka, O., Musci, T. S., and Capecchi, M. R. (1992). Developmental defects of the ear, cranial nerves and hindbrain resulting from targeted disruption of the mouse homeobox gene Hox-1.6. *Nature* **355,** 516–520.

Clarke, J. D., and Lumsden, A. (1993). Segmental repetition of neuronal phenotype sets in the chick embryo hindbrain. *Development* **118,** 151–162.

Dasen, J. S., Liu, J. P., and Jessell, T. M. (2003). Motor neuron columnar fate imposed by sequential phases of Hox-c activity. *Nature* **425,** 926–933.

Dasen, J. S., Tice, B. C., Brenner-Morton, S., and Jessell, T. M. (2005). A Hox regulatory network establishes motor neuron pool identity and target-muscle connectivity [see comment]. *Cell* **123,** 477–491.

Dasen, J. S., De Camilli, A., Wang, B., Tucker, P. W., and Jessell, T. M. (2008). Hox repertoires for motor neuron diversity and connectivity gated by a single accessory factor, foxp1. *Cell* **134,** 304–316.

Davenne, M., Maconochie, M. K., Neun, R., Pattyn, A., Chambon, P., Krumlauf, R., and Rijli, F. M. (1999). Hoxa2 and Hoxb2 control dorsoventral patterns of neuronal development in the rostral hindbrain. *Neuron* **22,** 677–691.

del Toro, E. D., Borday, V., Davenne, M., Neun, R., Rijli, F. M., and Champagnat, J. (2001). Generation of a novel functional neuronal circuit in *Hoxa1* mutant mice. *J. Neurosci.* **21,** 5637–5642.

Dessaud, E., McMahon, A. P., and Briscoe, J. (2008). Pattern formation in the vertebrate neural tube: A sonic hedgehog morphogen-regulated transcriptional network. *Development* **135,** 2489–2503.

Diaz, C., Puelles, L., Marin, F., and Glover, J. C. (1998). The relationship between rhombomeres and vestibular neuron populations as assessed in quail-chicken chimeras. *Dev. Biol.* **202,** 14–28.

Dickson, B. J. (2002). Molecular mechanisms of axon guidance. *Science* **298,** 1959–1964.

Dolle, P., Lufkin, T., Krumlauf, R., Mark, M., Duboule, D., and Chambon, P. (1993). Local alterations of Krox-20 and *Hox* gene expression in the hindbrain suggest lack of rhombomeres 4 and 5 in homozygote null Hoxa-1 (Hox-1.6) mutant embryos. *Proc. Natl. Acad. Sci. USA* **90,** 7666–7670.

Dupe, V., Davenne, M., Brocard, J., Dolle, P., Mark, M., Dierich, A., Chambon, P., and Rijli, F. M. (1997). *In vivo* functional analysis of the Hoxa-1 3′ retinoic acid response element (3′RARE). *Development* **124,** 399–410.

Eisen, J. S. (1999). Patterning motoneurons in the vertebrate nervous system. *Trends Neurosci.* **22,** 321–326.

Ericson, J., Rashbass, P., Schedl, A., Brenner-Morton, S., Kawakami, A., van Heyningen, V., Jessell, T. M., and Briscoe, J. (1997). Pax6 controls progenitor cell identity and neuronal fate in response to graded Shh signaling. *Cell* **90,** 169–180.

Farago, A. F., Awatramani, R. B., and Dymecki, S. M. (2006). Assembly of the brainstem cochlear nuclear complex is revealed by intersectional and subtractive genetic fate maps. *Neuron* **50,** 205–218.

Fortin, G., Jungbluth, S., Lumsden, A., and Champagnat, J. (1999). Segmental specification of gabaergic inhibition during development of hindbrain neural networks. *Nat. Neurosci.* **2,** 873–877.

Fortin, G., del Toro, E. D., Abadie, V., Guimaraes, L., Foutz, A. S., Denavit-Saubie, M., Rouyer, F., and Champagnat, J. (2000). Genetic and developmental models for the neural control of breathing in vertebrates. *Respir. Physiol.* **122,** 247–257.

Fraser, S., Keynes, R., and Lumsden, A. (1990). Segmentation in the chick embryo hindbrain is defined by cell lineage restrictions. *Nature* **344,** 431–435.

Gaufo, G. O., Flodby, P., and Capecchi, M. R. (2000). Hoxb1 controls effectors of sonic hedgehog and Mash1 signaling pathways. *Development* **127,** 5343–5354.

Gaufo, G. O., Thomas, K. R., and Capecchi, M. R. (2003). *Hox3* genes coordinate mechanisms of genetic suppression and activation in the generation of branchial and somatic motoneurons. *Development* **130,** 5191–5201.

Gaufo, G. O., Wu, S., and Capecchi, M. R. (2004). Contribution of *Hox* genes to the diversity of the hindbrain sensory system. *Development* **131,** 1259–1266.

Gavalas, A., Davenne, M., Lumsden, A., Chambon, P., and Rijli, F. M. (1997). Role of Hoxa-2 in axon pathfinding and rostral hindbrain patterning. *Development* **124,** 3693–3702.

Gavalas, A., Studer, M., Lumsden, A., Rijli, F. M., Krumlauf, R., and Chambon, P. (1998). Hoxa1 and Hoxb1 synergize in patterning the hindbrain, cranial nerves and second pharyngeal arch. *Development* **125,** 1123–1136.

Gavalas, A., Ruhrberg, C., Livet, J., Henderson, C. E., and Krumlauf, R. (2003). Neuronal defects in the hindbrain of *Hoxa1, Hoxb1* and *Hoxb2* mutants reflect regulatory interactions among these *Hox* genes. *Development* **130,** 5663–5679.

Geisen, M. J., Di Meglio, T., Pasqualetti, M., Ducret, S., Brunet, J. F., Chedotal, A., and Rijli, F. M. (2008). *Hox* paralogue group 2 genes control the migration of mouse pontine neurons through Slit-Robo Signaling. *PLoS Biol.* **6,** e142.

Gendron-Maguire, M., Mallo, M., Zhang, M., and Gridley, T. (1993). *Hoxa-2* mutant mice exhibit homeotic transformation of skeletal elements derived from cranial neural crest. *Cell* **75,** 1317–1331.

Glover, J. C. (2000a). Development of specific connectivity between premotor neurons and motoneurons in the brain stem and spinal cord. *Physiol. Rev.* **80,** 615–647.

Glover, J. C. (2000b). Neuroepithelial 'compartments' and the specification of vestibular projections. *Prog. Brain Res.* **124,** 3–21.

Glover, J. C. (2003). The development of vestibulo-ocular circuitry in the chicken embryo. *J. Physiol. (Paris)* **97,** 17–25.

Glover, J. C., Renaud, J. S., and Rijli, F. M. (2006). Retinoic acid and hindbrain patterning. *J. Neurobiol.* **66,** 705–725.

Goddard, J. M., Rossel, M., Manley, N. R., and Capecchi, M. R. (1996). Mice with targeted disruption of Hoxb-1 fail to form the motor nucleus of the VIIth nerve. *Development* **122,** 3217–3228.

Goridis, C., and Brunet, J. F. (1999). Transcriptional control of neurotransmitter phenotype. *Curr. Opin. Neurobiol.* **9,** 47–53.

Graham, A., Maden, M., and Krumlauf, R. (1991). The murine *Hox-2* genes display dynamic dorsoventral patterns of expression during central nervous system development. *Development* **112,** 255–264.

Grapin-Botton, A., Bonnin, M. A., McNaughton, L. A., Krumlauf, R., and Le Douarin, N. M. (1995). Plasticity of transposed rhombomeres: *Hox* gene induction is correlated with phenotypic modifications. *Development* **121,** 2707–2721.

Grapin-Botton, A., Bonnin, M. A., and Le Douarin, N. M. (1997). *Hox* gene induction in the neural tube depends on three parameters: Competence, signal supply and paralogue group. *Development* **124,** 849–859.

Guan, K. L., and Rao, Y. (2003). Signalling mechanisms mediating neuronal responses to guidance cues. *Nat. Rev. Neurosci.* **4,** 941–956.

Guidato, S., Prin, F., and Guthrie, S. (2003). Somatic motoneurone specification in the hindbrain: The influence of somite-derived signals, retinoic acid and Hoxa3. *Development* **130,** 2981–2996.

Gurdon, J. B. (1988). A community effect in animal development. *Nature* **336,** 772–774.

Guthrie, S., and Lumsden, A. (1992). Motor neuron pathfinding following rhombomere reversals in the chick embryo hindbrain. *Development* **114,** 663–673.

Hanneman, E., Trevarrow, B., Metcalfe, W. K., Kimmel, C. B., and Westerfield, M. (1988). Segmental pattern of development of the hindbrain and spinal cord of the zebrafish embryo. *Development* **103,** 49–58.

Helmbacher, F., Pujades, C., Desmarquet, C., Frain, M., Rijli, F. M., Chambon, P., and Charnay, P. (1998). Hoxa1 and Krox-20 synergize to control the development of rhombomere 3. *Development* **125,** 4739–4748.

Hunt, P., Gulisano, M., Cook, M., Sham, M. H., Faiella, A., Wilkinson, D., Boncinelli, E., and Krumlauf, R. (1991). A distinct Hox code for the branchial region of the vertebrate head. *Nature* **353,** 861–864.

Inoue, T., Chisaka, O., Matsunami, H., and Takeichi, M. (1997). Cadherin-6 expression transiently delineates specific rhombomeres, other neural tube subdivisions, and neural crest subpopulations in mouse embryos. *Dev. Biol.* **183,** 183–194.

Irving, C., and Mason, I. (2000). Signalling by FGF8 from the isthmus patterns anterior hindbrain and establishes the anterior limit of *Hox* gene expression. *Development* **127,** 177–186.

Itasaki, N., Sharpe, J., Morrison, A., and Krumlauf, R. (1996). Reprogramming Hox expression in the vertebrate hindbrain: Influence of paraxial mesoderm and rhombomere transposition. *Neuron* **16,** 487–500.

Jacob, J., Ferri, A. L., Milton, C., Prin, F., Pla, P., Lin, W., Gavalas, A., Ang, S. L., and Briscoe, J. (2007). Transcriptional repression coordinates the temporal switch from motor to serotonergic neurogenesis. *Nat. Neurosci.* **10,** 1433–1439.

Jacquin, T. D., Borday, V., Schneider-Maunoury, S., Topilko, P., Ghilini, G., Kato, F., Charnay, P., and Champagnat, J. (1996). Reorganization of pontine rhythmogenic neuronal networks in Krox-20 knockout mice. *Neuron* **17,** 747–758.

Jensen, P., Farago, A. F., Awatramani, R. B., Scott, M. M., Deneris, E. S., and Dymecki, S. M. (2008). Redefining the serotonergic system by genetic lineage. *Nat. Neurosci.* **11,** 417–419.

Jessell, T. M. (2000). Neuronal specification in the spinal cord: Inductive signals and transcriptional codes. *Nat. Rev. Genet.* **1,** 20–29.

Jungbluth, S., Bell, E., and Lumsden, A. (1999). Specification of distinct motor neuron identities by the singular activities of individual *Hox* genes. *Development* **126,** 2751–2758.

Kageyama, R., and Nakanishi, S. (1997). Helix-loop-helix factors in growth and differentiation of the vertebrate nervous system. *Curr. Opin. Genet. Dev.* **7,** 659–665.

Karis, A., Pata, I., van Doorninck, J. H., Grosveld, F., de Zeeuw, C. I., de Caprona, D., and Fritzsch, B. (2001). Transcription factor GATA-3 alters pathway selection of olivocochlear neurons and affects morphogenesis of the ear. *J. Comp. Neurol.* **429,** 615–630.

Kessel, M., and Gruss, P. (1991). Homeotic transformations of murine vertebrae and concomitant alteration of Hox codes induced by retinoic acid. *Cell* **67,** 89–104.

Kiecker, C., and Lumsden, A. (2005). Compartments and their boundaries in vertebrate brain development. *Nat. Rev. Neurosci.* **6,** 553–564.

Kontges, G., and Lumsden, A. (1996). Rhombencephalic neural crest segmentation is preserved throughout craniofacial ontogeny. *Development* **122,** 3229–3242.

Krumlauf, R. (1993a). *Hox* genes and pattern formation in the branchial region of the vertebrate head. *Trends Genet.* **9,** 106–112.

Krumlauf, R. (1993b). Mouse Hox genetic functions. *Curr. Opin. Genet. Dev.* **3,** 621–625.

Kuratani, S. C., and Eichele, G. (1993). Rhombomere transplantation repatterns the segmental organization of cranial nerves and reveals cell-autonomous expression of a homeodomain protein. *Development* **117,** 105–117.

Landsberg, R. L., Awatramani, R. B., Hunter, N. L., Farago, A. F., DiPietrantonio, H. J., Rodriguez, C. I., and Dymecki, S. M. (2005). Hindbrain rhombic lip is comprised of discrete progenitor cell populations allocated by Pax6. *Neuron* **48,** 933–947.

Lee, K. J., and Jessell, T. M. (1999). The specification of dorsal cell fates in the vertebrate central nervous system. *Annu. Rev. Neurosci.* **22,** 261–294.

Lu, Q. R., Sun, T., Zhu, Z., Ma, N., Garcia, M., Stiles, C. D., and Rowitch, D. H. (2002). Common developmental requirement for Olig function indicates a motor neuron/oligodendrocyte connection. *Cell* **109,** 75–86.

Lufkin, T., Dierich, A., LeMeur, M., Mark, M., and Chambon, P. (1991). Disruption of the Hox-1.6 homeobox gene results in defects in a region corresponding to its rostral domain of expression. *Cell* **66,** 1105–1119.

Lumsden, A. (1990). The cellular basis of segmentation in the developing hindbrain. *Trends Neurosci.* **13,** 329–335.

Lumsden, A., and Guthrie, S. (1991). Alternating patterns of cell surface properties and neural crest cell migration during segmentation of the chick hindbrain. *Development* (Suppl. 2), 9–15.

Lumsden, A., and Keynes, R. (1989). Segmental patterns of neuronal development in the chick hindbrain. *Nature* **337,** 424–428.

Lumsden, A., and Krumlauf, R. (1996). Patterning the vertebrate neuraxis. *Science* **274,** 1109–1115.

Machold, R., and Fishell, G. (2005). Math1 is expressed in temporally discrete pools of cerebellar rhombic-lip neural progenitors. *Neuron* **48,** 17–24.

Maconochie, M., Nonchev, S., Morrison, A., and Krumlauf, R. (1996). Paralogous *Hox* genes: Function and regulation. *Annu. Rev. Genet.* **30,** 529–556.

Maklad, A., and Fritzsch, B. (2003). Development of vestibular afferent projections into the hindbrain and their central targets. *Brain Res. Bull.* **60,** 497–510.

Marin, F., and Puelles, L. (1995). Morphological fate of rhombomeres in quail/chick chimeras: A segmental analysis of hindbrain nuclei. *Eur. J. Neurosci.* **7,** 1714–1738.

Mark, M., Lufkin, T., Vonesch, J. L., Ruberte, E., Olivo, J. C., Dolle, P., Gorry, P., Lumsden, A., and Chambon, P. (1993). Two rhombomeres are altered in *Hoxa-1* mutant mice. *Development* **119,** 319–338.

Matis, C., Oury, F., Remacle, S., Lampe, X., Gofflot, F., Picard, J. J., Rijli, F. M., and Rezsohazy, R. (2007). Identification of Lmo1 as part of a Hox-dependent regulatory network for hindbrain patterning. *Dev. Dyn.* **236,** 2675–2684.
McClintock, J. M., Kheirbek, M. A., and Prince, V. E. (2002). Knockdown of duplicated zebrafish *hoxb1* genes reveals distinct roles in hindbrain patterning and a novel mechanism of duplicate gene retention. *Development* **129,** 2339–2354.
McMahon, A. P. (2000). Neural patterning: The role of Nkx genes in the ventral spinal cord. *Genes Dev.* **14,** 2261–2264.
Metcalfe, W. K., Mendelson, B., and Kimmel, C. B. (1986). Segmental homologies among reticulospinal neurons in the hindbrain of the zebrafish larva. *J. Comp. Neurol.* **251,** 147–159.
Murphy, P., and Hill, R. E. (1991). Expression of the mouse labial-like homeobox-containing genes, *Hox 2.9* and *Hox 1.6*, during segmentation of the hindbrain. *Development* **111,** 61–74.
Niederlander, C., and Lumsden, A. (1996). Late emigrating neural crest cells migrate specifically to the exit points of cranial branchiomotor nerves. *Development* **122,** 2367–2374.
Novitch, B. G., Chen, A. I., and Jessell, T. M. (2001). Coordinate regulation of motor neuron subtype identity and pan-neuronal properties by the bhlh repressor Olig2. *Neuron* **31,** 773–789.
Osumi, N., Hirota, A., Ohuchi, H., Nakafuku, M., Iimura, T., Kuratani, S., Fujiwara, M., Noji, S., and Eto, K. (1997). Pax-6 is involved in the specification of hindbrain motor neuron subtype. *Development* **124,** 2961–2972.
Oury, F., Murakami, Y., Renaud, J. S., Pasqualetti, M., Charnay, P., Ren, S. Y., and Rijli, F. M. (2006). Hoxa2- and rhombomere-dependent development of the mouse facial somatosensory map. *Science* **313,** 1408–1413.
Pasqualetti, M., Diaz, C., Renaud, J. S., Rijli, F. M., and Glover, J. C. (2007). Fate-mapping the mammalian hindbrain: Segmental origins of vestibular projection neurons assessed using rhombomere-specific Hoxa2 enhancer elements in the mouse embryo. *J. Neurosci.* **27,** 9670–9681.
Pata, I., Studer, M., van Doorninck, J. H., Briscoe, J., Kuuse, S., Engel, J. D., Grosveld, F., and Karis, A. (1999). The transcription factor GATA3 is a downstream effector of Hoxb1 specification in rhombomere 4. *Development* **126,** 5523–5531.
Pattyn, A., Vallstedt, A., Dias, J. M., Samad, O. A., Krumlauf, R., Rijli, F. M., Brunet, J. F., and Ericson, J. (2003). Coordinated temporal and spatial control of motor neuron and serotonergic neuron generation from a common pool of CNS progenitors. *Genes Dev.* **17,** 729–737.
Pfaff, S., and Kintner, C. (1998). Neuronal diversification: Development of motor neuron subtypes. *Curr. Opin. Neurobiol.* **8,** 27–36.
Pierani, A., Brenner-Morton, S., Chiang, C., and Jessell, T. M. (1999). A sonic hedgehog-independent, retinoid-activated pathway of neurogenesis in the ventral spinal cord. *Cell* **97,** 903–915.
Popperl, H., Bienz, M., Studer, M., Chan, S. K., Aparicio, S., Brenner, S., Mann, R. S., and Krumlauf, R. (1995). Segmental expression of Hoxb-1 is controlled by a highly conserved autoregulatory loop dependent upon exd/pbx. *Cell* **81,** 1031–1042.
Rijli, F. M., Mark, M., Lakkaraju, S., Dierich, A., Dolle, P., and Chambon, P. (1993). A homeotic transformation is generated in the rostral branchial region of the head by disruption of Hoxa-2, which acts as a selector gene. *Cell* **75,** 1333–1349.
Rijli, F. M., Dolle, P., Fraulob, V., LeMeur, M., and Chambon, P. (1994). Insertion of a targeting construct in a Hoxd-10 allele can influence the control of Hoxd-9 expression. *Dev. Dyn.* **201,** 366–377.

Rijli, F. M., Gavalas, A., and Chambon, P. (1998). Segmentation and specification in the branchial region of the head: The role of the *Hox* selector genes. *Int. J. Dev. Biol.* **42,** 393–401.

Rodriguez, C. I., and Dymecki, S. M. (2000). Origin of the precerebellar system. *Neuron* **27,** 475–486.

Rossel, M., and Capecchi, M. R. (1999). Mice mutant for both Hoxa1 and Hoxb1 show extensive remodeling of the hindbrain and defects in craniofacial development. *Development* **126,** 5027–5040.

Samad, O. A., Geisen, M. J., Caronia, G., Varlet, I., Zappavigna, V., Ericson, J., Goridis, C., and Rijli, F. M. (2004). Integration of anteroposterior and dorsoventral regulation of Phox2b transcription in cranial motoneuron progenitors by homeodomain proteins. *Development* **131,** 4071–4083.

Sander, M., Paydar, S., Ericson, J., Briscoe, J., Berber, E., German, M., Jessell, T. M., and Rubenstein, J. L. (2000). Ventral neural patterning by Nkx homeobox genes: Nkx6.1 controls somatic motor neuron and ventral interneuron fates. *Genes Dev.* **14,** 2134–2139.

Santagati, F., and Rijli, F. M. (2003). Cranial neural crest and the building of the vertebrate head. *Nat. Rev. Neurosci.* **4,** 806–818.

Sechrist, J., Serbedzija, G. N., Scherson, T., Fraser, S. E., and Bronner-Fraser, M. (1993). Segmental migration of the hindbrain neural crest does not arise from its segmental generation. *Development* **118,** 691–703.

Serbedzija, G. N., Bronner-Fraser, M., and Fraser, S. E. (1992). Vital dye analysis of cranial neural crest cell migration in the mouse embryo. *Development* **116,** 297–307.

Sieber, M. A., Storm, R., Martinez-de-la-Torre, M., Muller, T., Wende, H., Reuter, K., Vasyutina, E., and Birchmeier, C. (2007). Lbx1 acts as a selector gene in the fate determination of somatosensory and viscerosensory relay neurons in the hindbrain. *J. Neurosci.* **27,** 4902–4909.

Simon, H., and Lumsden, A. (1993). Rhombomere-specific origin of the contralateral vestibulo-acoustic efferent neurons and their migration across the embryonic midline. *Neuron* **11,** 209–220.

Simon, H., Hornbruch, A., and Lumsden, A. (1995). Independent assignment of antero-posterior and dorso-ventral positional values in the developing chick hindbrain. *Curr. Biol.* **5,** 205–214.

Sotelo, C. (2004). Cellular and genetic regulation of the development of the cerebellar system. *Prog. Neurobiol.* **72,** 295–339.

Studer, M., Lumsden, A., Ariza-McNaughton, L., Bradley, A., and Krumlauf, R. (1996). Altered segmental identity and abnormal migration of motor neurons in mice lacking Hoxb-1. *Nature* **384,** 630–634.

Studer, M., Gavalas, A., Marshall, H., Ariza-McNaughton, L., Rijli, F. M., Chambon, P., and Krumlauf, R. (1998). Genetic interactions between Hoxa1 and Hoxb1 reveal new roles in regulation of early hindbrain patterning. *Development* **125,** 1025–1036.

Sugiyama, S., Di Nardo, A. A., Aizawa, S., Matsuo, I., Volovitch, M., Prochiantz, A., and Hensch, T. K. (2008). Experience-dependent transfer of Otx2 homeoprotein into the visual cortex activates postnatal plasticity. *Cell* **134,** 508–520.

Takahashi, M., and Osumi, N. (2002). Pax6 regulates specification of ventral neurone subtypes in the hindbrain by establishing progenitor domains. *Development* **129,** 1327–1338.

Tan, D. P., Ferrante, J., Nazarali, A., Shao, X., Kozak, C. A., Guo, V., and Nirenberg, M. (1992). Murine *Hox-1.11* homeobox gene structure and expression. *Proc. Natl. Acad. Sci. USA* **89,** 6280–6284.

Taneja, R., Thisse, B., Rijli, F. M., Thisse, C., Bouillet, P., Dolle, P., and Chambon, P. (1996). The expression pattern of the mouse receptor tyrosine kinase gene MDK1 is conserved through evolution and requires Hoxa-2 for rhombomere-specific expression in mouse embryos. *Dev. Biol.* **177,** 397–412.

Trainor, P. A., and Krumlauf, R. (2001). *Hox* genes, neural crest cells and branchial arch patterning. *Curr. Opin. Cell Biol.* **13,** 698–705.

Wang, V. Y., Rose, M. F., and Zoghbi, H. Y. (2005). Math1 expression redefines the rhombic lip derivatives and reveals novel lineages within the brainstem and cerebellum. *Neuron* **48,** 31–43.

Waskiewicz, A. J., Rikhof, H. A., and Moens, C. B. (2002). Eliminating zebrafish pbx proteins reveals a hindbrain ground state. *Dev. Cell* **3,** 723–733.

Wilkinson, D. G. (1995). Genetic control of segmentation in the vertebrate hindbrain. *Perspect. Dev. Neurobiol.* **3,** 29–38.

Wingate, R. J., and Hatten, M. E. (1999). The role of the rhombic lip in avian cerebellum development. *Development* **126,** 4395–4404.

Wingate, R. J., and Lumsden, A. (1996). Persistence of rhombomeric organisation in the postsegmental hindbrain. *Development* **122,** 2143–2152.

Wizenmann, A., and Lumsden, A. (1997). Segregation of rhombomeres by differential chemoaffinity. *Mol. Cell. Neurosci.* **9,** 448–459.

Wong, K., Park, H. T., Wu, J. Y., and Rao, Y. (2002). Slit proteins: Molecular guidance cues for cells ranging from neurons to leukocytes. *Curr. Opin. Genet. Dev.* **12,** 583–591.

Ye, W., Shimamura, K., Rubenstein, J. L., Hynes, M. A., and Rosenthal, A. (1998). FGF and Shh signals control dopaminergic and serotonergic cell fate in the anterior neural plate. *Cell* **93,** 755–766.

CHAPTER SIX

Hox Networks and the Origins of Motor Neuron Diversity

Jeremy S. Dasen* *and* Thomas M. Jessell[†]

Contents

* Smilow Neuroscience Program, Department of Physiology and Neuroscience, New York University School of Medicine, New York, NY, USA

† Departments of Neuroscience, and Biochemistry and Molecular Biophysics, Howard Hughes Medical Institute, Kavli Institute for Brain Science, Columbia University, New York, NY, USA

Current Topics in Developmental Biology, Volume 88
ISSN 0070-2153, DOI: 10.1016/S0070-2153(09)88006-X

Abstract

Motor behaviors are the primary means by which animals interact with their environment, forming the final output of most central nervous system (CNS) activity. The neural circuits that govern basic locomotor functions appear to be genetically hard wired and are comprised of discrete groups of neurons residing within the spinal cord. These local microcircuits coordinate simple reflexive behaviors in response to sensory stimuli and underlie the generation of rhythmic patterns of neural activity necessary for walking. In recent years there have been significant advances in understanding the genetic and molecular programs that determine the specificity of neural connections within the spinal cord that are critical for the emergence of coordinate motor behaviors.

The assembly of circuits within the spinal cord requires the generation of diverse cell types to accommodate the intricate sets of interconnections between motor neurons, sensory neurons, interneurons, and muscle. The first and most critical aspect of this process is that motor neurons select their appropriate muscle targets in the periphery with fidelity and precision. All of the subsequent steps in motor neuron connectivity, such as their descending inputs from higher brain centers, their circuits with sensory neurons and interneurons are constrained by the early connections formed between motor neurons and their muscle targets. The actions of a single family of transcription factors, encoded by the chromosomally clustered *Hox* genes, appear to have a central role in defining the specificity of motor neuron-muscle connectivity. The emerging logic of Hox protein function in motor neuron specification may provide more general insights into the programs that determine synaptic specificity in other CNS regions.

1. Introduction

Much of the computational power of vertebrate nervous systems is dedicated to the goal of controlling movement (Sherrington, 1906), and motor systems have necessarily evolved flexibility and adaptability to respond to the biomechanical challenges imposed by the outside world. Among the most sophisticated motor programs are those executed by the limbs—from the orderly recruitment of flexor and extensor muscles during locomotion to the higher level muscle synergies used in grasping and object manipulation. From a developmental perspective, the control of diverse motor behaviors presents a challenging problem in target specificity—demanding the coordinate activation of over a hundred distinct muscles, each by a dedicated set of spinal motor neurons. The selectivity with which motor neurons innervate their limb target muscles therefore represents an early and critical element in the assembly of vertebrate sensory-motor circuits.

The stereotypic nature of developing motor axonal projections within the limb led to the idea that motor neurons possess subtype identities that define their innervation patterns (Landmesser, 2001; Milner and Landmesser, 1999). The precision of nerve-muscle connectivity was argued to have its origins in the ability of motor neuron subtypes to send their axons into nerve branches that bring them to different muscle targets. The concept that motor neurons possess intrinsic features that direct selective patterns of axonal projection and target innervation received further support from studies showing that motor neurons are able to redirect their axons along new trajectories and find their correct muscle targets when forced to enter the limb from aberrant positions (Lance-Jones and Landmesser, 1980; Landmesser, 2001). Within the spinal cord, the cell bodies of motor neurons that project axons along a given peripheral pathway are grouped in discrete clusters—termed motor columns, divisions, and pools—and these neuronal subtypes occupy fixed positions within the spinal cord. These motor neuron subtypes appear to acquire an early identity that instructs their axons to grow along highly specific trajectories to their muscle targets.

Work over the past decade has begun to define the molecular programs that operate during embryonic development to determine motor neuron fate and associated patterns of connectivity. One key insight into these molecular programs is that core features of motor neuron identity that determine migratory routes, settling positions, patterns of axonal projections, and selection of synaptic targets are defined by the selective profile of transcription factor expression. The actions of one diverse family of transcription factors, encoded by the *Hox* genes, are central mediators of the intrinsic programs that shape motor neuron subtype identity and target muscle specificity.

The focus of this chapter is to outline recent studies that have helped to define the mechanisms by which a Hox-based transcriptional network controls motor neuron identity and connectivity in the developing spinal cord.

2. Spinal Motor Neuron Diversity

Terrestrial vertebrates possess hundreds of anatomically distinct muscle groups. The motor neurons that innervate these diverse targets are organized into discrete clusters within the spinal cord. The position that these groups of motor neurons occupy within the spinal cord is relatively fixed from animal to animal, and thus cell body position is often predictive of target innervation pattern (Hollyday and Jacobson, 1990; Landmesser, 1978b). Understanding the genetic programs that contribute to the formation of motor innervation maps has been a major challenge in the field. In this section we describe studies that have helped to define the early programs which establish motor neurons as a class and the emergence of motor neuron topographic maps.

2.1. Generation of generic motor neuron identity

Motor neurons and several classes of interneurons are generated in response to graded extrinsic signals acting along the dorsoventral axis of the neural tube. These secreted signals include sonic hedgehog (Shh) from the notochord and floor plate, and fibroblast growth factors (FGFs) and retinoic acid (RA) by the paraxial mesoderm. The detailed mechanisms through which Shh, RA, and FGFs define progenitor identities will not be addressed here since this topic has been the subject of several review articles (Dessaud *et al.*, 2008; Jessell, 2000; Shirasaki and Pfaff, 2002). In brief, Shh, FGF, and RA signaling induce the expression of distinct combinations of transcription factors in neural progenitors. These initial patterns are subsequently refined through the selective transcriptional cross-repressive interactions between transcription factors expressed at the boundaries between progenitor domains (Fig. 6.1A). Each progenitor domain in the neural tube expresses a unique profile of transcription factors, and these combinatorial patterns define progenitor fates (Briscoe *et al.*, 2000). Progenitors that give to motor neurons depend on the activities of the bHLH protein Olig2 and the homeodomain factors Pax6, Nkx6.1, and Nkx6.2 (Novitch *et al.*, 2001; Vallstedt *et al.*, 2001; Zhou and Anderson, 2002). Thus a major output of the dorsoventral signaling system is to define the identity of motor neurons as opposed to ventral interneurons.

Soon after their generation, at about embryonic day (*e*) 9.5 in mouse, spinal motor neurons express a set of homeodomain transcription factors (notably Hb9, Lhx3, Isl1, and Isl2), that that control features common to all spinal motor neurons as well as those that are involved in later aspects of subtype diversification (Fig. 6.1A) (Arber *et al.*, 1999; Pfaff *et al.*, 1996; Sharma *et al.*, 1998; Thaler *et al.*, 1999, 2004). These generic motor neurons characteristics include the projections of axons outside the spinal cord and the release of acetylcholine as the primary neurotransmitter. Although patterning events mediated by Shh, RA, and FGF signaling define how motor neurons as a class are specified, additional signaling pathways are presumably necessary for the further diversification of motor neurons into distinct subtypes.

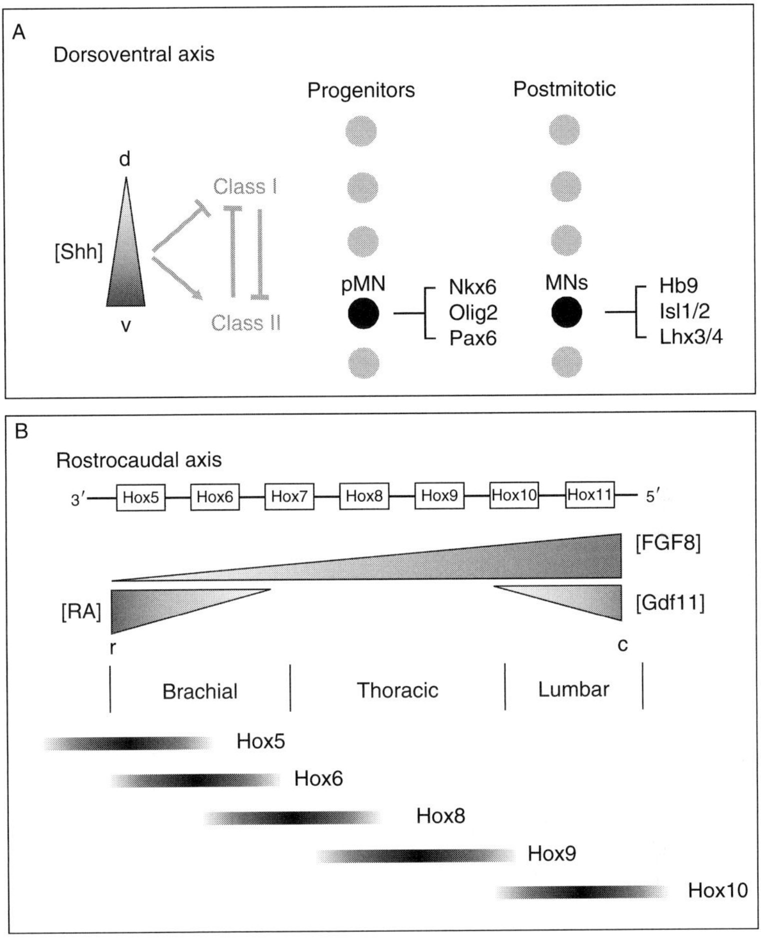

Figure 6.1 Patterning along the dorsoventral and rostrocaudal axes of the neural tube. (A) Motor neurons and ventral interneurons are generated along the dorsoventral (d–v) axis in response to the graded activities of sonic hedgehog (Shh) which induces the patterned expression of transcription factors in progenitor cells. Class I factors are induced by Shh while Class II transcription factors are repressed. Selective cross-repressive interactions between Class I and Class II transcription factor sharpen the boundaries between progenitor domains (Briscoe *et al.*, 2000). Retinoic acid (RA) from the paraxial mesoderm and fibroblast growth factor (FGF) signaling also influence the pattern of transcription factors in neural tube progenitors (not shown). (B) Along the rostrocaudal axis graded FGF signaling induces the expression of chromosomally linked *Hox* genes in the neural tube. *Hox* genes located at one end of the cluster are expressed more rostrally (r) while genes at the opposite end are expressed caudally (c) in response to higher levels of FGF. At more rostral levels *Hox* genes are regulated by graded RA signaling while at more caudal levels *Hox* genes are regulated by graded Gdf11.

2.2. Anatomical organization of spinal motor neuron subtypes

At the time of their birth, all motor neurons possess a set of core features that distinguish them from other classes of neurons, but rapidly diversify into subtype identities that allow them to form selective connections with target cells. Major distinctions in the identities of motor neurons have been defined through studies of their position, axon trajectory, and pattern of muscle innervation. One level of organization is the allocation of motor neurons to columnar groups, each column occupying a defined position along the rostrocaudal axis of the spinal cord. Four major columnar classes have been described, each innervating a unique set of peripheral target tissues (Fig. 6.2). The most prominent of these columnar groups are the lateral motor columns (LMCs) which are generated at limb levels of the spinal cord and innervate limb muscles. At thoracic levels visceral preganglionic column (PGC) motor neurons innervate sympathetic ganglia while hypaxial motor column (HMC) neurons innervate intercostal and abdominal wall musculature (Gutman *et al.*, 1993; Prasad and Hollyday, 1991). In contrast to these segmentally restricted motor columns, motor neurons in the median motor column (MMC) are present at all levels of the spinal cord and innervate dorsal axial musculature (Fetcho, 1987; Gutman *et al.*, 1993).

Additional layers of anatomical organization are present within these main columns, notably the segregation of the LMC into "divisions" and "pools." At both forelimb and hindlimb levels of the spinal cord, the LMC is split into two divisions: a medial division which contains neurons projecting axons ventrally within the limb mesenchyme and a lateral division which contains neurons that project dorsally (Fig. 6.2B) (Landmesser, 1978a; Tosney and Landmesser, 1985a,b). These two divisional subgroups have been linked to the particular mode of actions of their muscle targets: dorsally projecting lateral LMC axons frequently innervate extensor muscles, whereas medial LMC axons project ventrally and typically innervate flexor muscles. The functional relevance of this anatomical segregation is still unclear.

A third level of motor neuron diversity is evident in the segregation of motor neurons into motor pools (Romanes, 1942). Motor pools occupy distinct positions within the LMC, each pool innervating a dedicated target muscle (Hollyday and Jacobson, 1990; Landmesser, 1978b). Whereas forelimb and hindlimb motor neurons share similar columnar and divisional properties, the organization of motor pools between these two levels of the spinal cord is quite distinct, reflecting differences in the overall pattern of musculature between the forelimb and hindlimb. Nevertheless motor pools innervating forelimb and hindlimb form topographic maps which are similarly organized. At both levels of the spinal cord motor pools located more rostrally tend to innervate muscles located rostral and proximal within the limb, while caudal pools project to more caudal and distal regions (Hollyday and Jacobson, 1990; Landmesser, 1978b).

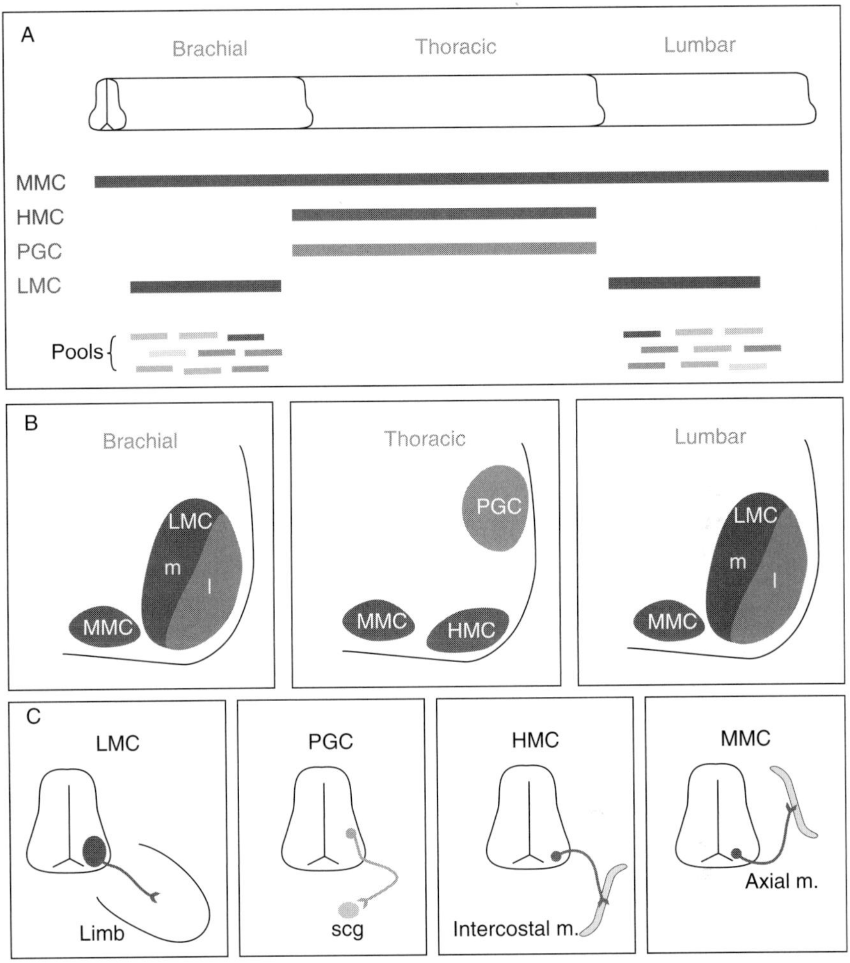

Figure 6.2 Motor neuron columnar, divisional and pool organization. (A) Motor columns and motor pools are generated at specific positions along the rostrocaudal axis. The cell bodies of motor neurons that send axons to the limb are contained within the lateral motor column (LMC) at brachial and lumbar levels of the spinal cord. Preganglionic column (PGC) motor neurons and hypaxial motor column (HMC) neurons are found at thoracic levels. Motor neurons within the medial motor column (MMC) are generated at all rostrocaudal levels of the spinal cord. Motor pools are generated at specific rostrocaudal positions within the LMC. (B) Schematic of cross sections of brachial, thoracic, and lumbar spinal cord showing the position of motor columns and divisions. Medial (m) and lateral (l) divisions of the LMC are indicated. (C) Projection patterns of motor neuron columnar subtypes. LMC neurons project to the limb, PGC neurons to sympathetic chain ganglia (scg), HMC neurons to intercostal and body wall muscles (m), MMC neurons to axial muscle.

3. Hox Expression in Developing Motor Neurons

The diversification of motor neurons into specific columnar, divisional, and pool subtypes relies on the acquisition of a positional identity along the rostrocaudal axis of the spinal cord. Like neuronal specification along the dorsoventral axis, motor neurons acquire positional information in response to graded signals which initiate transcriptional programs within progenitor and postmitotic cells. A critical output of the rostrocaudal signaling pathways acting on the neural tube is the establishment of selective patterns of *Hox* gene expression within specific motor neuron subtypes.

3.1. Rostrocaudal positional information in spinal cord development

Early insights into the role of rostrocaudal positional information in locomotor circuit assembly came from embryological manipulations in the chick embryo. One of the most dramatic set of experiments was performed by Victor Hamburger and colleagues who demonstrated that if the region of the spinal cord responsible for synchronous activation of wing muscles is grafted to the level of the hindlimb, chickens will synchronously activate muscle in the legs (Narayanan and Hamburger, 1971). Conversely, grafting hindlimb-level spinal cord to wing levels causes the chick to alternate wing movements, in a pattern similar to walking. These studies provide evidence that the intrinsic properties of neurons generated at specific rostrocaudal levels have critical roles in establishing the local circuitries controlling basic behavioral outputs.

More recent studies on rostrocaudal programming have demonstrated that positional identities are regulated by signals derived from axial structures, such as the node and notochord, as well as from paraxial tissues, the presomitic mesoderm and somites (Ensini *et al.*, 1998; Lance-Jones *et al.*, 2001; Liu *et al.*, 2001). Grafting experiments in chick indicated that regionally restricted signals govern the specification of motor neuron columnar subtypes. Transposition of thoracic and brachial levels of the neural tube during a critical period of development respecifies columnar fates—neurons derived from the former thoracic region of the neural tube acquire LMC identity, and conversely, motor neurons from the former brachial level acquire a CT identity (Ensini *et al.*, 1998; Shieh, 1951). The signals for columnar respecification derive, in part, from the adjacent paraxial mesoderm, since a similar switching of motor neuron columnar identity can be elicited by transposition of brachial and thoracic paraxial mesoderm (Ensini *et al.*, 1998).

3.2. Regulation of Hox expression by FGF, Retinoid, Wnt, and TGFβ signaling

While grafting studies in chick indicated that the regional identities of motor neurons are controlled through mesodermal signals, the identity of these signals and the intrinsic mediators of their actions were not known. One class of transcription factors with an evolutionarily conserved role in establishing differences in cell identity along the rostrocaudal axis are members of the chromosomally arrayed *Hox* gene family. The expression of *Hox* genes within the spinal cord is closely aligned with their position within the *Hox* cluster: genes located at the 3′ end of the cluster are expressed more anteriorly than genes at the 5′ end (Fig. 6.1B) (Kmita and Duboule, 2003; Lemons and McGinnis, 2006). The precise mechanism by which spatial colinear expression of *Hox* genes emerges in the neural tube is unknown; although gradients of signaling molecules appear to impart the initial profiles of *Hox* expression in most tissues where colinearity has been examined.

The expression of *Hox* genes within the CNS is controlled by multiple signaling molecules including FGFs, retinoids, Wnts, and members of the transforming growth factor (TGF) β superfamily (Bel-Vialar *et al.*, 2002; Diez del Corral and Storey, 2004; Liu, 2006; Liu *et al.*, 2001; Nordstrom *et al.*, 2006). Graded FGF signaling is involved in establishing the initial induction of *Hox* gene expression at brachial, thoracic, and lumbar levels of the spinal cord (Bel-Vialar *et al.*, 2002; Dasen *et al.*, 2003; Liu *et al.*, 2001). At the caudal end of the chick embryo, an organizing region called Hensen's node and the presomitic mesoderm are primary sources of FGF signals. As the tail bud regresses caudally during axis extension, more posterior regions of the spinal cord are exposed to FGF in higher concentration and over longer periods of time (Dubrulle and Pourquie, 2004; Liu *et al.*, 2001). Both *in vitro* and *in vivo* studies have shown that *Hox* genes located at the 3′ end of a cluster are induced by low levels of FGF while those at the 5′ end are induced by progressively higher FGF levels (Bel-Vialar *et al.*, 2002; Dasen *et al.*, 2003; Liu *et al.*, 2001). As a consequence, *Hox4-Hox8* paralog genes are expressed at brachial levels, *Hox8-Hox9* genes at thoracic, and *Hox10-Hox13* genes at lumbar levels of the spinal cord (Fig. 6.1B).

While graded FGF signals contribute to initial *Hox* patterns, other signaling systems participate in regulating subsets of *Hox* genes within a cluster. At more anterior levels RA signaling provided by paraxial mesoderm and somites has been shown to regulate Hox expression at brachial levels (Liu *et al.*, 2001). The action of retinoids is in part to antagonize the FGF gradient (Diez del Corral and Storey, 2004); although the mechanisms underlying this process are not known. At more caudal levels, the TGFβ family member Gdf11 has been shown to be essential both *in vitro* and *in vivo* in the regulation of *Hox8-Hox10* paralogs at thoracic and lumbar levels of the spinal cord (Liu, 2006; McPherron *et al.*, 1999), and this inductive factor appears to act in concert with high levels of FGF signaling (Liu *et al.*, 2001). Thus Hox expression in motor neurons relies on the convergent actions of multiple signaling pathways (Fig. 6.1B).

3.3. The emergence of definitive Hox patterns in motor neurons

Although graded signals are necessary to establish the initial pattern of *Hox* gene expression in motor neurons, the links between extrinsic signals and Hox protein expression in motor neurons are still unclear. While RA response elements have been characterized in the *Hox* genes controlling regional identity in the hindbrain (Glover *et al.*, 2006; Trainor and Krumlauf, 2000), similar control elements have yet to be described for *Hox* genes expressed within the spinal cord. How the neural tube interprets the FGF gradient is also unknown; although a class of homeodomain proteins related to the *Drosophila* gene *caudal* has been implicated. Manipulation of vertebrate *caudal* homeobox (*Cdx*) activities produces phenotypes which are similar to manipulations of FGF signaling. Misexpression of activated forms of Cdx can shift the patterns of Hox expression in the spinal cord (Bel-Vialar *et al.*, 2002), while in Zebrafish loss of Cdx activities deregulates Hox expression and the spinal cord acquires a hindbrain-like character (Shimizu *et al.*, 2006; Skromne *et al.*, 2007). The mechanisms by which Cdx proteins regulate *Hox* expression are not known, but may involve direct Cdx binding to individual *Hox* regulatory elements or interaction with locus control regions within the *Hox* clusters.

Another unresolved issue is the inordinate temporal delay between the time of neural tube exposure to graded FGF signals and the emergence of Hox expression in postmitotic motor neurons. FGF signaling appears to act on neural progenitors, and manipulation of FGF signals can rapidly switch *Hox* transcriptional profiles at early stages of development (Bel-Vialar *et al.*, 2002; Dasen *et al.*, 2003). However the expression of some Hox proteins by motor neurons is detected only in postmitotic cells (Dasen *et al.*, 2003). The mechanisms which introduce this apparent delay between Hox RNA and protein expression are not known; although *Hox* genes are under the control of several layers of posttranscriptional regulation, including silencing by microRNAs (Chopra and Mishra, 2006). Another possibility is that expression of *HoxB* cluster genes in progenitors, which are generally not detected in postmitotic motor neurons (Dasen *et al.*, 2005), prefigures progenitors to express *HoxA*, *HoxC*, and *HoxD* genes.

4. Hox Proteins Determine Motor Neuron Columnar Identity and Connectivity

Early insights into the role of *Hox* genes in CNS development came from studies in the vertebrate hindbrain. In the hindbrain the metameric organization of the rhombomeres provides anatomical landmarks for defining global aspects of Hox function, and several *Hox* gene mutants are characterized by homeotic transformation of rhombomeres identities and changes in neuronal identity (reviewed in Guthrie (2007)). In contrast, the relative morphological homogeneity and lack of columnar and pool specific molecular markers in the spinal cord created a challenge in linking Hox function to the specification of motor neuron subtypes. Although it had been recognized that different levels of the spinal cord express distinct *Hox* genes (Carpenter, 2002), the link to motor neuron columnar, divisional, and pool identities were unclear.

4.1. Specification of segmentally restricted columnar subtypes by *Hox* genes

Certain motor neuron populations can be defined by the combinatorial expression of LIM-homeodomain proteins (Tsuchida *et al.*, 1994); although no single LIM-homeodomain protein is specific for a single neuronal class in the spinal cord. The identification of molecular markers that are selectively expressed in two segmentally restricted motor columns, LMC and PGC neurons, provided a means to explore the signaling pathways that specify motor neuron subtypes. PGC neurons can be distinguished from other thoracic-level spinal motor neurons by expression of *BMP5*, a TGFβ family member (William *et al.*, 2003) as well as nuclear phospho-(p)-Smad1/5/8 (Dasen *et al.*, 2008). At limb levels LMC neurons can be defined by their selective expression of retinaldehyde dehydrogenase-2 (RALDH2), a key enzyme in retinoic acid synthesis (Sockanathan and Jessell, 1998).

Expression of Hox proteins is closely aligned with the position in which molecularly defined columnar subtypes are generated: expression of Hox6 proteins segregates with brachial (forelimb) LMC neurons, Hox9 proteins with thoracic PGC neurons, and Hox10 proteins with lumbar (hindlimb) LMC neurons (Fig. 6.3A) (Choe *et al.*, 2006; Dasen *et al.*, 2003; Lance-Jones *et al.*, 2001; Liu *et al.*, 2001). Consistent with the model where Hox protein expression is controlled by graded FGF signaling, elevation of FGF levels at brachial levels of the spinal cord induces the expression of Hoxc9, a *Hox* gene normally restricted to thoracic levels (Dasen *et al.*, 2003).

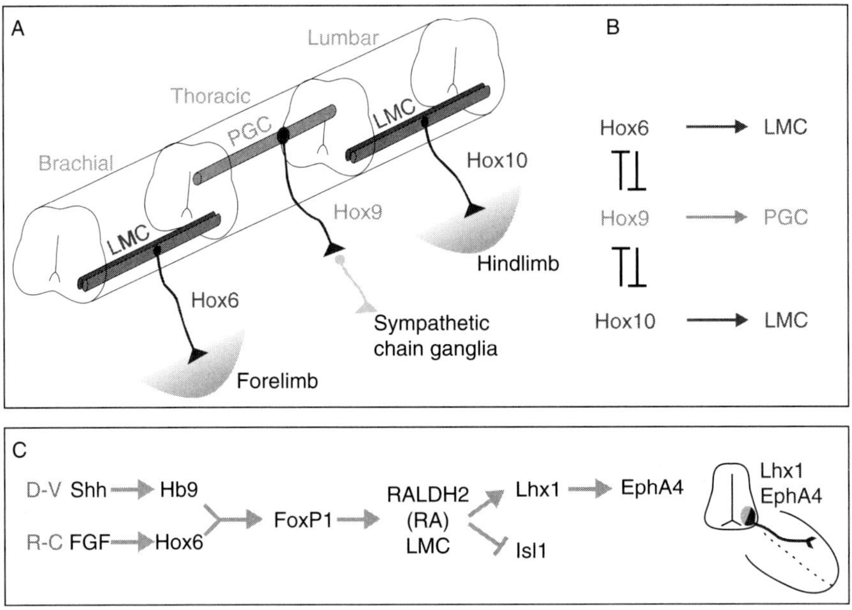

Figure 6.3 *Hox* genes and the specification of segmentally restricted motor columns. (A) Hox6, Hox9, and Hoxd10 are expressed in motor neurons at distinct rostrocaudal levels of the spinal cord and direct motor neuron identity and peripheral target connectivity. Hox6 activities control brachial LMC identity, Hox9 control PGC identity, and Hox10 lumbar LMC identity. (B) Regulatory interactions between *Hox* genes in motor neuron columnar fates. Cross-repressive interactions between Hox6, Hox9, and Hox10 proteins refine Hox profiles and Hox activator functions define LMC and PGC identities. (C) Role of LMC *Hox* genes in the program controlling motor neuron divisional identities and axonal projections along the dorsoventral axis of the limb. The convergent activities of motor neuron specific transcription factors (Hb9) and limb-level *Hox* genes (e.g., Hoxc6) direct expression of FoxP1 and RALDH2 in LMC neurons. RALDH2 creates a neuronal source of RA which leads to the induction of Lhx1 expression by lateral LMC neurons. Lhx1 directs expression of the guidance receptor EphA4, and EphA4 directs motor axons toward the dorsal limb.

This switch in Hox patterns is accompanied by the loss of brachial LMC neurons, characterized by the abolishment of Hoxc6 and RALDH2 expression. Motor neurons are instead converted to a PGC cell fate, defined by expression of *BMP5* and pSmad. The effects of elevated FGF are not restricted to changes in marker gene expression, but extend to multiple aspects of columnar identity including switches in the patterns of migration, peripheral connectivity, and the number of motor neurons generated at limb and thoracic levels (Dasen *et al.*, 2003).

Are the effects of elevating FGF signaling on motor neuron columnar identity mediated by changes in *Hox* gene expression? Consistent with a direct role in columnar specification, misexpression of Hoxc9 at brachial levels is sufficient to convert LMC to PGC motor neurons, while expression of Hoxc6 or Hoxd10 at thoracic levels can convert PGC and HMC to LMC neurons (Dasen *et al.*, 2003; Shah *et al.*, 2004). These actions of Hox proteins rely on their ability to cross-repress each others expression: Hox6 and Hox10 protein can repress Hox9 expression while Hox9 can repress Hox6 expression (Fig. 6.3B). Moreover, the actions of Hox proteins in inducing columnar fates and establishing Hox boundaries appear to be separable, as expression of a constitutive repressor forms block columnar inducing activities but retain their cross-repressive functions (Dasen *et al.*, 2003). Thus like many transcription factors Hox proteins possess intrinsic activator and repressor functions, and these functional differences serve distinct but coherent roles in motor neuron subtype specification (Fig. 6.3B).

The principles of Hox protein function along the rostrocaudal axis parallel those that operate along the dorsoventral axis which specify motor neurons as a class. Along both axes, the initial graded activity of a secreted signaling factor establishes broad domains of homeodomain protein expression that are subsequently refined through selective cross-repressive interactions. However, these two programs of transcriptional cross-repression appear to operate at different stages of neuronal specification. Dorsoventrally, homeodomain cross-repressive interactions are evident within neural progenitor cells (Briscoe *et al.*, 2000), whereas along the rostrocaudal axis Hox cross-repression occurs within postmitotic neurons (Dasen *et al.*, 2003). Nevertheless, the convergence of these two patterning programs ensures that Hox-directed features of columnar differentiation are confined to postmitotic motor neurons.

Aspects in the logic of Hox function in spinal motor neuron diversification are distinct from that used in rostrocaudal patterning of *Drosophila* larvae and in the vertebrate hindbrain. In these two systems, the actions of posteriorly expressed *Hox* genes typically dominate over those of more anteriorly expressed genes—a phenomenon termed posterior dominance (Duboule and Morata, 1994). The findings in the spinal cord argue against posterior prevalence of Hox function in postmitotic motor neuron specification, since ectopic caudal expression of Hoxc6 is as effective as ectopic rostral expression of Hoxc9. Several exceptions to the posterior prevalence rule of Hox function have been reported in both fly and vertebrate embryos (Duboule and Morata, 1994; Jegalian and De Robertis, 1992). Recent genetic analysis of axial skeleton patterning in *Hox* mutants provides further evidence against a global dominance of posterior *Hox* genes in somite derivatives (McIntyre *et al.*, 2007). Thus tissue context may influence functional dominances between *Hox* genes. In neural progenitors *Hox* genes are expressed in overlapping and nested patterns, one possibility is that posterior dominance operates at early stages of spinal cord development to help determine the final Hox pattern in postmitotic neurons.

4.2. Mechanisms of columnar Hox function in motor neuron connectivity

In addition to regulating the expression of columnar-specific genes, Hox proteins can also direct the peripheral connectivity of LMC and PGC neurons. At forelimb levels of the spinal cord the switching of LMC neurons to a PGC fate forces motor neurons to project their axons to the normal PGC targets, sympathetic chain ganglia. Conversely, expression of Hoxc6 or Hoxd10 at thoracic levels induces LMC fate and these neurons project into the limb (Dasen *et al.*, 2005; Shah *et al.*, 2004). Thus Hox proteins not only influence columnar identity at the level of molecular marker expression but also contribute to the initial specificity of motor axon projections in the periphery.

How might the activities of Hox proteins determine columnar-specific patterns of axonal innervation? Markers of newly generated LMC and PGC neurons, RALDH2 and BMP5/pSmad, are intimately involved in inductive signaling. Although the function of BMP5/pSmad signaling in PGC neurons is not known, RALDH2 activity appears to be required for the initial specificity in which LMC neurons project into the limb (Fig. 6.3C). The RALDH2-dependent synthesis of retinoids by LMC neurons is necessary for the specification of lateral LMC neuronal identity, in order to induce expression of Lhx1, a LIM-homeodomain protein (Kania *et al.*, 2000). Lhx1 expression has been shown to direct the dorsal projection of LMC motor axons in the developing limb through its ability to regulate EphA4 expression, a guidance receptor required for axons to avoid the ventral limb mesenchyme (Eberhart *et al.*, 2002; Kania and Jessell, 2003).

Thus an early step in the Hox-dependent specification of LMC identity is to direct RALDH2 expression, and trigger a series of downstream signaling events within postmitotic motor neurons that govern the pattern of motor neuron connectivity in the developing limb. More generally these findings suggest that a key step in the organization of columnar differentiation at different segmental levels of the spinal cord is the induction of signaling factors in different columnar subtypes of motor neurons, which in turn directs a molecular program for motor neuron connectivity.

5. Hox Transcriptional Networks and the Specification of Motor Pool Identities

Within LMC divisions, motor neurons are further subdivided into motor pools, each destined to innervate a single muscle target in the limb. A typical vertebrate limb contains over 50 muscle groups, requiring the generation of a diverse array of motor pool subtypes. Like columns, a motor pools occupy stereotypic rostrocaudal positions within the spinal cord and multiple pools can occupy a single segmental level (Fig. 6.4). As with the specification of segmentally restricted motor columns, Hox transcription factors appear to be critical determinants of pool identity and muscle target specificity.

The proposal that intrinsic motor pool identities direct target muscle connectivity emerged first through embryological manipulations which revealed that the axons of specific LMC neurons project to their limb muscle targets with high precision (Landmesser, 1978a), even when forced to enter the limb from innapropriate positions (Landmesser, 2001). Classical embryological studies have provided evidence that motor neurons within

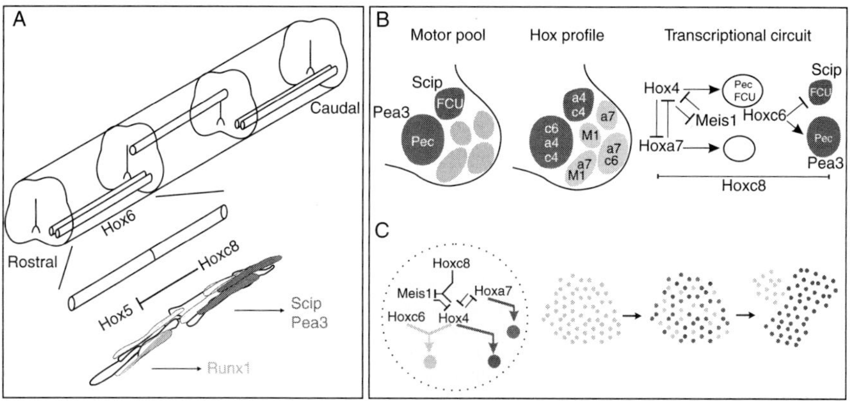

Figure 6.4 A *Hox* transcriptional network controls motor pool identity and connectivity. (A) *Hox* genes determine the rostrocaudal position of motor pools within the LMC. At brachial levels of the spinal cord cross-repressive interactions between Hox5 proteins and Hoxc8 establish the boundary between molecularly defined motor pools. Hox5 proteins (Hoxa5 and Hoxc5) are required to generate the motor pool that expresses the transcription factor Runx1 in rostral LMC neurons. Hoxc8 is required in caudal LMC neurons to generate the motor pools that express the transcription factors Pea3 and Scip. (B) Intrasegmental specification of motor pool identity. At a single segmental level of the spinal cord ~6–10 pools are generated. Motor pools projecting to the pectoralis (Pec) and flexor carpi ulnaris (FCU) can be molecularly defined by expression of the transcription factors Pea3 and Scip, respectively. Both Pec and FCU pools express unique profiles of Hox expression (M1:Meis1, Hoxc6 shown as c6, etc.). The patterns of Hox expression in the Pec and FCU pools are established through a transcriptional network which appears to be driven largely by Hox cross-repressive interactions. (C) Model for a Hox repressilator network in motor pool specification. Individual motor neurons initially inherit expression of multiple *Hox* genes as a function of their position along the rostrocaudal axis. These patterns are refined through repressive interactions on a cell by cell basis, giving rise to motor neurons with a specific Hox pattern (yellow, red, and green cells) that are scattered throughout the column. Subsequently, motor neurons cluster into discrete pools. Biases in the strength of repression may favor expression of one Hox protein over another giving rise to pools of different sizes. (See Color Insert.)

the LMC acquire aspects of their pool identity as their axons first invade the limb mesenchyme, well before approaching muscle targets (Hollyday, 1980a,b, 1995; Landmesser, 1978a,b). As we discuss below *Hox* genes appear to impart the initial selectivity of motor neuron connectivity during this early intrinsic phase of motor neuron differentiation and regulate a diverse repertoire of downstream transcriptional programs that control multiple characteristics of motor pool identity.

5.1. Assignment of motor pool identities by *Hox* genes

The establishment of diverse motor pool subtypes presumably requires the activities of a large number of transcriptional regulators. Systematic analysis of the expression of *Hox* gene expression in chick spinal cord revealed that nearly two dozen are expressed by motor neurons, in a manner consistent with a role in motor pool specification (Dasen *et al.*, 2005). Roles for Hox protein activities in pool specification have been most thoroughly investigated in the brachial level motor neurons that innervate forelimb musculature. Specific brachial motor pools can be molecularly defined by transcription factor expression (e.g., Runx1, Pea3, Scip, and Nkx6 proteins, see Fig. 6.4), and the combinatorial expression of Hox4, Hox5, Hox6, Hox7, and Hox8 proteins appear to define motor neuron pool fate. Experimental manipulation of the pattern of Hox expression in motor neurons leads to changes in motor pool identity, defined by a switch in the molecular profile of pool-specific transcription factors and a change in the pattern of peripheral connectivity of motor axons (Dasen *et al.*, 2005).

Two Hox-dependent programs appear to operate within LMC neurons to control pool fates, one assigning rostrocaudal motor pool position, and a second directing intrasegmental motor pool diversity (Fig. 6.4). The mechanisms by which the rostrocaudal positioning of motor pools is established by Hox proteins largely follow the strategy deployed in columnar specification. Graded FGF and RA signaling determines the initial pattern of *Hox3-Hox8* expression by brachial LMC pools, and rostrocaudal motor pool boundaries are established through selective cross-repressive interactions between pairs of Hox protein. One set of Hox interactions, exemplified by the activities of the Hox5 and Hox8 proteins, constrains motor pool specification to specific rostrocaudal levels of the LMC. At rostral levels of the brachial LMC two *Hox5* paralogs (*Hoxa5* and *Hoxc5*) define the position and identity of rostral motor pools whereas Hoxc8 defines caudal pools (Fig. 6.4A). Thus, the patterns of Hox expression that determine motor pool identity along the rostrocaudal axis of the LMC appear to be set by the same extrinsic signals that establish columnar identities, and are subsequently reinforced through selective cross-repressive interactions.

A single segmental level of the spinal cord contains as many as 10 motor pools (Fig. 6.4B). The neurons that populate these pools derive from progenitors that presumably have been exposed to the same level of rostro-caudal patterning signals, raising the issue of how this aspect of pool diversity is achieved. One model proposed for the intrasegmental diversification posits that all LMC neurons generated at a specific segmental level of the LMC initially express the same set of Hox proteins, as a reflection of their rostro-caudal position. But within this cohort of neurons, the expression of certain Hox proteins is favored over others, by virtue of their mutual repressive interactions (Fig. 6.4C). As a consequence, minor fluctuations in starting Hox conditions within individual motor neurons will result in the gradual extinction of expression of one or other of two opponent Hox proteins. The final complement of Hox proteins expressed within any given LMC neuron will therefore represent only a small subset of the starting repertoire. Consistent with this model expression of Hox4, Hox6, Hox7 proteins and a Hox cofactor called Meis1 are initially co-expressed by most motor neurons at caudal levels of the brachial LMC but eventually segregate in patterns that align with the expression of a set of pool specific transcription factors (Dasen *et al.*, 2005).

The existence of mechanisms that impart a bias to the outcome of Hox cross-regulatory interactions could account for the observation that motor neurons are allocated to distinct pools in different numbers, in anticipation of the size of their muscle target (Lin *et al.*, 1998). Mechanistically this process may involve asymmetries in the strength of Hox repression, or initial differences in the level or onset of Hox expression within individual neurons. This view shares elements in common with the workings of transcriptional repressor networks that have been engineered *de novo* in bacterial model systems (Elowitz and Leibler, 2000).

The analysis of motor pool specification reinforces the view that repressive interactions between Hox proteins direct motor neuron diversification within the spinal cord. But there are important distinctions in the interactions between Hox protein pairs. The Hox6/Hox9 protein pair exhibits mutual repressive interactions during motor neuron columnar specification (Dasen *et al.*, 2003), whereas the Hox5 and Hox8 interaction that occurs during motor pool specification is asymmetric (Dasen *et al.*, 2005). And in chick the repressive interactions between Hox4 and Hoxa7 proteins that occur during the intrasegmental diversification of pools appears not to be absolute, but rather unfolds gradually over several days of development. This reliance on Hox repressive interactions to allocate identities to spinal motor neurons is in apparent contrast with the Hox circuitry involved in hindbrain patterning, where positive auto-regulatory interactions have been observed (Gavalas *et al.*, 1997; Nonchev *et al.*, 1997).

5.2. *Hox* genes control the specificity of motor neuron-muscle connectivity

How might Hox activities in motor pools coordinate motor axon trajectory to specific muscle targets in the developing limb? On arriving at the base of the limb, the axons of LMC neurons select a ventral or dorsal trajectory in the limb mesenchyme, and then establish specific anteroposterior and proximo-distal trajectories which take them to the position of newly cleaved muscle masses (Tosney and Landmesser, 1985a). As developing axons project into the limb they navigate through a series of choice points en route to their synaptic targets. A major output of the Hox network in motor neurons is the control of downstream transcription factor expression, some of which are necessary for motor axon guidance decisions (Fig. 6.4). To what extent are the patterns of motor axon innervation driven through Hox-regulated intermediate transcription factors and to what extent might they be controlled directly by Hox targets?

As described earlier, aspects of limb innervation pattern can be linked to the program of columnar specification. The Hox6/10-activated program of LMC specification appears to direct axons toward the limb and determine a pattern of LIM-homeodomain protein expression that controls the dorsoventral trajectory of motor axons, through regulation of EphA4 expression (Kania and Jessell, 2003). The selection of certain muscle-specific nerve trajectories appears to be determined through activities of the Hox-induced transcription factors. Nkx6 homeodomain proteins are expressed by subsets of LMC neurons in a pool-specific pattern that is controlled by Hox proteins (Dasen *et al.*, 2003; De Marco Garcia and Jessell, 2008). In *Nkx6.1* mutant mice these motor neuron pools fail to innervate their normal target and invade foreign muscle targets (De Marco Garcia and Jessell, 2008). Another pool-specific Hox target, Pea3, is required for distinct aspects of motor neuron differentiation, including the clustering of neurons into pools, muscle-specific axonal arborization, and synaptic input onto motor neurons from sensory neurons (Livet *et al.*, 2002; Vrieseling and Arber, 2006). Thus multiple facets of the Hox programming of motor pool identities are regulated through intermediate transcription factors.

Could Hox proteins exert a more direct role in the control of axon guidance decisions within the limb? After making their initial dorsoventral choice at the base of the limb, motor axons follow cues that guide them along the anteroposterior and proximo-distal axes (Stirling and Summerbell, 1988). The basic pattern of muscle nerve branches is preserved in the absence of the target muscle itself (Lewis *et al.*, 1981; Phelan and Hollyday, 1990, 1991), implicating the limb mesenchyme as a source of cues that specify motor axon trajectories. Hox proteins are also expressed by the limb mesenchyme (Izpisua-Belmonte and Duboule, 1992) and could contribute to the establishment of axonal trajectory by positioning guidance cues at specialized decision regions (Tosney and Landmesser, 1985a). The existence of a topographic relationship between cell body position and projections along these axes raises the possibility that Hox proteins also exert direct roles in the control of specific guidance receptors, or regulate other determinants that direct innervation specificity within the limb.

5.3. Extrinsic and intrinsic programming of motor pool identities

While many aspects of motor pool identity appear to be programmed through cell-intrinsic Hox transcriptional networks, the expression of certain pool specific transcription factors relies on the presence extrinsic signals from the periphery. Expression of the ETS transcription factors Pea3 and Er81 in pools depends on target-derived neurotrophic signals provided by the limb mesoderm and muscle targets (Haase *et al.*, 2002; Lin *et al.*, 1998). These signals appear to be permissive rather than instructive and not all motor neurons are competent in their ability to respond to neurotrophins. In explants of spinal cord treated with glial-derived neurotrophic factor (GDNF), Pea3 is induced in a pattern approximating the normal number in vivo and confined to the level of the spinal cord which normally expresses Pea3 (Haase *et al.*, 2002). Thus not all motor neurons are equivalent in their ability to respond to GDNF.

The competence of motor neurons to activate ETS genes appears to be constrained by their pattern of Hox expression. Ectopic expression of Hoxc8 in LMC neurons is sufficient to expand the domain of Pea3 expression (Dasen *et al.*, 2005), suggesting the normal domain of Hoxc8 expression defines the region of GDNF competence. Consistent with this hypothesis, in *Hoxc8* mutants motor neurons fail to fully activate Pea3 expression (Vermot *et al.*, 2005), likely as a consequence of the inability of motor neurons to respond to GDNF. Hox proteins may therefore control the expression of targets that endow motor neurons with the ability to respond to peripheral cues. Hox targets may include receptors for neurotrophic factors or other components necessary for activation of the GDNF pathway.

These observations suggest that despite the importance of Hox-dependent steps in motor neuron specification, target-derived cues also contribute to the transcriptional programming of motor pools. Expression of the target-induced factor Pea3 is critical for later aspects of motor pool differentiation such as clustering of motor neurons into pools, muscle-specific patterns of axonal innervation, and sensory-motor connectivity (Livet *et al.*, 2002; Vrieseling and Arber, 2006). Thus, motor pool specification appears to unfold in two main phases: an early phase that confers aspects of motor neuron identity involved in the selection of target muscle connectivity (Landmesser, 2001; Milner and Landmesser, 1999), and a later phase, operating after motor axons have reached their muscle targets, that is associated with ETS gene expression and the clustering of motor neurons within the LMC (Livet *et al.*, 2002; Price *et al.*, 2002).

6. Restriction and Refinement of Hox Activities During Motor Neuron Differentiation

Although Hox protein activities appear to be critical in the generation of diverse motor neuron subtypes, several lines of evidence suggest that additional factors are necessary to restrict their functions. First, Hox proteins are broadly expressed throughout the embryo, and within the CNS they are expressed by multiple classes of neurons including interneurons and sensory neurons (Belting *et al.*, 1998; Dasen *et al.*, 2005; Ensini *et al.*, 1998). Second, within a given spinal segment the same Hox protein can be expressed by multiple columnar subtypes (Dasen *et al.*, 2008). These observations suggest the requirement for additional mechanisms to gate the actions of Hox proteins in motor neurons. This gating function may be controlled through transcriptional cofactors or through motor neuron-specific targets of Hox proteins.

Hox functions generally rely on interactions with a conserved family of DNA binding cofactors that refine and constrain their activities (Mann and Affolter, 1998). Two classes of canonical Hox cofactors, Meis and Pbx/Prep proteins (vertebrates homologs of the *Drosophila* Extradenticle and Homothorax proteins), have pervasive roles as regulators of Hox activity (Mann and Affolter, 1998; and Selleri, 2006). Within the spinal cord Meis and Pbx/Prep proteins display broad patterns of expression (Dasen *et al.*, 2005), and thus they are unlikely candidates as cell-type specific regulators of motor neuron Hox target specificities. Studies in *Drosophila* indicate that a distinct group of Hox accessory factors, the homeodomain protein Engrailed and the forkhead homeodomain proteins Sloppy-paired 1/2, functions with Extradenticle and Homothorax to modify Hox activities (Gebelein *et al.*, 2004). These factors have more restricted domains of expression and activity: Engrailed regulates Hox activity in posterior compartment cells whereas Slp1/2 regulate Hox activity in anterior cells (Gebelein *et al.*, 2004).

The vertebrate counterparts of these ancillary Hox cofactors, Engrailed and Fox proteins are expressed by subsets of spinal neurons, and a vertebrate FoxP protein, FoxP1, is selectively expressed by spinal motor neurons (Saueressig *et al.*, 1999; Shu *et al.*, 2001; Tamura *et al.*, 2003). Consistent with a role in gating Hox activities, expression of FoxP1 is restricted to two Hox sensitive columns (PGC and LMC neurons) while it is excluded from Hox-independent populations (HMC and MMC neurons). This exclusion from HMC and MMC motor neurons appears to be a function of the presence of inhibitory factors within these subtypes that prevent FoxP1 expression shortly after motor neurons are born (Dasen *et al.*, 2008). As described below, recent studies indicate that FoxP1 is an essential cofactor in the Hox-dependent program of differentiation in LMC and PGC neurons (Dasen *et al.*, 2008; Rousso *et al.*, 2008)

6.1. FoxP1: An accessory factor for Hox proteins in motor columns and pools

Although FoxP1 is expressed by LMC and PGC neurons, the levels of FoxP1 protein in these two columns differ dramatically: PGC neurons express low levels of FoxP1 and LMC neurons express high levels. These differences in FoxP1 levels appear to be set by the pattern of Hox expression as misexpression of Hox determinants of LMC identities such as Hoxc6 and Hoxd10 can switch HMC and PGC neurons to FoxP1 high LMC motor neurons at thoracic levels (Dasen *et al.*, 2008). In turn, the levels of FoxP1 protein appear to be critical in the specification of columnar fates as misexpression of FoxP1 in PGC and HMC neurons can convert these columns to LMC identities. These actions of FoxP1 appear to require continuous Hox function, since forced expression of FoxP1 is unable to induce LMC fates under conditions in which Hox activity is repressed (Dasen *et al.*, 2008). Thus FoxP1 appears to act jointly with Hox proteins rather than as a linear intermediary in the pathway of columnar specification (Fig. 6.5).

Consistent with a central role in the Hox-dependent steps of motor neuron differentiation, genetic inactivation of *Foxp1* does not effect the generation of motor neurons, or Hox expression, but nevertheless the number of LMC and PGC neurons is dramatically reduced (Dasen *et al.*, 2008; Rousso *et al.*, 2008). Instead prospective LMC and PGC motor neurons acquire molecular features of HMC neurons and the spinal cord consists almost entirely of two continuous columns (Fig. 6.5C). Strikingly all Hox-dependent steps of motor neuron differentiation, included the columnar, divisional and pool identities, are effectively erased. Molecular features of LMC divisional and pool identities are eroded in *Foxp1* mutants including the loss of specific LMC transcription factors (e.g., *Lhx1*, *Pea3*, *Nkx6.1*) as well as connectivity and synaptic specificity determinants (*EphA4*, *Sema3E*, *Cad20*) (Dasen *et al.*, 2008).

Although FoxP1 activities are essential for the transcriptional output of Hox proteins in LMC and PGC neurons, not all motor neuron Hox functions depend on FoxP1. Expression of FoxP1 is induced by Hox proteins, and thus this aspect of Hox function must be FoxP1 independent. In addition the repressive interactions between Hox paralogs are evident in MMC and HMC neurons (Dasen *et al.*, 2003), even though they lack FoxP1 expression. The activities of *Hox* genes involved in hindbrain motor neuron specification (Trainor and Krumlauf, 2000) also do not require FoxP1, since FoxP proteins are not expressed in motor neurons at

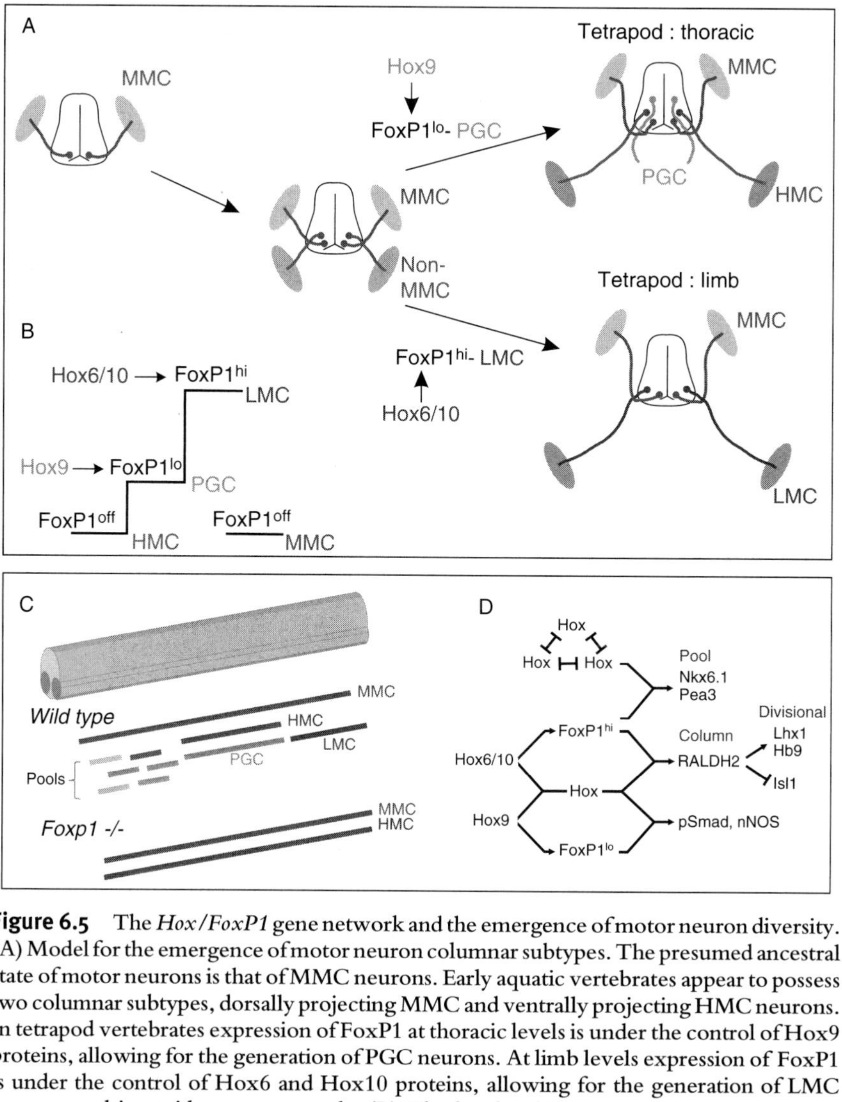

Figure 6.5 The *Hox/FoxP1* gene network and the emergence of motor neuron diversity. (A) Model for the emergence of motor neuron columnar subtypes. The presumed ancestral state of motor neurons is that of MMC neurons. Early aquatic vertebrates appear to possess two columnar subtypes, dorsally projecting MMC and ventrally projecting HMC neurons. In tetrapod vertebrates expression of FoxP1 at thoracic levels is under the control of Hox9 proteins, allowing for the generation of PGC neurons. At limb levels expression of FoxP1 is under the control of Hox6 and Hox10 proteins, allowing for the generation of LMC neurons and its resident motor pools. (B) The levels of FoxP1 are controlled by Hox proteins in PGC and LMC neurons. HMC but not MMC neurons are competent to respond the patterning activities of *Hox* genes, which control the level of FoxP1 expression and columnar fates. (C) Motor neuron columnar differentiation in wild type and *Foxp1* mutant mice, showing that FoxP1 controls the formation of PGC and LMC columns, as well as the diversification of motor pools within the LMC. In the absence of FoxP1, two continuous columns (MMC and HMC) are generated. (D) Interactions of FoxP1 and Hox proteins during the transcriptional programming of motor neuron columnar and pool fates. FoxP1 gates the activities of the two Hox networks involved in the assignment of columnar and pool identities. The Hox/FoxP1 network controls a repertoire of downstream genes including transcription factors, signaling molecules, and guidance receptors.

this level of the neuraxis. Meis and Pbx/Prep cofactors are expressed in the hindbrain, and Pbx proteins regulate cranial motor neuron identity (Moens and Selleri, 2006). Thus, the activities of FoxP1 as a mediator of Hox output during motor neuron differentiation may be exerted in a broader context of Meis and Pbx/Prep activities. Meis and Pbx/Prep cofactors may impart a first level of Hox specificity (Joshi *et al.*, 2007), with FoxP1 providing additional filters on target gene activation.

6.2. Coordinate control of motor axon targeting by FoxP1 and Hox factors

How does FoxP1/Hox network activity control the decisions taken by motor axons as they project to their targets? At thoracic levels of the spinal cord, the switch from PGC to HMC fate in *Foxp1* mutants is accompanied by a redirection of motor axons from sympathetic ganglia to body wall targets (Dasen *et al.*, 2008). In contrast, HMC-like motor neurons generated at brachial and lumbar levels in *Foxp1* mutants still project into the limb, following a path similar to that of LMC axons. This observation is consistent with studies showing that after transplantation of thoracic spinal cord to limb levels, motor axons are capable of innervating the limb (O'Brien *et al.*, 1990; O'Brien and Oppenheim, 1990). Further evidence that this is the predicted trajectory of HMC neurons generated at limb levels comes from more recent studies showing that ectopic limbs induced adjacent to thoracic spinal cord are innervated selectively by axons of HMC neurons (Turney *et al.*, 2003). Thus, HMC and LMC neurons appear to be similar in their initial pursuit of a distal trajectory that takes them to body wall or limb targets.

In the limb, LMC axons establish stereotypic projections to individual muscles (Landmesser, 2001). The establishment of these precise patterns of connectivity requires the actions of a multitude of Hox-dependent transcription factors, recognition molecules and guidance receptors. Despite the loss of expression of these factors in *Foxp1* mutants, nerve branching and limb muscle innervation patterns are remarkably well preserved (Dasen *et al.*, 2008; Rousso *et al.*, 2008). These findings fit best with the view that the overall pattern of motor nerve branching within the limb is determined by preestablished permissive or inhibitory domains within the limb mesenchyme, with the FoxP1/Hox program providing LMC neurons with identities that enable axons to respond to local cues that promote the selection of just one of many available conduits. In this view, the trajectory of motor neurons deprived of FoxP1 activity will still be constrained by the existence of preordained paths, but without Hox-dependent intrinsic molecular programming, individual axons are reduced to choosing haphazardly among their potential options. Consistent with this model, in *Foxp1* mutants the topographic map of motor projections is lost, as motor neurons no longer project into the limb in an organized manner (Dasen *et al.*, 2008).

6.3. Hox/FoxP1 interactions and the origins of motor neuron diversity

Analysis of the Hox/FoxP1 transcriptional network has provided clues into the possible mechanisms by which the tetrapod motor system emerged during evolution. Aquatic vertebrates with behaviors driven by axial and hypaxial muscles lack PGC and LMC neurons but possess neurons that are similar to MMC and HMC neurons in character (Fetcho, 1992; Kusakabe and Kuratani, 2005). The appearance of LMC and PGC neurons is linked to the formation of paired appendages (lateral fins and limbs) and a sympathetic nervous system—structures that emerged later in vertebrate evolution (Fetcho, 1992; Freitas *et al.*, 2006; Funakoshi and Nakano, 2007). The Hox/FoxP1 dependent program of motor neuron diversification may therefore have evolved to meet the demands of a new and diverse set of peripheral target tissues, and to generate more elaborate set of motor behaviors.

How did the Hox/FoxP1 transcriptional network emerge within motor neurons? The ancestral state of spinal motor neurons appears to be marked by expression of Lhx3, Isl1/2, and Hb9, a set of transcription factors conserved in the motor neurons of invertebrates (Landgraf and Thor, 2006). This transcriptional profile also defines early-born primary motor neurons of Zebrafish and Xenopus embryos (Appel *et al.*, 1995; Borodinsky *et al.*, 2004), presumed counterparts of the motor neurons of jawless vertebrates (Fetcho, 1992), as well as the MMC neurons of birds and mammals (Jessell, 2000). The conserved transcriptional profile of this ancestral set of motor neurons may reflect common patterns of connectivity—the innervation of segmentally arrayed muscles involved in undulatory locomotor behaviors. The diversification of columnar subtypes from this ancestral group requires relief from the confining influence of the LIM-homeodomain factor Lhx3, as Lhx3 exerts a dominant activity in specifying MMC fates over other columnar subtypes (Sharma *et al.*, 2000; William *et al.*, 2003). This evasive step may have involved a decrease in the strength of the Wnt signaling component of the dorsoventral inductive pathway, since reducing Wnt4/5 activity in mice promotes the generation of HMC neurons at the expense of MMC neurons (Agalliu *et al.*, 2009). Thus the basic spinal motor system induced by the dorsoventral signaling pathway comprises MMC and HMC neurons, arrayed in coextensive columns.

The induction of *Foxp1* expression under the regulatory control of Hox proteins may have permitted the twinned columnar organization of limbless vertebrates to diversify and generate PGC and LMC neurons. The formation of HMC neurons from an ancestral MMC-like group was a crucial step in this diversification program—creating a malleable population of Lhx3-negative motor neurons to serve as the substrate for the FoxP1/Hox program of columnar and pool specification. Prospective HMC neurons, freed from the dominant actions of Lhx3, became competent to respond to the rostrocaudal patterning activities of Hox proteins, allowing them to acquire PGC or LMC fates. FoxP1, through its dose-dependent inductive activity, is a key mediator of the Hox program of columnar specification (Figure 6.5). Under the influence of Hox9 activity, newly generated Lhx3-negative thoracic motor neurons acquire the capacity for low-level FoxP1 expression and appear to resolve their bi-potential HMC or PGC fate through mutual repressive interactions between FoxP1 and Hb9 (Dasen *et al.*, 2008). In the limb-level domains of Hox6/10 expression, FoxP1 is induced at high levels which allow motor neurons to acquire an LMC fate.

The mechanisms by which the FoxP1/Hox network was recruited to the task of motor neuron columnar diversification remains unclear. The absence of PGC and LMC neurons from early vertebrates could have its basis in changes in the cis-regulatory elements that control *Foxp1* expression (Prud'homme *et al.*, 2007; Shubin *et al.*, 1997), such that Hox-sensitive elements responsible for expression in spinal motor neurons were configured only at the time of formation of the sympathetic nervous system and paired appendages. Alternatively, the rostrocaudal pattern of expression of *Hox* genes in the spinal cord of jawless vertebrates may differ from that in birds and mammals (Force *et al.*, 2002; Takio *et al.*, 2007), and thus may fail to produce a productive Hox code capable of activating FoxP1 expression. Further investigation into the regulatory elements controlling the *Foxp1* and *Hox* genes and their targets in motor neurons may provide insights into how columnar subtypes arose in diverse vertebrate species.

6.4. Coordinate regulation of neuronal and mesodermal Hox programs

The same extrinsic signaling systems which pattern Hox profiles in the CNS also control Hox expression within the mesoderm, raising the possibility that regulation of Hox expression in these distinct tissues may be involved in coordinating the position of motor neuron subtypes with the location and identity of their peripheral target fields. One aspect of this program is the establishment of a register between the rostrocaudal positions motor columns and their targets. Application of FGFs to thoracic mesoderm has been shown to induce the formation of an ectopic limb in place of body wall mesoderm (Cohn *et al.*, 1995), and the regulation of *Hox9* gene paralog expression in lateral plate mesoderm in response to FGFs has been implicated to determine limb bud position (Cohn *et al.*, 1997). These observations support the idea that exposure of neural and lateral plate mesodermal cells to a common source of FGFs establishes distinct profiles of *Hox* gene expression in these two tissues, profiles that in turn control the alignment of LMC and limb position.

Similarly, the muscle targets of LMC motor neurons within the limb appear to acquire aspects of their identity through Hox-dependent programs.

Limb musculature is derived from a population of migratory muscle precursors that is generated selectively at limb levels through the actions of the LIM-homeodomain protein Lbx1 (Brohmann *et al.*, 2000; Gross *et al.*, 2000). Interestingly, a Hox determinant of LMC identity, Hoxa10, has been shown to be sufficient to induce Lbx1 expression in muscle precursors at thoracic levels and to reprogram nonmigratory myoblasts into migratory muscle cells (Alvares *et al.*, 2003). Moreover, *Hox* genes control patterning within the limb mesenchyme and may therefore exert additional roles in the guidance of axons along the anterior-posterior and proximal-distal axes of the limb (Izpisua-Belmonte and Duboule, 1992), prior to establishment of the mature pattern of musculature. Thus, the coordinate regulation of Hox expression patterns during vertebrate evolution may provide a plausible basis for linking the formation and diversification of LMC motor neuron subtypes to the appearance and patterning of paired appendages.

7. Conclusions

The task of specifying hundreds distinct motor neuron subtypes, each projecting to a specific target cell group, appears to have been met by deploying a regulatory network of nearly two dozen *Hox* genes. The selective connections formed between motor neurons and muscle are however just one aspect of circuit assembly in the spinal cord. Additional components of motor neuron connectivity, such as the specificity of synaptic inputs from sensory neurons and interneurons, may also rely on the actions of the FoxP1/Hox transcriptional network. Hox and Fox proteins are expressed by other classes of spinal neurons, suggesting these transcription factor families could have a more extensive role in the assembly oflocomotor circuits. The self-organizing features inherent in the FoxP1/Hox network may have therefore helped endow developing motor circuits with their apparent high degree of genetic determination.

Studies on the transcriptional programs that control diversity and synaptic specificity in the spinal cord have provided a valuable model system for understanding the mechanisms of neuronal specification throughout the CNS. Emerging work on retinal and cortical neurons suggest an equivalent high degree of subtype diversity, even within a single class of neurons (Klausberger and Somogyi, 2008; Kong *et al.*, 2005). Although retinal and cortical neurons are devoid of chromosomally clustered *Hox* gene expression, it is likely that similar cell-intrinsic transcriptional networks control neuronal specification throughout the CNS. Whether other CNS circuits utilize coherent gene families to generate highly diverse neuronal subtypes remains to be determined.

REFERENCES

Agalliu, D., Takada, S., Agalliu, I., McMahon, A. P., and Jessell, T. M. (2009). Motor neurons with axial muscle projections specified by Wnt4/5 signaling. *Neuron.* **61,** 708–720.

Alvares, L. E., Schubert, F. R., Thorpe, C., Mootoosamy, R. C., Cheng, L., Parkyn, G., Lumsden, A., and Dietrich, S. (2003). Intrinsic, Hox-dependent cues determine the fate of skeletal muscle precursors. *Dev. Cell* **5,** 379–390.

Appel, B., Korzh, V., Glasgow, E., Thor, S., Edlund, T., Dawid, I. B., and Eisen, J. S. (1995). Motoneuron fate specification revealed by patterned LIM homeobox gene expression in embryonic zebrafish. *Development* **121,** 4117–4125.

Arber, S., Han, B., Mendelsohn, M., Smith, M., Jessell, T. M., and Sockanathan, S. (1999). Requirement for the homeobox gene Hb9 in the consolidation of motor neuron identity. *Neuron* **23,** 659–674.

Belting, H. G., Shashikant, C. S., and Ruddle, F. H. (1998). Multiple phases of expression and regulation of mouse Hoxc8 during early embryogenesis. *J. Exp. Zool.* **282,** 196–222.

Bel-Vialar, S., Itasaki, N., and Krumlauf, R. (2002). Initiating *Hox* gene expression: In the early chick neural tube differential sensitivity to FGF and RA signaling subdivides the HoxB genes in two distinct groups. *Development* **129,** 5103–5115.

Borodinsky, L. N., Root, C. M., Cronin, J. A., Sann, S. B., Gu, X., and Spitzer, N. C. (2004). Activity-dependent homeostatic specification of transmitter expression in embryonic neurons. *Nature* **429,** 523–530.

Briscoe, J., Pierani, A., Jessell, T. M., and Ericson, J. (2000). A homeodomain protein code specifies progenitor cell identity and neuronal fate in the ventral neural tube. *Cell* **101,** 435–445.

Brohmann, H., Jagla, K., and Birchmeier, C. (2000). The role of Lbx1 in migration of muscle precursor cells. *Development* **127,** 437–445.

Carpenter, E. M. (2002). *Hox* genes and spinal cord development. *Dev. Neurosci.* **24,** 24–34.

Choe, A., Phun, H. Q., Tieu, D. D., Hu, Y. H., and Carpenter, E. M. (2006). Expression patterns of Hox10 paralogous genes during lumbar spinal cord development. *Gene Expr. Patterns* **6,** 730–737.

Chopra, V. S., and Mishra, R. K. (2006). "Mir"acles in *Hox* gene regulation. *Bioessays* **28,** 445–448.

Cohn, M. J., Izpisua-Belmonte, J. C., Abud, H., Heath, J. K., and Tickle, C. (1995). Fibroblast growth factors induce additional limb development from the flank of chick embryos. *Cell* **80,** 739–746.

Cohn, M. J., Patel, K., Krumlauf, R., Wilkinson, D. G., Clarke, J. D., and Tickle, C. (1997). Hox9 genes and vertebrate limb specification. *Nature* **387,** 97–101.

Dasen, J. S., Liu, J. P., and Jessell, T. M. (2003). Motor neuron columnar fate imposed by sequential phases of Hox-c activity. *Nature* **425,** 926–933.

Dasen, J. S., Tice, B. C., Brenner-Morton, S., and Jessell, T. M. (2005). A Hox regulatory network establishes motor neuron pool identity and target-muscle connectivity. *Cell* **123,** 477–491.

Dasen, J. S., De Camilli, A., Wang, B., Tucker, P. W., and Jessell, T. M. (2008). Hox repertoires for motor neuron diversity and connectivity gated by a single accessory factor, FoxP1. *Cell* **134,** 304–316.

De Marco Garcia, N. V., and Jessell, T. M. (2008). Early motor neuron pool identity and muscle nerve trajectory defined by postmitotic restrictions in Nkx6.1 activity. *Neuron* **57,** 217–231.

Dessaud, E., McMahon, A. P., and Briscoe, J. (2008). Pattern formation in the vertebrate neural tube: A sonic hedgehog morphogen-regulated transcriptional network. *Development* **135,** 2489–2503.

Diez del Corral, R., and Storey, K. G. (2004). Opposing FGF and retinoid pathways: A signalling switch that controls differentiation and patterning onset in the extending vertebrate body axis. *Bioessays* **26,** 857–869.

Duboule, D., and Morata, G. (1994). Colinearity and functional hierarchy among genes of the homeotic complexes. *Trends Genet.* **10,** 358–364.

Dubrulle, J., and Pourquie, O. (2004). fgf8 mRNA decay establishes a gradient that couples axial elongation to patterning in the vertebrate embryo. *Nature* **427,** 419–422.

Eberhart, J., Swartz, M. E., Koblar, S. A., Pasquale, E. B., and Krull, C. E. (2002). EphA4 constitutes a population-specific guidance cue for motor neurons. *Dev. Biol.* **247,** 89–101.

Elowitz, M. B., and Leibler, S. (2000). A synthetic oscillatory network of transcriptional regulators. *Nature* **403,** 335–338.

Ensini, M., Tsuchida, T. N., Belting, H. G., and Jessell, T. M. (1998). The control of rostrocaudal pattern in the developing spinal cord: Specification of motor neuron subtype identity is initiated by signals from paraxial mesoderm. *Development* **125,** 969–982.

Fetcho, J. R. (1987). A review of the organization and evolution of motoneurons innervating the axial musculature of vertebrates. *Brain Res.* **434,** 243–280.

Fetcho, J. R. (1992). The spinal motor system in early vertebrates and some of its evolutionary changes. *Brain Behav. Evol.* **40,** 82–97.

Force, A., Amores, A., and Postlethwait, J. H. (2002). Hox cluster organization in the jawless vertebrate *Petromyzon marinus*. *J. Exp. Zool.* **294,** 30–46.

Freitas, R., Zhang, G., and Cohn, M. J. (2006). Evidence that mechanisms of fin development evolved in the midline of early vertebrates. *Nature* **442,** 1033–1037.

Funakoshi, K., and Nakano, M. (2007). The sympathetic nervous system of anamniotes. *Brain Behav. Evol.* **69,** 105–113.

Gavalas, A., Davenne, M., Lumsden, A., Chambon, P., and Rijli, F. M. (1997). Role of Hoxa-2 in axon pathfinding and rostral hindbrain patterning. *Development* **124,** 3693–3702.

Gebelein, B., McKay, D. J., and Mann, R. S. (2004). Direct integration of Hox and segmentation gene inputs during *Drosophila* development. *Nature* **431,** 653–659.

Glover, J. C., Renaud, J. S., and Rijli, F. M. (2006). Retinoic acid and hindbrain patterning. *J. Neurobiol.* **66,** 705–725.

Gross, M. K., Moran-Rivard, L., Velasquez, T., Nakatsu, M. N., Jagla, K., and Goulding, M. (2000). Lbx1 is required for muscle precursor migration along a lateral pathway into the limb. *Development* **127,** 413–424.

Guthrie, S. (2007). Patterning and axon guidance of cranial motor neurons. *Nat. Rev. Neurosci.* **8,** 859–871.

Gutman, C. R., Ajmera, M. K., and Hollyday, M. (1993). Organization of motor pools supplying axial muscles in the chicken. *Brain Res.* **609,** 129–136.

Haase, G., Dessaud, E., Garces, A., de Bovis, B., Birling, M., Filippi, P., Schmalbruch, H., Arber, S., and deLapeyriere, O. (2002). GDNF acts through PEA3 to regulate cell body positioning and muscle innervation of specific motor neuron pools. *Neuron* **35,** 893–905.

Hollyday, M. (1980a). Motoneuron histogenesis and the development of limb innervation. *Curr. Top. Dev. Biol.* **15,** 181–215.

Hollyday, M. (1980b). Organization of motor pools in the chick lumbar lateral motor column. *J. Comp. Neurol.* **194,** 143–170.

Hollyday, M. (1995). Chick wing innervation. I. Time course of innervation and early differentiation of the peripheral nerve pattern. *J. Comp. Neurol.* **357,** 242–253.

Hollyday, M., and Jacobson, R. D. (1990). Location of motor pools innervating chick wing. *J. Comp. Neurol.* **302,** 575–588.

Izpisua-Belmonte, J. C., and Duboule, D. (1992). Homeobox genes and pattern formation in the vertebrate limb. *Dev. Biol.* **152,** 26–36.

Jegalian, B. G., and De Robertis, E. M. (1992). Homeotic transformations in the mouse induced by overexpression of a human Hox3.3 transgene. *Cell* **71,** 901–910.

Jessell, T. M. (2000). Neuronal specification in the spinal cord: Inductive signals and transcriptional codes. *Nat. Rev. Genet.* **1,** 20–29.

Joshi, R., Passner, J. M., Rohs, R., Jain, R., Sosinsky, A., Crickmore, M. A., Jacob, V., Aggarwal, A. K., Honig, B., and Mann, R. S. (2007). Functional specificity of a Hox protein mediated by the recognition of minor groove structure. *Cell* **131,** 530–543.

Kania, A., and Jessell, T. M. (2003). Topographic motor projections in the limb imposed by LIM homeodomain protein regulation of ephrin-A:EphA interactions. *Neuron* **38,** 581–596.

Kania, A., Johnson, R. L., and Jessell, T. M. (2000). Coordinate roles for LIM homeobox genes in directing the dorsoventral trajectory of motor axons in the vertebrate limb. *Cell* **102,** 161–173.

Klausberger, T., and Somogyi, P. (2008). Neuronal diversity and temporal dynamics: The unity of hippocampal circuit operations. *Science* **321,** 53–57.

Kmita, M., and Duboule, D. (2003). Organizing axes in time and space; 25 years of colinear tinkering. *Science* **301,** 331–333.

Kong, J. H., Fish, D. R., Rockhill, R. L., and Masland, R. H. (2005). Diversity of ganglion cells in the mouse retina: Unsupervised morphological classification and its limits. *J. Comp. Neurol.* **489,** 293–310.

Kusakabe, R., and Kuratani, S. (2005). Evolution and developmental patterning of the vertebrate skeletal muscles: Perspectives from the lamprey. *Dev. Dyn.* **234,** 824–834.

Lance-Jones, C., and Landmesser, L. (1980). Motoneurone projection patterns in embryonic chick limbs following partial deletions of the spinal cord. *J. Physiol.* **302,** 559–580.

Lance-Jones, C., Omelchenko, N., Bailis, A., Lynch, S., and Sharma, K. (2001). Hoxd10 induction and regionalization in the developing lumbosacral spinal cord. *Development* **128,** 2255–2268.

Landgraf, M., and Thor, S. (2006). Development of *Drosophila motoneurons*: Specification and morphology. *Semin. Cell Dev. Biol.* **17,** 3–11.

Landmesser, L. (1978a). The development of motor projection patterns in the chick hind limb. *J. Physiol.* **284,** 391–414.

Landmesser, L. (1978b). The distribution of motoneurones supplying chick hind limb muscles. *J. Physiol.* **284,** 371–389.

Landmesser, L. T. (2001). The acquisition of motoneuron subtype identity and motor circuit formation. *Int. J. Dev. Neurosci.* **19,** 175–182.

Lemons, D., and McGinnis, W. (2006). Genomic evolution of *Hox* gene clusters. *Science* **313,** 1918–1922.

Lewis, J., Chevallier, A., Kieny, M., and Wolpert, L. (1981). Muscle nerve branches do not develop in chick wings devoid of muscle. *J. Embryol. Exp. Morphol.* **64,** 211–232.

Lin, J. H., Saito, T., Anderson, D. J., Lance-Jones, C., Jessell, T. M., and Arber, S. (1998). Functionally related motor neuron pool and muscle sensory afferent subtypes defined by coordinate ETS gene expression. *Cell* **95,** 393–407.

Liu, J. P. (2006). The function of growth/differentiation factor 11 (Gdf11) in rostrocaudal patterning of the developing spinal cord. *Development* **133,** 2865–2874.

Liu, J. P., Laufer, E., and Jessell, T. M. (2001). Assigning the positional identity of spinal motor neurons: Rostrocaudal patterning of Hox-c expression by FGFs, Gdf11, and retinoids. *Neuron* **32,** 997–1012.

Livet, J., Sigrist, M., Stroebel, S., De Paola, V., Price, S. R., Henderson, C. E., Jessell, T. M., and Arber, S. (2002). ETS gene Pea3 controls the central position and terminal arborization of specific motor neuron pools. *Neuron* **35,** 877–892.

Mann, R. S., and Affolter, M. (1998). Hox proteins meet more partners. *Curr. Opin. Genet Dev.* **8,** 423–429

McIntyre, D. C., Rakshit, S., Yallowitz, A. R., Loken, L., Jeannotte, L., Capecchi, M. R., and Wellik, D. M. (2007). Hox patterning of the vertebrate rib cage. *Development* **134,** 2981–2989.

McPherron, A. C., Lawler, A. M., and Lee, S. J. (1999). Regulation of anterior/posterior patterning of the axial skeleton by growth/differentiation factor 11. *Nat. Genet.* **22,** 260–264.

Milner, L. D., and Landmesser, L. T. (1999). Cholinergic and GABAergic inputs drive patterned spontaneous motoneuron activity before target contact. *J. Neurosci.* **19,** 3007–3022.

Moens, C. B., and Selleri, L. (2006). Hox cofactors in vertebrate development. *Dev. Biol.* **291,** 193–206.

Narayanan, C. H., and Hamburger, V. (1971). Motility in chick embryos with substitution of lumbosacral by brachial and brachial by lumbosacral spinal cord segments. *J. Exp. Zool.* **178,** 415–431.

Nonchev, S., Maconochie, M., Gould, A., Morrison, A., and Krumlauf, R. (1997). Cross-regulatory interactions between *Hox* genes and the control of segmental expression in the vertebrate central nervous system. *Cold Spring Harb. Symp. Quant. Biol.* **62,** 313–323.

Nordstrom, U., Maier, E., Jessell, T. M., and Edlund, T. (2006). An early role for WNT signaling in specifying neural patterns of *Cdx* and *Hox* gene expression and motor neuron subtype identity. *PLoS Biol.* **4,** e252.

Novitch, B. G., Chen, A. I., and Jessell, T. M. (2001). Coordinate regulation of motor neuron subtype identity and pan-neuronal properties by the bHLH repressor Olig2. *Neuron* **31,** 773–789.

O'Brien, M. K., and Oppenheim, R. W. (1990). Development and survival of thoracic motoneurons and hindlimb musculature following transplantation of the thoracic neural tube to the lumbar region in the chick embryo: Anatomical aspects. *J. Neurobiol.* **21,** 313–340.

O'Brien, M. K., Landmesser, L., and Oppenheim, R. W. (1990). Development and survival of thoracic motoneurons and hindlimb musculature following transplantation of the thoracic neural tube to the lumbar region in the chick embryo: Functional aspects. *J. Neurobiol.* **21,** 341–355.

Pfaff, S. L., Mendelsohn, M., Stewart, C. L., Edlund, T., and Jessell, T. M. (1996). Requirement for LIM homeobox gene Isl1 in motor neuron generation reveals a motor neuron-dependent step in interneuron differentiation. *Cell* **84,** 309–320.

Phelan, K. A., and Hollyday, M. (1990). Axon guidance in muscleless chick wings: The role of muscle cells in motoneuronal pathway selection and muscle nerve formation. *J. Neurosci.* **10,** 2699–2716.

Phelan, K. A., and Hollyday, M. (1991). Embryonic development and survival of brachial motoneurons projecting to muscleless chick wings. *J. Comp. Neurol.* **311,** 313–320.

Prasad, A., and Hollyday, M. (1991). Development and migration of avian sympathetic preganglionic neurons. *J. Comp. Neurol.* **307,** 237–258.

Price, S. R., De Marco Garcia, N. V., Ranscht, B., and Jessell, T. M. (2002). Regulation of motor neuron pool sorting by differential expression of type II cadherins. *Cell* **109,** 205–216.

Prud'homme, B., Gompel, N., and Carroll, S. B. (2007). Emerging principles of regulatory evolution. *Proc. Natl. Acad. Sci. USA* **104**(Suppl 1), 8605–8612.

Romanes, G. J. (1942). The development and significance of the cell columns in the ventral horn of the cervical and upper thoracic spinal cord of the rabbit. *J. Anat. Lond.* **76,** 112–130.

Rousso, D. L., Gaber, Z. B., Wellik, D., Morrisey, E. E., and Novitch, B. G. (2008). Coordinated actions of the forkhead protein Foxp1 and Hox proteins in the columnar organization of spinal motor neurons. *Neuron* **59,** 226–240.

Saueressig, H., Burrill, J., and Goulding, M. (1999). Engrailed-1 and netrin-1 regulate axon pathfinding by association interneurons that project to motor neurons. *Development* **126,** 4201–4212.

Shah, V., Drill, E., and Lance-Jones, C. (2004). Ectopic expression of Hoxd10 in thoracic spinal segments induces motoneurons with a lumbosacral molecular profile and axon projections to the limb. *Dev. Dyn.* **231,** 43–56.

Sharma, K., Sheng, H. Z., Lettieri, K., Li, H., Karavanov, A., Potter, S., Westphal, H., and Pfaff, S. L. (1998). LIM homeodomain factors Lhx3 and Lhx4 assign subtype identities for motor neurons. *Cell* **95,** 817–828.

Sharma, K., Leonard, A. E., Lettieri, K., and Pfaff, S. L. (2000). Genetic and epigenetic mechanisms contribute to motor neuron pathfinding. *Nature* **406,** 515–519.

Sherrington, C. S. (1906). "The Integrative Action of the Nervous System." C. Scribner's sons, New York.

Shieh, P. (1951). The neofomation of cells of preganglionic type in the cervical spinal cord of the chick embryo following its transplantation to the thoracic level. *J. Exp. Zool.* **117,** 359–395.

Shimizu, T., Bae, Y. K., and Hibi, M. (2006). Cdx-Hox code controls competence for responding to Fgfs and retinoic acid in zebrafish neural tissue. *Development* **133,** 4709–4719.

Shirasaki, R., and Pfaff, S. L. (2002). Transcriptional codes and the control of neuronal identity. *Annu. Rev. Neurosci.* **25,** 251–281.

Shu, W., Yang, H., Zhang, L., Lu, M. M., and Morrisey, E. E. (2001). Characterization of a new subfamily of winged-helix/forkhead (Fox) genes that are expressed in the lung and act as transcriptional repressors. *J. Biol. Chem.* **276,** 27488–27497.

Shubin, N., Tabin, C., and Carroll, S. (1997). Fossils, genes and the evolution of animal limbs. *Nature* **388,** 639–648.

Skromne, I., Thorsen, D., Hale, M., Prince, V. E., and Ho, R. K. (2007). Repression of the hindbrain developmental program by Cdx factors is required for the specification of the vertebrate spinal cord. *Development* **134,** 2147–2158.

Sockanathan, S., and Jessell, T. M. (1998). Motor neuron-derived retinoid signaling specifies the subtype identity of spinal motor neurons. *Cell* **94,** 503–514.

Stirling, R. V., and Summerbell, D. (1988). Specific guidance of motor axons to duplicated muscles in the developing amniote limb. *Development* **103,** 97–110.

Takio, Y., Kuraku, S., Murakami, Y., Pasqualetti, M., Rijli, F. M., Narita, Y., Kuratani, S., and Kusakabe, R. (2007). *Hox* gene expression patterns in Lethenteron japonicum embryos-insights into the evolution of the vertebrate Hox code. *Dev. Biol.* **308,** 606–620.

Tamura, S., Morikawa, Y., Iwanishi, H., Hisaoka, T., and Senba, E. (2003). Expression pattern of the winged-helix/forkhead transcription factor Foxp1 in the developing central nervous system. *Gene Expr. Patterns* **3,** 193–197.

Thaler, J., Harrison, K., Sharma, K., Lettieri, K., Kehrl, J., and Pfaff, S. L. (1999). Active suppression of interneuron programs within developing motor neurons revealed by analysis of homeodomain factor HB9. *Neuron* **23,** 675–687.

Thaler, J. P., Koo, S. J., Kania, A., Lettieri, K., Andrews, S., Cox, C., Jessell, T. M., and Pfaff, S. L. (2004). A postmitotic role for Isl-class LIM homeodomain proteins in the assignment of visceral spinal motor neuron identity. *Neuron* **41,** 337–350.

Tosney, K. W., and Landmesser, L. T. (1985a) Development of the major pathways for neurite outgrowth in the chick hindlimb. *Dev. Biol.* **109,** 193–214.

Tosney, K. W., and Landmesser, L. T. (1985b). Growth cone morphology and trajectory in the lumbosacral region of the chick embryo. *J. Neurosci.* **5,** 2345–2358.

Trainor, P. A., and Krumlauf, R. (2000). Patterning the cranial neural crest: Hindbrain segmentation and *Hox* gene plasticity. *Nat. Rev. Neurosci.* **1,** 116–124.

Tsuchida, T., Ensini, M., Morton, S. B., Baldassare, M., Edlund, T., Jessell, T. M., and Pfaff, S. L. (1994). Topographic organization of embryonic motor neurons defined by expression of LIM homeobox genes. *Cell* **79,** 957–970.

Turney, B. W., Rowan-Hull, A. M., and Brown, J. M. (2003). The innervation of FGF-induced additional limbs in the chick embryo. *J. Anat.* **202,** 83–92.

Vallstedt, A., Muhr, J., Pattyn, A., Pierani, A., Mendelsohn, M., Sander, M., Jessell, T. M., and Ericson, J. (2001). Different levels of repressor activity assign redundant and specific roles to Nkx6 genes in motor neuron and interneuron specification. *Neuron* **31,** 743–755.

Vermot, J., Schuhbaur, B., Le Mouellic, H., McCaffery, P., Garnier, J. M., Hentsch, D., Brulet, P., Niederreither, K., Chambon, P., and Dolle, P. (2005). Retinaldehyde dehydrogenase 2 and Hoxc8 are required in the murine brachial spinal cord for the specification of Lim1+ motoneurons and the correct distribution of Islet1+ motoneurons. *Development* **132,** 1611–1621.

Vrieseling, E., and Arber, S. (2006). Target-induced transcriptional control of dendritic patterning and connectivity in motor neurons by the ETS gene Pea3. *Cell* **127,** 1439–1452.

William, C. M., Tanabe, Y., and Jessell, T. M. (2003). Regulation of motor neuron subtype identity by repressor activity of Mnx class homeodomain proteins. *Development* **130,** 1523–1536.

Zhou, Q., and Anderson, D. J. (2002). The bHLH transcription factors OLIG2 and OLIG1 couple neuronal and glial subtype specification. *Cell* **109,** 61–73.

CHAPTER SEVEN

Establishment of Hox Vertebral Identities in the Embryonic Spine Precursors

Tadahiro Iimura,* Nicolas Denans,† *and* Olivier Pourquié†,‡,§

Contents

Abstract

The vertebrate spine exhibits two striking characteristics. The first one is the periodic arrangement of its elements—the vertebrae—along the anteroposterior axis. This segmented organization is the result of somitogenesis, which takes place during organogenesis. The segmentation machinery involves a molecular oscillator—the segmentation clock—which delivers a periodic signal controlling somite production. During embryonic axis elongation, this signal is displaced posteriorly by a system of traveling signaling gradients—the

* Tokyo Medical and Dental University, Global Center of Excellence (GCOE) Program, International Research Center for Molecular Science in Tooth and Bone Diseases, Tokyo, Japan
† Stowers Institute for Medical Research, Kansas City, Missouri, USA
‡ Howard Hughes Medical Institute, Kansas City, Missouri, USA
§ Department of Anatomy and Cell Biology, The University of Kansas School of Medicine, Kansas City, Kansas, USA

Current Topics in Developmental Biology, Volume 88
ISSN 0070-2153, DOI: 10.1016/S0070-2153(09)88007-1

wavefront—which depends on the Wnt, FGF, and retinoic acid pathways. The other characteristic feature of the spine is the subdivision of groups of vertebrae into anatomical domains, such as the cervical, thoracic, lumbar, sacral, and caudal regions. This axial regionalization is controlled by a set of transcription factors called *Hox* genes. *Hox* genes exhibit nested expression domains in the somites which reflect their linear arrangement along the chromosomes—a property termed colinearity. The colinear disposition of *Hox* genes expression domains provides a blueprint for the regionalization of the future vertebral territories of the spine. In amniotes, *Hox* genes are activated in the somite precursors of the epiblast in a temporal colinear sequence and they were proposed to control their progressive ingression into the nascent paraxial mesoderm. Consequently, the positioning of the expression domains of *Hox* genes along the anteroposterior axis is largely controlled by the timing of Hox activation during gastrulation. Positioning of the somitic Hox domains is subsequently refined through a crosstalk with the segmentation machinery in the presomitic mesoderm. In this review, we focus on our current understanding of the embryonic mechanisms that establish vertebral identities during vertebrate development.

1. Introduction

The vertebrate spine is formed by the periodic series of vertebrae distributed along the anteroposterior (AP) body axis (Hirsinger *et al.*, 2000). This repetitive pattern is established during somitogenesis, a process by which segmental structures called somites are produced in the embryo. In vertebrates, somites are epithelial blocks of mesodermal cells that contain the precursors of vertebrae, skeletal muscles and dorsal dermis (Hirsinger *et al.*, 2000). They form as the tissue pinches off from the anterior tip of the presomitic mesoderm (PSM) in a rhythmic fashion. Somitogenesis begins anteriorly at the level immediately caudal to the otic vesicle (Hinsch and Hamilton, 1956; Huang *et al.*, 1997) and proceeds posteriorly on both sides of the neural tube and notochord to the caudal tip of the embryo. Although the number of somites that forms in a given species is highly constrained, it varies widely between species, ranging from approximately 30 somite pairs in some fish to several hundred pairs in snakes (Gomez *et al.*, 2008; Richardson *et al.*, 1998). In amniotes, such as birds or mammals, the anterior-most somites contribute to the baso-occipital bone at the base of the skull (Christ *et al.*, 2007; Couly *et al.*, 1993; Noden, 1986); whereas, the more posterior somites form the vertebral column. The paraxial tissue anterior to the otic vesicle is called the cephalic or head mesoderm. It does not segment into somites and gives rise to bones and muscles of the head. Together with the somitic series, the cephalic mesoderm forms the paraxial mesoderm.

Although all newly formed somites exhibit very similar morphology (Fig. 7.1A), they eventually differentiate into very distinct anatomical structures depending on their position along the AP axis (Fig. 7.1B). The vertebrate spine is partitioned into domains that exhibit different identities, such as cervical, thoracic, lumbar, sacral, and caudal. Within a given species, the number of each type of vertebrae—the vertebral formula—is usually fixed. The acquisition of vertebral identities is controlled by the Hox transcription factors (Krumlauf, 1994; Wellik, 2007), which are arranged in clusters in the genome (Duboule, 2007). Mammals have 39 *Hox* genes organized into four clusters, and birds share a very similar set of *Hox* genes (Richardson *et al.*, 2007). In each cluster, the genes are arranged on the chromosome in a sequence that reflects the timing of their expression during embryogenesis (temporal colinearity) and the position of their expression domain along the AP axis (spatial colinearity) (Dolle *et al.*, 1989; Gaunt *et al.*, 1988; Graham *et al.*, 1989). Thus, *Hox* genes exhibit nested expression domains in the vertebral precursors along the AP axis. The spatial colinearity of the expression domains of *Hox* genes results in a specific combination of genes to be expressed in each somite (Kessel and Gruss, 1991). This combination of *Hox* genes is involved in the control of

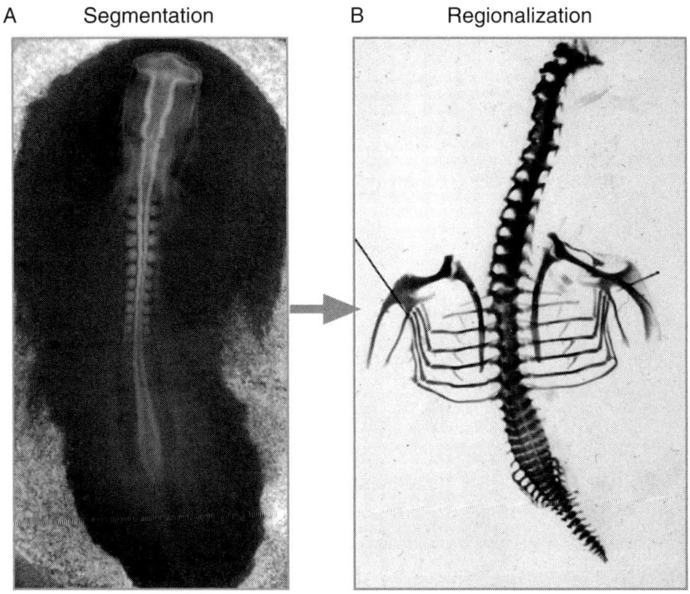

Figure 7.1 Segmentation and regional patterning of somitic derivatives. (A) 2-day-old chicken embryo in the process of forming the embryonic somites. (B) Skeletal preparation of a 10-day-old chicken embryo illustrating the different anatomical regions along the anteroposterior axis of the spine. (See Color Insert.)

the specification of vertebral identities (Carapuco *et al.*, 2005; Deschamps and van Nes, 2005; Kessel and Gruss, 1991; Wellik, 2007 and references therein). However, whereas the expression domain of many *Hox* genes extends from the posterior end of the embryo to a defined anterior level in the somites, their action is essentially restricted to their anterior-most expression domain. Thus, the identity of a segment is controlled only by the posterior-most *Hox* genes that are expressed in this segment (Burke *et al.*, 1995; Duboule and Morata, 1994). This property is termed posterior prevalence in vertebrates (Burke *et al.*, 1995; Duboule and Morata, 1994).

In the vertebral precursors of amniotes, *Hox* genes are first activated prior to gastrulation in the superficial epiblast in a colinear manner (Deschamps *et al.*, 1999; Forlani *et al.*, 2003; Gaunt and Strachan, 1994; 1996; Iimura and Pourquié, 2006). *Hox* genes were proposed to control the timing of ingression of mesodermal precursors in the primitive streak (Iimura and Pourquié, 2006). Thus, epiblast precursors expressing more 5′ genes ingress later than those expressing more 3′ genes and, hence, are positioned more posteriorly along the AP axis because of the progressive mode of cell deposition involved in the formation of the embryonic axis. This mechanism was proposed to be involved in the conversion of the temporal colinearity into the colinear or nested arrangement of the expression domain of these genes along the AP axis (Iimura and Pourquié, 2007). In this review, we will first discuss the establishment of the colinear Hox expression domains in the somites of amniote embryos.

2. Initial *Hox* Gene Activation in Paraxial Mesoderm Precursors in the Epiblast

At the beginning of gastrulation, during the early stages of primitive streak formation, the presumptive territory of the paraxial mesoderm in the epiblast is located bilaterally to the forming primitive streak (Fig. 7.2A) (Bortier and Vakaet, 1992; Hatada and Stern, 1994; Iimura *et al.*, 2007; Lawson *et al.*, 1991; Nicolet, 1971; Rosenquist, 1966; Schoenwolf *et al.*, 1992; Tam and Beddington, 1987; Tam and Trainor, 1994; Waddington, 1952). During gastrulation, these domains converge toward the streak where they undergo an epithelium-to-mesenchyme transition, which allows their ingression to form the paraxial mesoderm (Hatada and Stern, 1994; Psychoyos and Stern, 1996; Schoenwolf *et al.*, 1992; Selleck and Stern, 1991). The first paraxial mesoderm precursors to ingress form the head mesoderm (Lawson *et al.*, 1991; Nicolet, 1971; Psychoyos and Stern, 1996). This tissue lies at the anterior tip of the embryo and does not express *Hox* genes (Iimura and Pourquié, 2006; Kuratani *et al.*, 1997; Prince and Lumsden, 1994). After the primitive streak reaches a maximal size

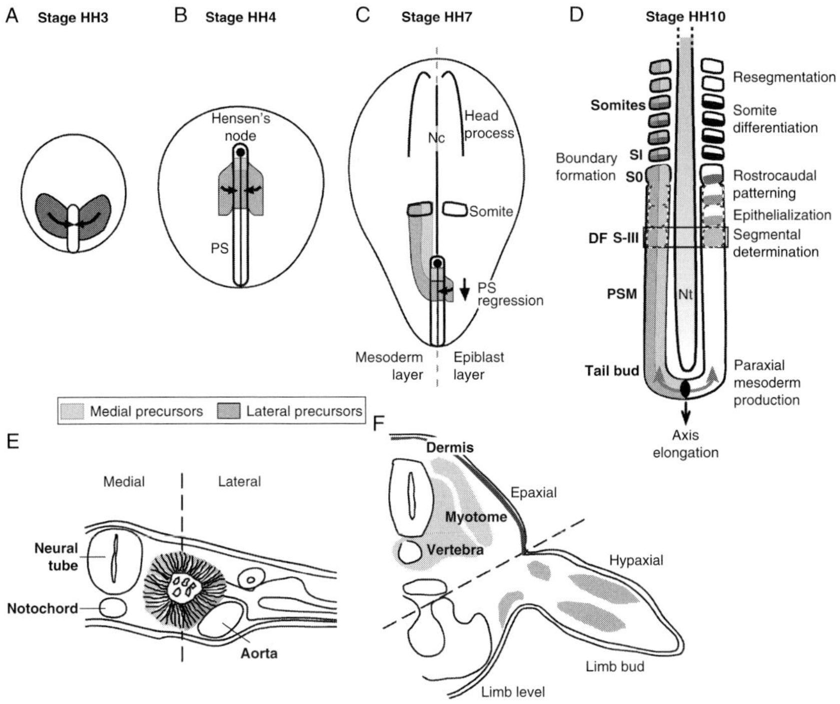

Figure 7.2 Paraxial mesoderm formation and segmentation in the chicken embryo. (A–D) Dorsal views of chicken embryos. (A) Gastrulating chicken embryo at stage 3 Hamburger and Hamilton (stage HH3) (Hamburger and Hamilton, 1992). The presumptive territory of the paraxial mesoderm (red), which contains the precursors of vertebrae and skeletal muscles, converge toward the primitive streak (PS). (B) Stage HH4 chicken embryo. At this stage, the PS has reached its maximal length. Presumptive territories of the paraxial mesoderm are located in the superficial epiblast just below Hensen's node (medial precursor population, green) and in two symmetrical domains located on both sides of the PS (lateral precursor population, purple). These cells are ingressing (arrows) through the PS to form the paraxial mesoderm. (C) Stage HH7 chicken embryo. The PS and node have begun their posterior regression (arrow), leaving in their wake the embryonic axis comprising the head process anteriorly and the notochord (Nc) axially. Epiblast cells (purple) continue to ingress in the PS (arrow) and join the descendents of a population of resident stem cells located in the anterior primitive streak (green) to generate the paraxial mesoderm. The mesodermal layer is represented on the left side without the superficial epiblast. (D) Posterior region of a stage HH10 chicken embryo. Somitogenesis progresses posteriorly on both sides of the neural tube (Nt) in concert with axis elongation (arrow). Paraxial mesoderm cells are produced at the tail bud level and undergo a maturation process in the presomitic mesoderm (PSM), leading to the periodic formation of new pairs of somites. Segmental determination occurs at the level of the determination front (DF, black rectangle). Presumptive somite nomenclature according to Pourquié and Tam (2001). (E, F) Transverse sections showing the fate of medial and lateral paraxial mesoderm precursors (only the right side of the embryos are shown). The hatched line separates the medial and the lateral domains. (E) Epithelial somite in a 2-day-old chicken embryo.

(Fig. 7.2B), it begins to shrink and regress, and its anterior tip—which corresponds to the Spemann organizer of amniotes (called the Hensen's node or the Node)—begins to move posteriorly (Spratt, 1947) (Fig. 7.2B–D). This regression movement lays in its wake the forming body axis as cells become progressively deposited by the regressing primitive streak. Fate maps of early chicken and mouse embryos show that paraxial mesoderm precursors are located in the anterior portion of the streak (medial precursors, green) and in the adjacent epiblast (lateral precursors, purple) (Fig. 7.2B and C) (Forlani *et al.*, 2003; Hatada and Stern, 1994; Iimura *et al.*, 2007; Lawson *et al.*, 1991; Nicolet, 1971; Psychoyos and Stern, 1996; Rosenquist, 1966; Schoenwolf *et al.*, 1992; Selleck and Stern, 1991; Tam and Beddington, 1987; Tam and Trainor, 1994; Waddington, 1952).

The onset of Hox activation in vertebral precursors has been examined in detail for several genes in chicken and mouse embryos (Fig. 7.3A–C). *Hox* gene transcription is first initiated in the epiblast at the level of the posterior streak, a region that gives rise to mainly the lateral plate and extraembryonic mesoderm (Fig. 7.3B and C) (Deschamps and Wijgerde, 1993; Forlani *et al.*, 2003; Frohman *et al.*, 1990; Gaunt and Strachan, 1994, 1996; Gaunt *et al.*, 1986; Iimura and Pourquié, 2006). *Hox* gene expression initially appears as a salt-and-pepper pattern (Iimura and Pourquié, 2006) and progressively extends anteriorly in the epiblast along the primitive streak (Fig. 7.3B and C) (Forlani *et al.*, 2003; Gaunt and Strachan, 1994, 1996; Iimura and Pourquié, 2006). Then, expression of *Hox* genes progressively spreads to the surrounding cells, resulting in their expression in virtually all of the epiblast cells along the streak up to the node level (Deschamps *et al.*, 1999; Forlani *et al.*, 2003; Gaunt and Strachan, 1994, 1996; Iimura and Pourquié, 2006). Next, *Hox* gene expression subsequently spreads even further anterior to the node in the posterior neural plate (Deschamps *et al.*, 1999; Forlani *et al.*, 2003; Gaunt and Strachan, 1994, 1996; Iimura and Pourquié, 2006). This progressive mode of expression has been described as anterior propagation, forward spreading, or rostral expansion (Deschamps and van Nes, 2005; Gaunt and Strachan, 1994).

In chicken embryos, spreading occurs across an implanted glass barrier (Gaunt and Strachan, 1994). In contrast, in mouse embryos, cultures of isolated anterior primitive streak region taken prior to activation of *Hoxb1* or *Hoxb8* activate these genes if they are taken from embryos less than 12 h

---

Medial (green) and lateral (purple) somitic cells are indicated according to their origins in panels (B) and (C). (F) Differentiation of the somitic derivatives in a 5-day-old chicken embryo at the limb level. Medial somitic cells (green) contribute to epaxial body structures, such as vertebra, muscle (myotome), and dermis; whereas, lateral somitic cells (purple) give rise to hypaxial muscles in the ventral body and limbs. PS, primitive streak; Nc, notochord; Nt, neural tube; PSM, presomitic mesoderm. (See Color Insert.)

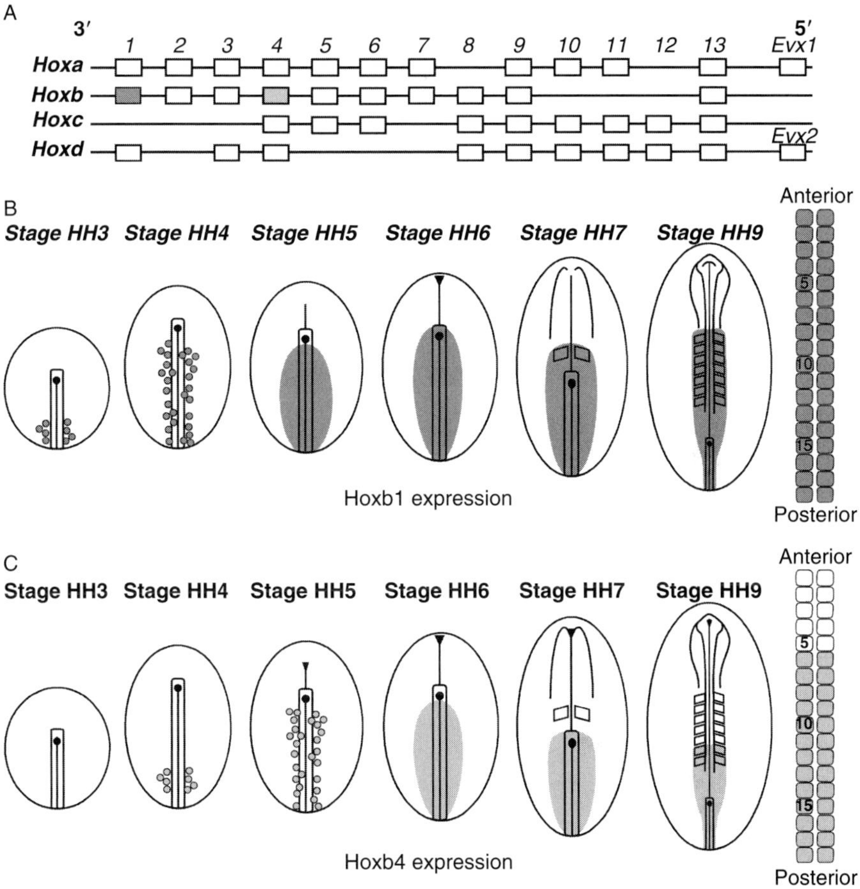

Figure 7.3 Onset of *Hox* gene activation in the chicken embryo. (A) Chicken Hox clusters showing the location of *Hoxb1* (purple) and *Hoxb4* (green). (B) Activation of *Hoxb1* in the chicken embryo. *Hoxb1* (purple) is first expressed in a salt-and-pepper pattern in the posterior part of the primitive streak at stage HH3 and its expression subsequently spreads anteriorly along the streak. At stage HH4, only precursors of the lateral somitic cells express the gene. Expression continues to spread until all cells of the epiblast adjacent to the streak, as well as cells entering the primitive streak, activate *Hoxb1* at stage HH4. From this stage onward, *Hoxb1* is expressed by the precursors of the medial and the lateral somites in the epiblast and the streak, and then this expression is maintained in the descendents of these cells entering the posterior presomitic mesoderm. While the embryonic axis forms, paraxial mesoderm cells continue to express *Hoxb1*, leading to the formation of the columns of *Hoxb1*-expressing somites. This strategy results in the position of the *Hoxb1* expression domain in the paraxial mesoderm to be roughly defined by the timing of Hox activation in paraxial mesoderm precursors in the epiblast. (C) Activation of *Hoxb4* (green). Similar expression kinetics to that of *Hoxb1* is observed, albeit starting a bit later in the posterior streak, at stage HH4. As a result, the presumptive territory of the paraxial mesoderm is reached later by *Hoxb4* at stage HH5, after the cells that will form the anterior-most somites expressing

before activation of these genes or if they are cocultured with the posterior primitive streak region (Forlani *et al.*, 2003). Therefore, Hox activation in the anterior primitive streak requires a signal from the posterior primitive streak. However, the necessary molecular events for the activation of a given *Hox* gene along the epiblast appear to have already taken place in the epiblast precursors at least 12 h before the gene is actually detected, thus potentially providing an explanation as to why the glass barrier does not inhibit the spreading in the chicken embryo. The molecular basis underlying this spreading of Hox expression is not understood. This anterior spreading movement of Hox expression continues for some time in the neural tissue until the level of the anterior boundary is reached; whereas in the paraxial mesoderm, no significant forward spreading is observed in the PSM after ingression of the descendents of the Hox-expressing epiblast cells (Forlani *et al.*, 2003; Iimura and Pourquié, 2006). Consequently, boundaries in the paraxial mesoderm and neural tube are usually located at different AP levels, and the regulation of their positioning obeys different rules (Bel-Vialar *et al.*, 2002). Here, we essentially discuss the positioning of boundaries in the mesoderm.

The first *Hoxb* gene to be expressed, *Hoxb1*, is activated early during gastrulation before complete extension of the primitive streak, and it follows the expression kinetics described above (Forlani *et al.*, 2003; Frohman *et al.*, 1990; Iimura and Pourquié, 2006) (Fig. 7.3A and B). All *Hoxb* genes examined follow similar activation kinetics but their expression is initiated in the posterior primitive streak at progressively later times as one considers genes located more 5′ along the cluster (Forlani *et al.*, 2003; Iimura and Pourquié, 2006) (Fig. 7.3A–C). Thus, the activation of *Hox* genes in the epiblast follows a temporal colinear sequence (Forlani *et al.*, 2003; Gaunt and Strachan, 1996; Iimura and Pourquié, 2006). Soon after the expression domain of a given *Hox* gene reaches the node level, all cells in the paraxial mesoderm territory of the epiblast now express this gene (as well as the more 3′ genes). Descendents of the epiblast cells that express this *Hox* gene enter the posterior PSM at the primitive streak level and maintain expression of this gene. Thus, the anterior boundary of a given *Hox* gene in the paraxial mesoderm is roughly defined by the moment when the expression domain of this gene reaches the paraxial mesoderm territory of the epiblast (Forlani *et al.*, 2003; Iimura and Pourquié, 2006). The temporal colinearity of *Hox* gene expression ensures that this territory is reached progressively later by

---

Hoxb1 have been produced by the primitive streak. Consequently, the first cells expressing *Hoxb4* will be positioned more posteriorly (at the level of somites 5–6) than the *Hoxb1*-expressing cells in the somitic series. In this way, the temporal colinear activation of *Hox* genes in the posterior primitive streak is translated into nested colinear domains along the somitic series. Expression domains of *Hox* genes in other tissues such the neural plate are not represented. (See Color Insert.)

successively more 5′ *Hox* genes. Hence, cells that express progressively more posterior *Hox* genes sequentially enter into the posterior PSM. Due to the posterior elongation of the embryo resulting from primitive streak regression, this leads to the establishment of the nested Hox expression domains along the AP axis.

Whereas the primitive streak contributes to the formation of the most anterior somites; more posteriorly, the paraxial mesoderm is formed from the tail bud (Fig. 7.2D). In this structure, gastrulation movements, similar to those of the primitive streak, are still observed (Cambray and Wilson, 2002, 2007; Catala *et al.*, 1995; Davis and Kirschner, 2000; Gont *et al.*, 1993; Kanki and Ho, 1997; Knezevic *et al.*, 1998; Pasteels, 1937; Stern, 2006). In the tail bud, precursors of the paraxial mesoderm are located immediately posterior to the region where the territory of the notochord and of the neural tube become fused (the chordoneural hinge) (Cambray and Wilson, 2007; Charrier *et al.*, 1999; McGrew *et al.*, 2008). During axis elongation, newly ingressed paraxial mesoderm cells are deposited bilaterally to the chordoneural hinge, resulting in the continuous formation of the two strips of paraxial mesoderm tissue. The sequence of *Hox* gene activation in the primitive streak appears to continue in the tail bud with the colinear activation of the posterior-most *AbdB*-related genes (paralogs 10–13) (Izpisua-Belmonte *et al.*, 1991). However, the cellular details of their expression kinetics in the tail bud have not been reported. Nevertheless, there are very significant differences in the timing of the activation of the 3′ and 5′ genes. Whereas the time between activation of *Hoxb1* and *Hoxb9* in the streak is very short in mouse and chicken embryos (~10 h) (Forlani *et al.*, 2003; Gaunt and Strachan, 1996; Iimura and Pourquié, 2006), the activation time between the paralogs 10–13 is much slower (Izpisua-Belmonte *et al.*, 1991).

Recent observations support a conservation of the mode of mesoderm formation among vertebrates (Cambray and Wilson, 2007; Iimura *et al.*, 2007; Solnica-Krezel, 2005). A *Hox* gene activation pattern similar to that reported in the paraxial mesoderm precursors in chicken and mouse embryos is observed in the Xenopus marginal zone (Wacker *et al.*, 2004b). In the frog, temporal colinear activation of *Hox* genes is observed in the entire marginal zone, except in the organizer region. Colinear *Hox* gene activation proceeds normally in ultraviolet (UV) ventralized Xenopus embryos, suggesting that organizer signals are not required for the temporal colinearity of Hox activation during gastrulation (Wacker *et al.*, 2004b). The organizer is required for the spreading of the colinear activation of *Hox* genes to the neural plate (Wacker *et al.*, 2004b). These experiments in frogs confirm that the colinear activation of *Hox* genes is first initiated in the mesodermal precursors prior to their internalization and then spreads to the neural plate.

3. Molecular Control of Temporal Colinearity

In vertebrates, *Hox* genes are arranged colinearly in intact clusters of ~100 kb in length (Duboule, 2007). The striking conservation of the integrity of vertebrate Hox clusters has led to propose a functional link between the tight colinear arrangement of *Hox* genes in the genome and their temporal activation during embryogenesis (Iimura and Pourquié, 2007; Kmita and Duboule, 2003). Thus far, such a temporal colinearity has been observed only in species that display intact Hox clusters, suggesting that this process relies on some higher-order properties of the organization of the clusters. Animals that show temporal colinearity also exhibit a progressive mode of axis formation and segmentation. For instance, short-germ band insects such as Tribolium or cephalochordates such as Amphioxus, which elongate and segment their body axis progressively, exhibit an intact Hox cluster (Putnam *et al.*, 2008; Richards *et al.*, 2008). In contrast, flies or ascidians, in which the Hox cluster is broken into several pieces, do not show temporal colinearity and form their axis and segments in a highly derived fashion (Dehal *et al.*, 2002; Duboule, 2007; Seo *et al.*, 2004). A segmentation machinery involving dynamic expression of the Notch pathway and a posterior Wnt gradient also appears to be a shared feature of animals showing a progressive mode of axis formation and segmentation (Angelini and Kaufman, 2005; Bolognesi *et al.*, 2008; Chipman and Akam, 2008; McGregor *et al.*, 2008; Stollewerk *et al.*, 2003). Therefore, the progressive mode of segmentation strategy coupled to the temporal colinearity of *Hox* gene expression in the segment precursors might reflect an ancestral patterning strategy of Bilateria.

The molecular mechanism governing the sequential activation of *Hox* genes in the epiblast is unknown. In vertebrates, it has been proposed that temporal colinearity reflects the progressive opening of chromatin in a 3′-to-5′ direction, thus providing a temporally controlled access of *Hox* genes to the transcription machinery (Kmita and Duboule, 2003). This conclusion was initially suggested by relocation experiments of *Hoxd9/LacZ* and *Hoxd11/LacZ* transgenes at the 5′ extremity of the *Hoxd* complex (between *Hoxd13* and *Evx2*). This resulted in a delayed activation of these transgenes in the tail bud, which behaved like *Hoxd13* in the transgenic animals (van der Hoeven *et al.*, 1996). Insertion of a *Hoxd9/LacZ* transgene upstream of the *Hoxd* cluster followed by a series of deletions identified a global regulatory element in the 5′ end of the *Hoxd* complex (Kondo and Duboule, 1999). This element is essential for the early colinear activation of the cluster. Together, these observations led to propose a silencing mechanism originating at the 5′ end of the cluster (Kmita and Duboule, 2003). Progressive opening of the chromatin (or derepression) in a 3′-to-5′

direction would make genes progressively available for transcription. The sequential activation of genes along the *Hoxb* cluster was imaged *in vivo* in mouse embryonic stem (ES) cells (Chambeyron and Bickmore, 2004) and in the primitive streak of early mouse embryos (Chambeyron *et al.*, 2005) using fluorescent in situ hybridization. Progressive chromatin decondensation was observed for the mouse *Hoxb* and *Hoxd* clusters in differentiating ES cells and in the primitive streak/tail bud of mouse embryos (Chambeyron *et al.*, 2005; Morey *et al.*, 2007). Chromatin decondensation, accompanied by successive looping out of the chromosomal territories (CT), is concomitant with gene transcription both *in vitro* and in the primitive streak context (Chambeyron *et al.*, 2005; Morey *et al.*, 2007). Although a functional link between the looping out of genes and transcription remains to be established, extrusion of a locus from the CT probably represents a poised state for transcription. However, no extrusion of the *Hoxd* genes CT was observed in the limb bud despite chromatin decondensation and transcription of the genes, indicating that both processes can occur also independently (Morey *et al.*, 2007).

The colinearity has also been proposed to reflect the action of graded signaling pathways, such as retinoic acid (RA) and FGF (and their target genes Cdx), acting on *cis*-regulatory elements dispersed in the Hox clusters during gastrulation (Duboule, 1998). The sequence of the human genome shows an extremely low density of interspersed repeats (as *Alu* sequence) in the four Hox clusters, suggesting that *cis*-regulatory elements in each cluster are a strong selective constraint that cannot tolerate being interrupted by insertions (Lander *et al.*, 2001). Several experimental configurations, in which *cis*-regulatory sequences within the cluster have been mutated, have been shown to alter the temporal colinearity of *Hox* gene activation during gastrulation (Gerard *et al.*, 1996; Juan and Ruddle, 2003). In mutant mice, in which a *Hoxb1/LacZ* transgene was relocated in the 5′ end of the *Hoxd* cluster, not only was the transgene expressed early in the mesoderm but it also induced ectopic expression of *Hoxd13* in the primitive streak in a pattern reminiscent of *Hoxb1* (Kmita *et al.*, 2000). In this transgenic mouse, ectopic chromatin decondensation and looping out of the CT resulting in a breaking of *Hoxd* colinearity are observed (Morey *et al.*, 2008). Thus, the *Hoxb1* locus contains *cis*-regulatory elements that can initiate local chromatin rearrangement, leading to gene activation. Ectopic expression of *Hoxb1* and *Hoxb2* in response to RA depends on whether the genes are within or outside the cluster (Roelen *et al.*, 2002). A randomly integrated *Hoxb1/LacZ* transgene that contains retinoic acid response elements (RARE) faithfully mimics the endogenous spatial expression pattern of *Hoxb1*; however, unlike the endogenous locus, it cannot be prematurely activated by short RA exposure (Roelen *et al.*, 2002). While able to prematurely activate *Hoxb1* and *Hoxb2*, RA is unlikely to control temporal colinearity because roughly normal activation of *Hox* genes is observed in mouse *raldh2*$^{-/-}$ mutants in which RA synthesis is blocked

(Niederreither *et al.*, 1999). Other candidates, such as Wnt, BMP, and FGF signaling, have been proposed to play a role in the control of the onset of *Hox* gene activation; however, there is no conclusive evidence implicating them in the control of temporal colinearity thus far (Deschamps and van Nes, 2005). Therefore, the timing of activation of a *Hox* gene in the cluster depends on its position within the cluster and on local *cis*-regulatory sequences. Interaction between *cis*-regulatory sequences and the chromatin remodeling machinery might lead to control the progressive availability of *Hox* genes loci for transcription.

The temporal colinearity also involves a temporal regulation of transcription by global enhancers. In vertebrates, temporal and spatial transcriptional regulation of *Hoxd* genes in the early developing mouse limb was shown to depend on interactions between two opposite global regulatory regions located on both sides of the *Hoxd* cluster (Deschamps, 2007). Analysis of an inversion and a series of internal deletions in the *Hoxd* cluster showed that the posterior genes, now positioned more 3′ in the cluster due to the deletion, were activated earlier in the limb bud (Tarchini and Duboule, 2006; Zakany *et al.*, 2004). This work identified an enhancer located in the 3′ region upstream of the cluster called early limb control region (ELCR). The ELCR controls progressive activation of genes in a 3′-to-5′ sequence, while a repressor element in the 5′ region controls the spatial restriction of Hox expression in the limb bud. Since the lateral plate mesoderm that forms the limb bud mesenchyme in which the ELCR is acting is derived from the epiblast, the colinear expression of *Hoxd* genes reported during the early phase of limb bud development likely reflects the earlier temporal colinearity observed in the mesodermal precursors in the epiblast described above. Examination of the same deletion mutants at an earlier stage should reveal whether *Hox* genes that are precociously activated in the limb bud are also precociously activated in the epiblast. This would indicate whether the ELCR enhancer is also operating at the trunk level to control temporal Hox activation in the epiblast. These observations suggest that the initial transcriptional availability of *Hox* genes depends both on their position within the cluster and on specific regulatory elements located outside of the cluster.

4. Converting Temporal into Spatial Colinearity

In vertebrates, the expression dynamics of *Hox* genes described above follows the gastrulation movements of cells from the epiblast to the mesoderm, suggesting that *Hox* genes could play a role in mesoderm formation through gastrulation. This hypothesis was tested in chicken embryos by electroporation of Hox-expressing constructs containing a fluorescent reporter in epiblast cells

(Iimura and Pourquié, 2006). In this way, the fate of the descendants of Hox-overexpressing cells can be easily tracked in the embryo. Overexpression of *Hoxb* genes in the paraxial mesoderm territory in the epiblast at the gastrula stage resulted in colinear phenotypes in terms of AP distribution of Hox-expressing cells (Fig. 7.4A–F). In other words, cells overexpressing more 3′ *Hox* genes contributed to more anterior portions of the somitic columns; whereas, cells overexpressing more 5′ *Hox* genes contributed to more caudal domains. When posterior *Hox* genes, such as *Hoxb7* or *Hoxb9*, were expressed prior to their normal activation, the Hox-expressing cells retained their epithelial structure and remained in the epiblast longer than cells expressing either control GFP or anterior *Hox* genes, such as *Hoxb1* and *Hoxb4* (Fig. 7.4G–L). A deletion mutant of the third helix of the *Hoxb9* homeodomain did not show this delay effect on epiblast ingression, suggesting that DNA binding is essential for this function. Similar results were observed in Xenopus, when the B1 blastomere was injected with *Hoxb9* (*XlhBox6*) at the 32-cell stage (Niehrs and De Robertis, 1991). Involution of the expressing mesodermal cells appeared delayed compared to control cells. The injected cells changed their fate and became incorporated into more posterior territories in the embryo. Therefore, the different *Hox* genes appear to exert a graded (colinear) control over the properties involved in epiblast cell ingression, such as adhesion of epithelial cells and epithelium-to-mesenchyme transition. Consistently, in the developing limb bud mesenchyme, *Hox* genes elicit changes in homophilic cell adhesion properties, possibly through controlling cell adhesion molecules, such as EphA7 receptor (Salsi and Zappavigna, 2006; Stadler *et al.*, 2001; Yokouchi *et al.*, 1995). Strikingly, a similar role for *C. elegans Hox* genes in the control of cell migration has been demonstrated (Kenyon *et al.*, 1997). In the worm, the QL neuroblast, which is located on the left side of the embryo, expresses the *mab-5 Hox* gene (*AntP-AbdA* homolog) and migrates posteriorly; whereas, the QR neuroblast located on the right side expresses the *lin-39* (*Pb-Scr* homolog) and migrates anteriorly.

These data suggest that the sequential activation of *Hox* genes controls the timing of cell ingression during gastrulation. This mechanism has been proposed to control the translation of the early temporal colinearity into the colinear positioning of the *Hox* gene expression domains along the AP axis (Iimura and Pourquié, 2006).

5. Posterior Prevalence is Required for the Establishment of Spatial Colinearity

In the model described above, the colinear activation of *Hox* genes in the epiblast controls the sequential ingression of mesodermal cells in the embryo (Iimura and Pourquié, 2006). Progressively more 5′ paralog genes

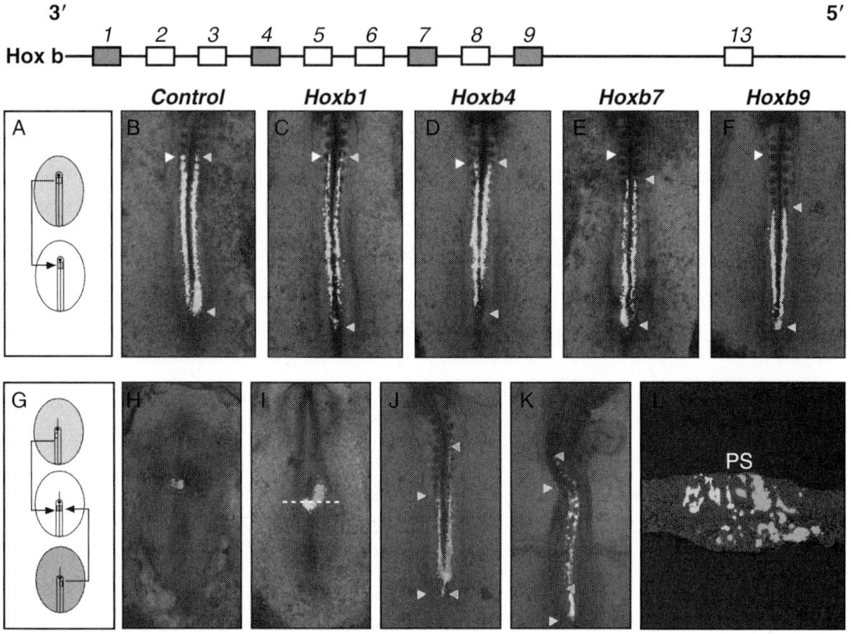

Figure 7.4 *Hox* genes control the timing of ingression of epiblast cells into the primitive streak. (A) Schematic representation of the homotopic and homochronic grafts of fragments of the 80% level of the primitive streak electroporated with Hox-expressing constructs. (B–F) Contributions of the electroporated somitic precursors expressing control (B) or *Hoxb1* (C), *Hoxb4* (D), *Hoxb7* (E), and *Hoxb9* (F). Expression vectors driven by a ubiquitous CAGGS promoter with *IRES2-ZsGreen* were observed following a reincubation period of 16 h. White arrowheads denote the fifth somite level, and green arrowheads denote the anterior limit of the clones overexpressing the constructs. Note the colinear distribution of the anterior limit of the green clones shifting more posteriorly as progressively more 5′ *Hox* genes are overexpressed. (G) Schematic representation of homotopic and homochronic double graft of fragments of the 80% level of the primitive streak from embryos electroporated with *Hoxb4-IRES2-DsRed* and *Hoxb9-IRES2-ZsGreen*, respectively. The grafted embryo is shown just before reincubation (H), and 6 h (I), 16 h (J), and 40 h (K) after reincubation. Green and red arrowheads denote the anterior and posterior extension of the descendants of the grafted labeled cells by each reporter along the AP axis (H–K). The descendents of the cells expressing *Hoxb4* (red) are located more anteriorly than those expressing *Hoxb9* (J, K). (L) Transverse section of an embryo grafted as in (G) and incubated for 6 h at the level indicated in (I) (white hatched line). Note that the cells expressing *Hoxb4* (red) have entered the mesodermal layer; whereas, *Hoxb9*-expressing cells (green) remain in the surface epiblast layer. Ventral views, anterior to the top. PS, primitive streak (panels H–K). (See Color Insert.)

are successively activated in the paraxial mesoderm territory in the epiblast. Activation of these genes begins in a salt-and-pepper fashion in cells that already express more anterior *Hox* genes (Fig. 7.3). In the chicken embryo, when 5′ *Hox* genes are overexpressed in epiblast cells fated to contribute to more anterior territories, these cells alter their migratory properties, remain longer in the epiblast, and contribute to more posterior derivatives (Fig. 7.4). Furthermore, when a 3′ and a 5′ *Hox* gene are co-overexpressed in the same cells in the chicken embryo epiblast, the resulting phenotype is similar to that elicited when only the most-posterior gene is overexpressed (Iimura and Pourquié, 2006). Therefore, in the epiblast, the function of the 5′ *Hox* genes is dominant over that of the 3′ *Hox* genes. This well-established property of functional dominance of the posterior *Hox* genes is termed posterior prevalence in vertebrates (Duboule and Morata, 1994). In the absence of such a mechanism, one might expect that the temporal colinearity of *Hox* gene expression along the primitive streak would result in mixed populations of cells expressing consecutive *Hox* paralog genes to enter simultaneously the posterior PSM and to become located at similar AP levels. Thus, the control of *Hox* genes over epiblast cell ingression might provide a mechanism to ensure a separation of *Hox* gene expression domains by controlling the serial ingression of cells expressing progressively more 5′ *Hox* genes. This is particularly important for the segregation of cells in the anterior-most domain of expression of *Hox* genes, which is the domain where the genes exert their patterning role (Duboule and Morata, 1994). Hence, in this model, the posterior prevalence of *Hox* genes is required for translating the temporal colinearity of their activation in their spatial colinear expression along the somitic columns.

Posterior prevalence or phenotypic suppression was initially characterized in flies mutant for the Polycomb gene *extra-sex combs* (*esc*) in which all *Hox* genes are expressed ubiquitously along the AP axis (Struhl, 1983). In these mutants, all segments acquire the identity of the posterior-most A8 segment, which is controlled by the most 5′ *Hox* gene *AbdB*. Also, ubiquitous overexpression of *Ubx* in the fly embryo overrides the effect of *Hox* genes expressed more anteriorly, such as *Sex comb reduced* (*Scr*) but not that of more 5′ genes such as *AntP* (Gonzalez-Reyes and Morata, 1990). Such observations led to the idea that posterior *Hox* genes are functionally dominant over the anterior ones (Duboule and Morata, 1994). In mouse loss-of-function mutants for *Hox* genes, the actual vertebral phenotypes are usually restricted to a narrow window in the anterior-most expression domain of the mutated genes (Horan *et al.*, 1995a; van den Akker *et al.*, 2001; Wellik and Capecchi, 2003). These mutant mice usually exhibit a normal morphology posteriorly, where more posterior *Hox* genes are normally expressed but the mutated gene is lacking. In transgenic mice that overexpress *Hox* genes anterior to their normal domains, vertebrae that ectopically express the transgene undergo posterior homeotic transformations (Kessel *et al.*, 1990; Lufkin *et al.*, 1992). Thus, only the posterior-most *Hox* genes expressed in a somite play a functional role in specifying its future vertebral identity.

In vertebrates, the ancestral *Hox* gene cluster has been repeatedly duplicated, resulting in seven sets of Hox clusters in zebrafish and four sets in birds and mammals (Duboule, 2007). These duplications generated groups of paralog genes that share high similarities in sequence, expression pattern and function. Paralog swapping experiments in the mouse using a knock-in strategy, in which *Hoxa3* was substituted by *Hoxd3*, or *Hoxa1* by *Hoxb1*, show that these paralog genes carry out similar, although not fully identical functions (Greer *et al.*, 2000; Tvrdik and Capecchi, 2006). Combinations of *Hox4* paralog mutants show a dose-dependent increase in the number of affected vertebrae in mutants lacking more paralogs (Horan *et al.*, 1995b). Furthermore, null mutants for all paralog genes in a group show a much more severe phenotype than mutants retaining one single, wild-type allele (McIntyre *et al.*, 2007; van den Akker *et al.*, 2001; Wellik and Capecchi, 2003).

Whereas the posterior region is normal in mouse compound mutants for complete paralog groups as predicted by posterior prevalence, the posterior part of the affected domain overlaps with the functional domain of the next paralog group (McIntyre *et al.*, 2007; Wellik, 2007). Thus, the axial identity of a vertebra is not strictly controlled by the posterior-most *Hox* genes expressed in its precursors but requires input from adjacent paralogs. Therefore, a combination of adjacent paralog *Hox* genes is required for the specification of vertebral identities. Thus, posterior prevalence probably reflects a graded trend of functional dominance of posterior-over-anterior *Hox* genes, rather than an absolute functional suppression of *Hox* genes by genes located more 5′ in the clusters.

Experiments in the chicken embryo further support this idea. Following premature expression of a given *Hoxb* gene in the epiblast, cells do not end up exactly in the normal expression domain of this gene as one might have predicted in a strict interpretation of posterior prevalence (Iimura and Pourquié, 2006). For example, precocious expression of *Hoxb9* in the epiblast leads the expressing cells to delay their ingression and become positioned at least five to six somites more posteriorly than control cells electroporated with a GFP vector (Fig. 7.4). However, the anterior-most location of *Hoxb9*-overexpressing cells does not correspond to the normal boundary of *Hoxb9* in the somites, which is located at the level of somites 22–23. In fact, cells electroporated with *Hoxb9* in the epiblast before its normal activation, end up being located posterior to somite 11, hence, much more anterior than their endogenous expression domain. Varying the amount of a given *Hox* gene overexpressed in the epiblast by using different promoters does not alter the definitive positioning of the over-expressing cells along the AP axis. Thus, posterior prevalence appears to rely on qualitative rather than quantitative differences between *Hox* genes (Iimura and Pourquié, 2006).

The molecular mechanism underlying posterior prevalence remains unclear. *Hox* gene expression domains largely overlap along the body axis, indicating that posterior genes do not repress transcription of anterior genes. miRNA-mediated, posttranscriptional regulation of *Hox* gene expression has recently been proposed to play a role in the control of posterior prevalence (Yekta *et al.*, 2008). Three miRNAs—*miR-196a-1, miR-196a-2*, and *miR-196b*—can direct cleavage of *Hoxb8* mRNA *in vitro* through their complementary sequence to this gene (Yekta *et al.*, 2004). *Hoxb, Hoxc*, and *Hoxa* clusters, respectively, encode these miRNAs between *Hox9* and *Hox10* genes, at a very similar position to the miRNA *mir-iab4* in the *Drosophila* HOM complex, which lies between *AbdA* and *AbdB* genes (Ronshaugen *et al.*, 2005). These miRNAs are extensively conserved among vertebrates and potentially bind to the 3′UTR sequence of *Hoxb8, Hoxc8, Hoxd8*, and *Hoxa7* (Yekta *et al.*, 2008). These miRNAs were shown to be involved in targeting *Hoxb8* 3′UTR during posterior limb bud patterning, thus providing a safeguard mechanism against inappropriate Hox activity (Hornstein *et al.*, 2005). Target sequences for these miRNAs are enriched in *Hox* genes predominantly located more 3′ than *miR-196*, consistent with their potential role in posterior prevalence (Ronshaugen *et al.*, 2005; Yekta *et al.*, 2008). Two other miRNA—*miR-10a* and *miR-10b*—were identified 5′ to the group 4 genes in the *Hoxb* and *Hoxd* complexes (Woltering and Durston, 2008). Blocking their function leads to extensive vertebral defects, including homeotic transformation.

Posttranscriptional mechanisms other than miRNAs are also likely to be involved in the control of posterior prevalence. Gain-of-function experiments in which *Hox* genes coding sequences were overexpressed from ubiquitous promoters in transgenic flies or mice show specific defects only in the region anterior to the normal expression domain of these genes (Gonzalez-Reyes and Morata, 1990; Kessel *et al.*, 1990; Lufkin *et al.*, 1992). Also, in the overexpression experiments in the chicken epiblast, the Hox expression constructs only contain the coding sequence driven by a ubiquitous chicken *β-actin* promoter (Iimura and Pourquié, 2006). Thus, in these experiments, the functional suppression is neither acting transcriptionally nor through the 3′UTR of *Hox* genes. This raises the possibility that the suppression of the anterior gene function by posterior genes occurs at the protein level.

6. Spatial Dissociation of Segmentation and *Hox* Gene Activation Programs

The origin of the paraxial mesoderm during gastrulation can be traced to two distinct populations of precursors (Fig. 7.2B and C) (Bellairs, 1986; Iimura *et al.*, 2007; Nicolet, 1970; Rosenquist, 1966; Selleck and Stern, 1991).

These two populations exhibit different origins and fates, and contribute to the future medial and lateral somitic compartments (Fig. 7.2E). The medial compartment gives rise to the epaxial muscles, the vertebral column, and the dermis of the back; whereas, the lateral compartment produces essentially the ribs and the hypaxial muscles that include intercostals and limb muscles (Olivera-Martinez *et al.*, 2000; Ordahl and Le Douarin, 1992) (Fig. 7.2E and F).

In the chicken embryo, medial somites derive from a population of precursors that exhibit a stem cell-like behavior and are located in the anterior-most primitive streak and Hensen's node, while the lateral somite precursors derive from the epiblast adjacent to the anterior primitive streak that continues to ingress during axis formation (Bellairs, 1986; Iimura *et al.*, 2007; Selleck and Stern, 1991). A stem cell-like population, located in the primitive streak and contributing to somites, has also been identified in the mouse (Cambray and Wilson, 2002; Eloy-Trinquet and Nicolas, 2002; McGrew *et al.*, 2008; Nicolas *et al.*, 1996). The self-renewal capacity of these stem cells has been revealed by lineage analysis using various cell-labeling strategies and by serial transplants in chicken and mouse embryos (Cambray and Wilson, 2002, 2007; McGrew *et al.*, 2008; Nicolas *et al.*, 1996; Selleck and Stern, 1991). This dual origin of paraxial mesoderm precursors is likely conserved across vertebrates (Iimura *et al.*, 2007).

The distinction between medial and lateral precursors is important because the two populations are involved in different aspects of spine patterning. Microsurgically separating the medial from the lateral PSM results in a segmented medial PSM, while the lateral one remains unsegmented (Freitas *et al.*, 2001). Furthermore, transplantation of the region containing the anterior primitive streak to different regions of the primitive streak of a stage-matched host shows that the descendents of the grafted cells form ectopic rows of somites; whereas, no such effect is observed with other epiblast regions (Schoenwolf, 1992). These lines of evidence suggest that the segmentation program is controlled by the medial cells precursors, which eventually give rise to most of the segmental derivatives of the body axis.

The dynamics of the posterior-to-anterior waves of activation of *Hox* genes along the epiblast described earlier imply that the territory fated to give rise to the lateral portion of the somite in the epiblast first activates expression of *Hox* genes. Due to forward spreading, the stem cell-containing territory located in the anterior streak will be reached by the Hox expression wave only later. Therefore, Hox activation (initiated in the posterior primitive streak by temporal colinearity) and segmentation (initiated in the anterior primitive streak) are spatially segregated.

A conserved feature of the tetrapod axis is the order of morphologically distinct regions along the AP axis level (e.g., cervical, thoracic, lumbar, sacral, and caudal vertebrae). Yet, different numbers of segments contribute to various regions, resulting in the diversity of vertebral formulae between vertebrate species. For instance, mice have 14 rib-bearing thoracic vertebrae, while chickens have seven and corn snakes have 225. Mutation of an entire paralog series of *Hox* genes in the mouse does not affect the total number of segments in the mutants, suggesting that *Hox* genes are not directly involved in control of the total segment number (Wellik, 2007). While *Hox* gene expression has not been investigated in detail in snakes, the large expansion of the thoracic and caudal regions observed (225 thoracic and 65 caudal vertebrae) was shown to be associated with an acceleration of the segmentation clock relative to the axis growth (Gomez *et al.*, 2008). This acceleration was proposed to lead to the formation of smaller segments, hence allowing the formation of more segments for the same amount of growth compared to mouse and chicken embryos. Such an uncoupling between the processes of segmentation and axial regionalization is consistent with their spatial dissociation. It further suggests that their control is likely to be largely independent, hence allowing the huge diversity of vertebral formulae observed across vertebrates.

7. Definitive Positioning of *Hox* Gene Boundaries in the Somites

The precise timing of *Hox* gene activation serves a critical function in defining the final spatial colinear distribution of *Hox* gene expression domains along the body axis, which ultimately determines axial identity of the different vertebrae (Forlani *et al.*, 2003; Juan and Ruddle, 2003; Wacker *et al.*, 2004a). Importantly, however, this initial blueprint does not entirely match the definitive *Hox* gene expression (Forlani *et al.*, 2003). For instance, in the mouse embryo, *Hoxb8* expression is initiated at the head-fold stage (day 7.5 *dpc*) in cells located in the anterior primitive streak (Forlani *et al.*, 2003) and the most anterior localization of descendents of these cells is found to be mainly at the level of somites 6/7 (Forlani *et al.*, 2003). However, the final expression boundary of *Hoxb8* will be more posterior, at the level of somites 10/11. Thus, the position of the definitive *Hox* gene expression boundaries is not solely imposed by the colinear initiation phase, but in addition, requires modification of *Hox* gene expression domains after cells exit the primitive streak and before they are incorporated into a somite, hence, while located in the PSM.

Cells of the posterior PSM experience transcriptional oscillations of the cyclic genes that are driven by the segmentation clock (Fig. 7.5A) (Dequeant *et al.*, 2008). Signaling components, including many negative feedback inhibitors of the Notch, Wnt, and Fgf pathways, are involved in the segmentation clock and have been proposed to control the periodic

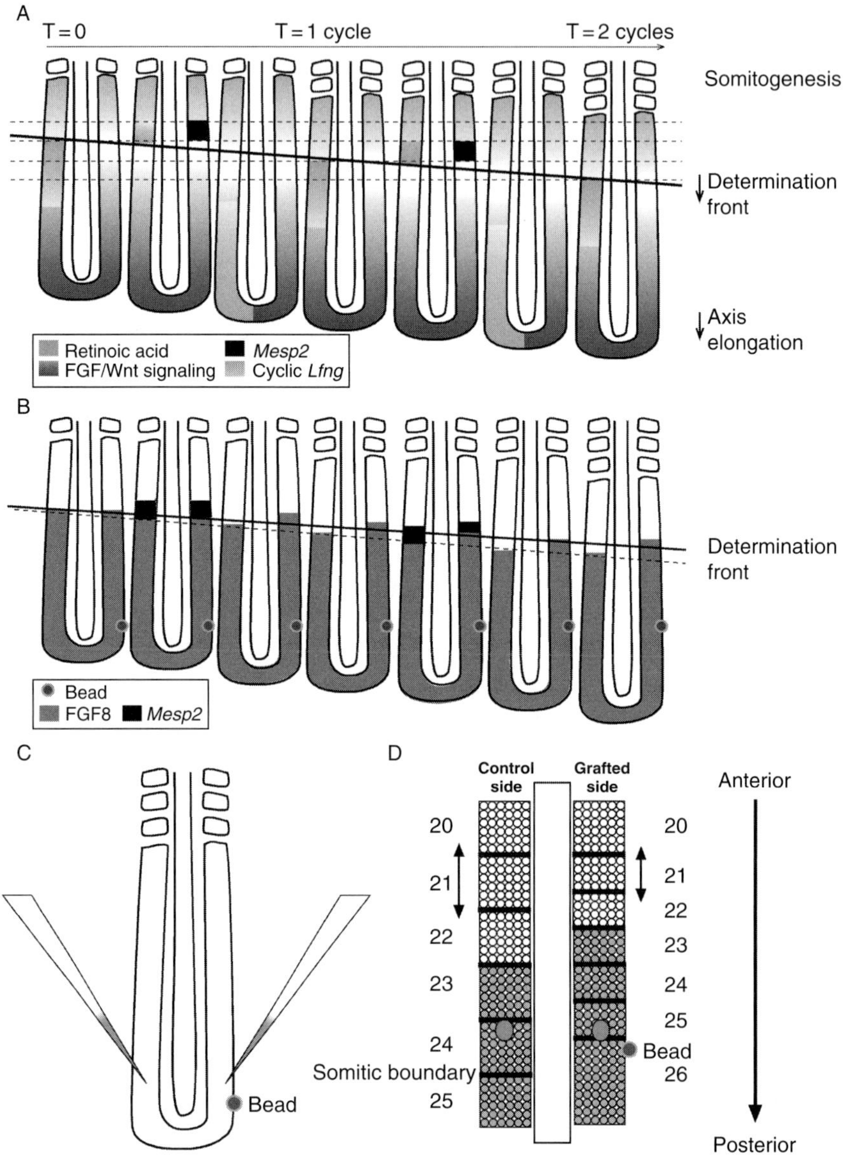

Figure 7.5 Interaction between Hox patterning and the segmentation machinery. (A) The Clock and Wavefront model for somitic segment determination. Antagonistic gradients of FGF/Wnt signaling (purple) and retinoic acid signaling (green) position the determination front (thick, black line). The periodic wave of cyclic gene expression reflecting the segmentation clock signal is shown in orange (represented on the left side only). As the embryo extends posteriorly, the determination front moves caudally. Cells that pass the determination front are exposed to the periodic clock signal, initiating the segmentation program and activating expression of genes, such as *Mesp2*

activation of the system. The segmentation clock oscillator generates a complex periodic pulse of signaling in the PSM, which was proposed to set the rhythm of somite production. Molecular evidence for the existence of such an oscillator has now been obtained in the chicken, mouse, frog, and zebrafish, suggesting that it is a conserved feature of vertebrates (Dequeant *et al.*, 2008 and references therein). Posterior-to-anterior gradients of Fgf and Wnt signaling within the PSM were shown to define a signaling threshold where PSM cells become competent to respond to this periodic clock signal. Due to axis elongation, this threshold (termed the determination front or wavefront) constantly regresses posteriorly while maintaining its relative position in the PSM at the same level. This displacement controls the spacing of the clock response along the AP axis and the conversion of the clock pulse into the spatial periodicity of somites (Dequeant *et al.*, 2008). Therefore, the size of a somite is determined by the distance traveled by the wavefront during one period of the segmentation clock oscillation. Reciprocal inhibitory interactions between RA produced by the anterior PSM and FGF produced by the posterior PSM provide a fine-tuning mechanism involved in refining the positioning of the determination front (Diez del Corral *et al.*, 2003; Moreno and Kintner, 2004). In addition to the important function of controlling differentiation and defining the somite-forming unit, the gradient system of Wnt/Fgf and RA signaling is also involved in the control of *Hox* gene expression in the PSM.

Local modification of the FGF signaling gradient by grafting a bead of Fgf8 in the posterior PSM was shown to result in the formation of smaller somites anterior to the bead (Dubrulle *et al.*, 2001) (Fig. 7.5B). This effect was interpreted as a slowing down of the determination front progression (due to the local FGF concentration increase). As a result, less competent cells are released at the front level during one period of the clock oscillation,

(black squares, represented on the right side only), in a stripe that prefigures the future segment. This establishes the segmental pattern of the presumptive somites. (B) The graft of an Fgf8 bead in the posterior part of the embryo leads to an anterior extension of the FGF gradient (purple), corresponding to a slowing down of the determination front regression. As a result, less competent cells pass the determination front during one oscillation of the clock, leading to a smaller segment. Ultimately, this segment will form a smaller somite compared to the control side (shown in D). (C, D) Labeling cells at the same axial position with DiI in embryos grafted with a Fgf8 bead (C) shows that cells remain at the same axial position but that the position of the somite boundaries is changed on the grafted side due to the formation of smaller somites (Dubrulle *et al.*, 2001). (D) As a result, cells from the same axial level become incorporated into differently numbered somites. Strikingly, the expression of *Hox* genes (here, *Hoxb9*, green) is maintained at the appropriate somitic boundary rather than at the appropriate axial level, providing evidence for a crosstalk between the segmentation machinery and Hox patterning. Red spots show the DiI-labeled cells marked in (C). Dorsal views, anterior to the top. (See Color Insert.)

hence, resulting in smaller somite formation (Fig. 7.5B). Remarkably, the position of the *Hoxb9* and *Hoxa10* anterior boundary at the level of these smaller somites become located more anteriorly on the grafted side, remaining at the appropriate somitic level (Alvares *et al.*, 2003; Dubrulle *et al.*, 2001). In other words, the *Hox* gene expression boundary remains located at the appropriate somite number (somite 23 for *Hoxb9*) rather than at the absolute axial position (as defined by the position of the Hox boundary on the unoperated contralateral side) (Fig. 7.5C and D). Cells that ectopically alter their Hox expression in response to the bead graft also alter their axial properties. For instance, in response to ectopic *Hoxa10* expression triggered by the Fgf8 bead graft, cells begin to activate the muscle migratory program downstream of *Lbx1* (Alvares *et al.*, 2003; Dubrulle *et al.*, 2001). Thus, in these experiments, a gain of function of FGF signaling in the posterior PSM results in a posterior transformation of cells in the smaller somites. Therefore, *Hox* gene expression in the PSM remains plastic and can still be altered by manipulating the segmentation machinery (Dubrulle *et al.*, 2001; Zakany *et al.*, 2001). Additional evidence indicates that Notch signaling, which shows oscillating activity in the posterior PSM (Huppert *et al.*, 2005; Morimoto *et al.*, 2005), likewise is involved in the control of *Hox* gene expression in the PSM (Cordes *et al.*, 2004).

Various perturbations of FGF and RA signaling have been shown to cause changes in *Hox* gene expression and homeotic transformations in mouse embryos. The effect of gain- or loss-of-function mutations of the FGF pathway on *Hox* gene expression is consistent with the results of the FGF bead graft experiments described above. For instance, Fgfr1 gain-of-function mutations introduced into the mouse embryo led to posterior homeotic transformation; whereas, hypomorphic mutations led to anterior transformation (Partanen *et al.*, 1998). FGF signaling and RA signaling were shown to exert mutually antagonistic actions in the PSM (Diez del Corral *et al.*, 2003; Moreno and Kintner, 2004; Vermot and Pourquié, 2005). Hence, a RA gain of function leads to an FGF loss of function in the PSM. The modulation of RA signaling levels in developing mouse embryos causes homeotic transformations which, depending on the time of treatment, cause vertebrae to acquire either more anterior or more posterior identities (Kessel and Gruss, 1991). RA treatment of mice at 8.5 *dpc* causes a posterior shift of 5′ *Hox* genes, consistent with the anterior homeotic transformations observed in the FGF hypomorphic mutants. In contrast, the same experiment at gastrulation stage (7.0 *dpc*) induces an anterior shift of 3′ *Hox* genes (Kessel and Gruss, 1991), consistent with their ability to precociously activate *Hoxb1* in the epiblast (Marshall *et al.*, 1992). These important functions of RA signaling in specifying vertebral axial identity and the control of *Hox* genes were also confirmed through genetic deletion of RA signaling components (Lohnes *et al.*, 1994).

Finally, Wnt signaling has also been connected directly to the process of axial specification. In *Wnt3a* mutant mouse embryos (Ikeya and Takada, 2001) or in Wnt3a hypomorphic embryos *vestigial tail* (*vt*), homeotic transformations and corresponding changes in *Hox* gene expression are observed (Greco *et al.*, 1996). In these mutants, *Fgf8* expression is downregulated in the PSM (Aulehla *et al.*, 2003), suggesting that the effect on *Hox* gene expression could result from the downregulation of FGF signaling in the PSM. The caudal transcription factors (*Cdx*) are targets of the Wnt, FGF, and RA pathways, which play a conserved role in Hox regulation during AP axis formation (Chawengsaksophak *et al.*, 2004; Copf *et al.*, 2004; Houle *et al.*, 2003a; Pilon *et al.*, 2007; van den Akker *et al.*, 2002). In mice, *Cdx* gene expression (*Cdx1, 2*, and *4*) is initiated at the primitive streak stage and later, is restricted to the posterior part of the embryo (Bel-Vialar *et al.*, 2002; Deschamps and van Nes, 2005; Houle *et al.*, 2000, 2003b; Ikeya and Takada, 2001; Isaacs *et al.*, 1994; Lohnes, 2003; Pilon *et al.*, 2006, 2007; Pownall *et al.*, 1996; Prinos *et al.*, 2001). Cdx transcription factors act by regulating a subset of *Hox* genes in a dose-dependent manner (van den Akker *et al.*, 2002). Together, these data suggest that the FGF, Wnt, and RA pathways control the expression of *Hox* gene domains in the PSM.

Several other factors involved in the regulation of *Hox* genes and in the patterning of the axial skeleton have been identified. In the mouse, the TGF-β family secreted factor, Gdf11 (which is expressed in the tail bud) acts upstream of *Hox* genes. Mice mutant for *Gdf11* show homeotic transformation of the vertebral column and tail truncation, correlating with disruption of Hox expression domains (McPherron *et al.*, 1999; Nakashima *et al.*, 1999). These patterning defects are partially phenocopied in mice mutant for the Gdf11 proprotein convertase *Pcsk5*, specifically expressed in the PSM, as well as in mutants for the TGF-β type I receptor (*Alk5*) and type II receptors (*ActRIIA/B* [*Acvr2/b*]) (Andersson *et al.*, 2006; Essalmani *et al.*, 2008; Oh and Li, 1997; Oh *et al.*, 2002; Rancourt and Rancourt, 1997; Szumska *et al.*, 2008). How and if these genes interact with the pathways described above remain to be investigated.

8. Positioning of *Hox* Gene Boundaries in the Forming Segments

After passing through the determination front, competent cells that receive the clock signal simultaneously activate the transcription factor *Mesp2*, which defines the future segmental domain and positions the future somitic boundary (Fig. 7.5A, black) (Dequeant *et al.*, 2008; Morimoto *et al.*, 2005). Based on mathematical modeling, it was proposed that the system of opposing FGF and RA gradients defines a bistability window in the PSM, where cells can adopt either of two distinct steady states: FGF dominant or RA dominant (Goldbeter *et al.*, 2007). Cells in the FGF-dominant state can abruptly switch to the RA-dominant stable state, enabling a group of cells that will form the future segment to be exposed simultaneously to RA signaling. This switch was proposed to be triggered by the periodic signaling pulse delivered by the segmentation clock.

The resulting segmental prepattern, which appears as bilateral stripes of *Mesp2* located immediately anterior to the determination front, provides the blueprint for the morphological segments. The morphological boundaries form when the epithelial somite separates from the anterior tip of the PSM, and subsequently, somitic tissue begins to differentiate as directed by the surrounding tissues (Hirsinger *et al.*, 2000). During this step, somites subdivide into the ventral sclerotome (which contains the skeletal precursors) and the dorsal dermomyotome (which contributes to the myotome and dermatome, giving rise to the skeletal muscles of the body and the dorsal dermis, respectively).

Mesp2 also plays a key role in establishing the rostrocaudal polarity of future somites (Takahashi *et al.*, 2000). This rostrocaudal subdivision of somites imposes the segmental patterning of the nervous system by restricting migration of neural crest cells and axonal guidance of motor neurons within the rostral half of the somite (Bronner-Fraser, 2000). This rostrocaudal subdivision is also critical for the formation of vertebrae, which develop when the caudal half of the skeletal precursors (the sclerotome) of one somite fuses to the rostral half of the sclerotome of the following somite during a process called resegmentation (Christ *et al.*, 2007).

The spatial regulation of *Hoxd* genes in the anterior PSM is controlled by the segmentation machinery (Zakany *et al.*, 2001). Replacing the entire *Hoxd* cluster with a *LacZ* reporter showed periodic stripes of gene expression in the anterior PSM, suggesting that a global enhancer outside of the cluster is responsible for the segmental expression in the PSM. Strikingly, *Hoxd1* is expressed in the same striped domain as *Mesp2* but slightly later in time, suggesting that *Hoxd1* activation might lie downstream of *Mesp2*. Subsequently, *Hoxd1* expression, like *Mesp2*, becomes restricted to the anterior compartment of the forming somite. A similar expression pattern is observed for other *Hox* genes (e.g., *Hoxd3*). This regulatory expression is under the influence of the Notch pathway, since the striped expression is lost in the mouse *Rbpjk*-null mutants in which Notch signaling is abolished. The role of *Hox* genes during later stages of somite patterning remains to be investigated.

9. Conclusion: Determination of the Axial Fate of Vertebral Precursors

Due to posterior prevalence, only the anterior expression domain of *Hox* genes appears to be functionally relevant for patterning the vertebral column. Thus, the positioning of the anterior boundaries of Hox expression domains is of key importance for the establishment of axial identity along the vertebrate spine. Data discussed in this review indicate that the positioning of Hox anterior boundaries is essentially defined by the timing of their activation during gastrulation. The importance of the precise temporal activation of *Hox* genes is illustrated by mutations that change the timing of Hox activation in mouse embryos (Juan and Ruddle, 2003). Mutation of the *cis*-regulatory region controlling *Hoxc8* expression, which causes an initial delay in expression, affects skeletal patterning by largely phenocopying the *Hoxc8*-null mutant, even if the normal somitic expression appeared to be recovered at later stages (Juan and Ruddle, 2003).

Grafting experiments in the chicken embryo demonstrated that Hox expression and axial identity of vertebrae are already fixed in the PSM (Kieny *et al.*, 1972; Nowicki and Burke, 2000). Furthermore, overexpression of *Hox* genes driven by a PSM-specific promoter in transgenic mice leads to severe skeletal patterning defects; whereas, overexpression from a promoter driving expression in somites does not elicit such phenotypes (Carapuco *et al.*, 2005). Thus, the appropriate expression of *Hox* genes early in the development of paraxial mesoderm is critical for the establishment of the future vertebral identity.

In fact, Hox expression appears to be fixed extremely early in the paraxial mesoderm precursors—soon after their initial activation at the epiblast stage (Iimura and Pourquié, 2006). Heterochronic transplantation of epiblast cells in the chicken embryo showed that groups of transplanted cells maintain their endogenous Hox expression schedule. Strikingly, however, in grafts of very small fragments that contain the paraxial mesoderm precursors or in isolated cells that detach from the graft, the original *Hox* gene expression is not maintained, and cells adapt to their new location (McGrew *et al.*, 2008; O. Pourquié and T. Iimura, unpublished observations). Such a situation is also observed in the nervous system (Trainor and Krumlauf, 2000), suggesting that whereas the axial fate appears to be determined very early at the tissue level, small group of cells can be reprogrammed.

Together, theses experiments suggest that the axial identity of paraxial mesoderm precursors is sealed very early in the differentiation of this lineage, possibly even before their ingression during gastrulation, in response to *Hox* gene expression. Therefore, although the position of the anterior boundaries of Hox expression domains somehow becomes subsequently refined in the PSM resulting in their definitive positioning along the AP axis, the timing of *Hox* gene activation during gastrulation largely controls vertebral identity.

ACKNOWLEDGMENTS

The current work was supported by the Stowers Institute for Medical Research and NIH grant R01 HD043158 to OP, and by the grant from the Japanese Ministry of Education, Global Center of Excellence (GCOE) Program, "International Research Center for Molecular Science in Tooth and Bone Diseases" to TI. We thank Pourquié lab members for critical reading of the manuscript and S. Esteban for artwork. OP is a Howard Hughes Medical Institute Investigator.

REFERENCES

Alvares, L. E., Schubert, F. R., Thorpe, C., Mootoosamy, R. C., Cheng, L., Parkyn, G., Lumsden, A., and Dietrich, S. (2003). Intrinsic, Hox-dependent cues determine the fate of skeletal muscle precursors. *Dev. Cell* **5,** 379–390.

Andersson, O., Reissmann, E., and Ibanez, C. F. (2006). Growth differentiation factor 11 signals through the transforming growth factor-beta receptor ALK5 to regionalize the anterior-posterior axis. *EMBO Rep.* **7,** 831–837.

Angelini, D. R., and Kaufman, T. C. (2005). Functional analyses in the milkweed bug *Oncopeltus fasciatus* (Hemiptera) support a role for Wnt signaling in body segmentation but not appendage development. *Dev. Biol.* **283,** 409–423.

Aulehla, A., Wehrle, C., Brand-Saberi, B., Kemler, R., Gossler, A., Kanzler, B., and Herrmann, B. G. (2003). Wnt3a plays a major role in the segmentation clock controlling somitogenesis. *Dev. Cell* **4,** 395–406.

Bellairs, R. (1986). The primitive streak. *Anat. Embryol. (Berl.)* **174,** 1–14.

Bel-Vialar, S., Itasaki, N., and Krumlauf, R. (2002). Initiating Hox gene expression: In the early chick neural tube differential sensitivity to FGF and RA signaling subdivides the HoxB genes in two distinct groups. *Development* **129,** 5103–5115.

Bolognesi, R., Farzana, L., Fischer, T. D., and Brown, S. J. (2008). Multiple Wnt genes are required for segmentation in the short-germ embryo of *Tribolium castaneum*. *Curr. Biol.* **18,** 1624–1629.

Bortier, H., and Vakaet, L. C. (1992). Fate mapping the neural plate and the intraembryonic mesoblast in the upper layer of the chicken blastoderm with xenografting and time-lapse videography. *Dev. Suppl.* 93–97.

Bronner-Fraser, M. (2000). Rostrocaudal differences within the somites confer segmental pattern to trunk neural crest migration. *Curr. Top. Dev. Biol.* **47,** 279–296.

Burke, A. C., Nelson, C. E., Morgan, B. A., and Tabin, C. (1995). Hox genes and the evolution of vertebrate axial morphology. *Development* **121,** 333–346.

Cambray, N., and Wilson, V. (2002). Axial progenitors with extensive potency are localised to the mouse chordoneural hinge. *Development* **129,** 4855–4866.

Cambray, N., and Wilson, V. (2007). Two distinct sources for a population of maturing axial progenitors. *Development* **134,** 2829–2840.

Carapuco, M., Novoa, A., Bobola, N., and Mallo, M. (2005). Hox genes specify vertebral types in the presomitic mesoderm. *Genes Dev.* **19,** 2116–2121.

Catala, M., Teillet, M. A., and Le Douarin, N. M. (1995). Organization and development of the tail bud analyzed with the quail-chick chimaera system. *Mech. Dev.* **51,** 51–65.

Chambeyron, S., and Bickmore, W. A. (2004). Chromatin decondensation and nuclear reorganization of the HoxB locus upon induction of transcription. *Genes Dev.* **18,** 1119–1130.

Chambeyron, S., Da Silva, N. R., Lawson, K. A., and Bickmore, W. A. (2005). Nuclear re-organisation of the Hoxb complex during mouse embryonic development. *Development* **132,** 2215–2223.

Charrier, J. B., Teillet, M. A., Lapointe, F., and Le Douarin, N. M. (1999). Defining subregions of Hensen's node essential for caudalward movement, midline development and cell survival. *Development* **126,** 4771–4783.

Chawengsaksophak, K., de Graaff, W., Rossant, J., Deschamps, J., and Beck, F. (2004). Cdx2 is essential for axial elongation in mouse development. *Proc. Natl. Acad. Sci. USA* **101,** 7641–7645.

Chipman, A. D., and Akam, M. (2008). The segmentation cascade in the centipede *Strigamia maritima*: Involvement of the Notch pathway and pair-rule gene homologues. *Dev. Biol.* **319,** 160–169.

Christ, B., Huang, R., and Scaal, M. (2007). Amniote somite derivatives. *Dev. Dyn.* **236,** 2382–2396.

Copf, T., Schroder, R., and Averof, M. (2004). Ancestral role of caudal genes in axis elongation and segmentation. *Proc. Natl. Acad. Sci. USA* **101,** 17711–17715.

Cordes, R., Schuster-Gossler, K., Serth, K., and Gossler, A. (2004). Specification of vertebral identity is coupled to Notch signalling and the segmentation clock. *Development* **131,** 1221–1233.

Couly, G., Coltey, P., and Le Douarin, N. M. (1993). The triple origin of skull in higher vertebrates: A study in quail-chick chimeras. *Dev. Suppl.* **117,** 409–429.

Davis, R. L., and Kirschner, M. W. (2000). The fate of cells in the tailbud of *Xenopus laevis*. *Development* **127,** 255–267.

Dehal, P., Satou, Y., Campbell, R. K., Chapman, J., Degnan, B., De Tomaso, A., Davidson, B., Di Gregorio, A., Gelpke, M., Goodstein, D. M., Harafuji, N., Hastings, K. E., *et al.* (2002). The draft genome of *Ciona intestinalis*: Insights into chordate and vertebrate origins. *Science* **298,** 2157–2167.

Dequeant, M. L., Ahnert, S., Edelsbrunner, H., Fink, T. M., Glynn, E. F., Hattem, G., Kudlicki, A., Mileyko, Y., Morton, J., Mushegian, A. R., Pachter, L., Rowicka, M., *et al.* (2008). Comparison of pattern detection methods in microarray time series of the segmentation clock. *PLoS ONE* **3,** e2856.

Deschamps, J. (2007). Ancestral and recently recruited global control of the Hox genes in development. *Curr. Opin. Genet. Dev.* **17,** 422–427.

Deschamps, J., and van Nes, J. (2005). Developmental regulation of the Hox genes during axial morphogenesis in the mouse. *Development* **132,** 2931–2942.

Deschamps, J., and Wijgerde, M. (1993). Two phases in the establishment of HOX expression domains. *Dev. Biol.* **156,** 473–480.

Deschamps, J., van den Akker, E., Forlani, S., De Graaff, W., Oosterveen, T., Roelen, B., and Roelfsema, J. (1999). Initiation, establishment and maintenance of Hox gene expression patterns in the mouse. *Int. J. Dev. Biol.* **43,** 635–650.

Diez del Corral, R., Olivera-Martinez, I., Goriely, A., Gale, E., Maden, M., and Storey, K. (2003). Opposing FGF and retinoid pathways control ventral neural pattern, neuronal differentiation, and segmentation during body axis extension. *Neuron* **40,** 65–79.

Dolle, P., Izpisua-Belmonte, J. C., Falkenstein, H., Renucci, A., and Duboule, D. (1989). Coordinate expression of the murine Hox-5 complex homoeobox-containing genes during limb pattern formation. *Nature* **342,** 767–772.

Duboule, D. (1998). Vertebrate hox gene regulation: Clustering and/or colinearity? *Curr. Opin. Genet. Dev.* **8,** 514–518.

Duboule, D. (2007). The rise and fall of Hox gene clusters. *Development* **134,** 2549–2560.

Duboule, D., and Morata, G. (1994). Colinearity and functional hierarchy among genes of the homeotic complexes. *Trends Genet.* **10,** 358–364.

Dubrulle, J., McGrew, M. J., and Pourquié, O. (2001). FGF signaling controls somite boundary position and regulates segmentation clock control of spatiotemporal Hox gene activation. *Cell* **106,** 219–232.

Eloy-Trinquet, S., and Nicolas, J. F. (2002). Cell coherence during production of the presomitic mesoderm and somitogenesis in the mouse embryo. *Development* **129,** 3609–3619.

Essalmani, R., Zaid, A., Marcinkiewicz, J., Chamberland, A., Pasquato, A., Seidah, N. G., and Prat, A. (2008). *In vivo* functions of the proprotein convertase PC5/6 during mouse development: Gdf11 is a likely substrate. *Proc. Natl. Acad. Sci. USA* **105,** 5750–5755.

Forlani, S., Lawson, K. A., and Deschamps, J. (2003). Acquisition of Hox codes during gastrulation and axial elongation in the mouse embryo. *Development* **130,** 3807–3819.

Freitas, C., Rodrigues, S., Charrier, J. B., Teillet, M. A., and Palmeirim, I. (2001). Evidence for medial/lateral specification and positional information within the presomitic mesoderm. *Development* **128,** 5139–5147.

Frohman, M. A., Boyle, M., and Martin, G. R. (1990). Isolation of the mouse Hox-2.9 gene; analysis of embryonic expression suggests that positional information along the anterior-posterior axis is specified by mesoderm. *Development* **110,** 589–607.

Gaunt, S. J., and Strachan, L. (1994). Forward spreading in the establishment of a vertebrate Hox expression boundary: The expression domain separates into anterior and posterior zones, and the spread occurs across implanted glass barriers. *Dev. Dyn.* **199,** 229–240.

Gaunt, S. J., and Strachan, L. (1996). Temporal colinearity in expression of anterior Hox genes in developing chick embryos. *Dev. Dyn.* **207,** 270–280.

Gaunt, S. J., Miller, J. R., Powell, D. J., and Duboule, D. (1986). Homoeobox gene expression in mouse embryos varies with position by the primitive streak stage. *Nature* **324,** 662–664.

Gaunt, S. J., Sharpe, P. T., and Duboule, D. (1988). Spatially restricted domains of homeo-gene transcripts in mouse embryos: Relation to a segmented body plan. *Development* **104,** 169–179.

Gerard, M., Chen, J. Y., Gronemeyer, H., Chambon, P., Duboule, D., and Zakany, J. (1996). *In vivo* targeted mutagenesis of a regulatory element required for positioning the Hoxd-11 and Hoxd-10 expression boundaries. *Genes Dev.* **10,** 2326–2334.

Goldbeter, A., Gonze, D., and Pourquié, O. (2007). Sharp developmental thresholds defined through bistability by antagonistic gradients of retinoic acid and FGF signaling. *Dev. Dyn.* **236,** 1495–1508.

Gomez, C., Ozbudak, E. M., Wunderlich, J., Baumann, D., Lewis, J., and Pourquié, O. (2008). Control of segment number in vertebrate embryos. *Nature* **454,** 335–339.

Gont, L. K., Steinbeisser, H., Blumberg, B., and de Robertis, E. M. (1993). Tail formation as a continuation of gastrulation: The multiple cell populations of the Xenopus tailbud derive from the late blastopore lip. *Development* **119,** 991–1004.

Gonzalez-Reyes, A., and Morata, G. (1990). The developmental effect of overexpressing a Ubx product in Drosophila embryos is dependent on its interactions with other homeotic products. *Cell* **61,** 515–522.

Graham, A., Papalopulu, N., and Krumlauf, R. (1989). The murine and Drosophila homeobox gene complexes have common features of organization and expression. *Cell* **57,** 367–378.

Greco, T. L., Takada, S., Newhouse, M. M., McMahon, J. A., McMahon, A. P., and Camper, S. A. (1996). Analysis of the vestigial tail mutation demonstrates that Wnt-3a gene dosage regulates mouse axial development. *Genes Dev.* **10,** 313–324.

Greer, J. M., Puetz, J., Thomas, K. R., and Capecchi, M. R. (2000). Maintenance of functional equivalence during paralogous Hox gene evolution. *Nature* **403,** 661–665.

Hamburger, V., and Hamilton, H. L. (1992). A series of normal stages in the development of the chick embryo (1951). *Dev. Dyn.* **195,** 231–272.

Hatada, Y., and Stern, C. D. (1994). A fate map of the epiblast of the early chick embryo. *Development* **120,** 2879–2889.

Hinsch, C. W., and Hamilton, H. L. (1956). The developmental fate of the first somite in the chick embryo. *Anat. Rec.* **125,** 225–246.

Hirsinger, E., Jouve, C., Dubrulle, J., and Pourquié, O. (2000). Somite formation and patterning. *Int. Rev. Cytol.* **198,** 1–65.

Horan, G. S., Kovacs, E. N., Behringer, R. R., and Featherstone, M. S. (1995a). Mutations in paralogous Hox genes result in overlapping homeotic transformations of the axial skeleton: Evidence for unique and redundant function. *Dev. Biol.* **169,** 359–372.

Horan, G. S., Ramirez-Solis, R., Featherstone, M. S., Wolgemuth, D. J., Bradley, A., and Behringer, R. R. (1995b). Compound mutants for the paralogous hoxa-4, hoxb-4, and hoxd-4 genes show more complete homeotic transformations and a dose-dependent increase in the number of vertebrae transformed. *Genes Dev.* **9,** 1667–1677.

Hornstein, E., Mansfield, J. H., Yekta, S., Hu, J. K., Harfe, B. D., McManus, M. T., Baskerville, S., Bartel, D. P., and Tabin, C. J. (2005). The microRNA miR-196 acts upstream of Hoxb8 and Shh in limb development. *Nature* **438,** 671–674.

Houle, M., Prinos, P., Iulianella, A., Bouchard, N., and Lohnes, D. (2000). Retinoic acid regulation of Cdx1: An indirect mechanism for retinoids and vertebral specification. *Mol. Cell. Biol.* **20,** 6579–6586.

Houle, M., Allan, D., Lohnes, D., and Lufkin, T. (2003a). "Cdx Homeodomain Proteins in Vertebral Patterning." Advances in Developmental Biology and Biochemistry vol. 13, pp. 69–105. Elsevier, Amsterdam.

Houle, M., Sylvestre, J. R., and Lohnes, D. (2003b). Retinoic acid regulates a subset of Cdx1 function *in vivo*. *Development* **130,** 6555–6567.

Huang, R., Zhi, Q., Ordahl, C. P., and Christ, B. (1997). The fate of the first avian somite. *Anat. Embryol. (Berl.)* **195,** 435–449.

Huppert, S. S., Ilagan, M. X., De Strooper, B., and Kopan, R. (2005). Analysis of Notch function in presomitic mesoderm suggests a gamma-secretase-independent role for presenilins in somite differentiation. *Dev. Cell* **8,** 677–688.

Iimura, T., and Pourquié, O. (2006). Collinear activation of Hoxb genes during gastrulation is linked to mesoderm cell ingression. *Nature* **442,** 568–571.

Iimura, T., and Pourquié, O. (2007). Hox genes in time and space during vertebrate body formation. *Dev. Growth Differ.* **49,** 265–275.

Iimura, T., Yang, X., Weijer, C. J., and Pourquié, O. (2007). Dual mode of paraxial mesoderm formation during chick gastrulation. *Proc. Natl. Acad. Sci. USA* **104,** 2744–2749.

Ikeya, M., and Takada, S. (2001). Wnt-3a is required for somite specification along the anteroposterior axis of the mouse embryo and for regulation of cdx-1 expression. *Mech. Dev.* **103,** 27–33.

Isaacs, H. V., Pownall, M. E., and Slack, J. M. (1994). eFGF regulates Xbra expression during Xenopus gastrulation. *EMBO J.* **13,** 4469–4481.

Izpisua-Belmonte, J. C., Falkenstein, H., Dolle, P., Renucci, A., and Duboule, D. (1991). Murine genes related to the Drosophila AbdB homeotic genes are sequentially expressed during development of the posterior part of the body. *EMBO J.* **10,** 2279–2289.

Juan, A. H., and Ruddle, F. H. (2003). Enhancer timing of Hox gene expression: Deletion of the endogenous Hoxc8 early enhancer. *Development* **130,** 4823–4834.
Kanki, J. P., and Ho, R. K. (1997). The development of the posterior body in zebrafish. *Development* **124,** 881–893.
Kenyon, C. J., Austin, J., Costa, M., Cowing, D. W., Harris, J. M., Honigberg, L., Hunter, C. P., Maloof, J. N., Muller-Immergluck, M. M., Salser, S. J., Waring, D. A., Wang, B. B., *et al.* (1997). The dance of the Hox genes: Patterning the anteroposterior body axis of *Caenorhabditis elegans*. *Cold Spring Harb. Symp. Quant. Biol.* **62,** 293–305.
Kessel, M., and Gruss, P. (1991). Homeotic transformations of murine vertebrae and concomitant alteration of Hox codes induced by retinoic acid. *Cell* **67,** 89–104.
Kessel, M., Balling, R., and Gruss, P. (1990). Variations of cervical vertebrae after expression of a Hox-1.1 transgene in mice. *Cell* **61,** 301–308.
Kieny, M., Mauger, A., and Sengel, P. (1972). Early regionalization of somitic mesoderm as studied by the development of axial skeleton of the chick embryo. *Dev. Biol.* **28,** 142–161.
Kmita, M., and Duboule, D. (2003). Organizing axes in time and space; 25 years of colinear tinkering. *Science* **301,** 331–333.
Kmita, M., van Der Hoeven, F., Zakany, J., Krumlauf, R., and Duboule, D. (2000). Mechanisms of Hox gene colinearity: Transposition of the anterior Hoxb1 gene into the posterior HoxD complex. *Genes Dev.* **14,** 198–211.
Knezevic, V., De Santo, R., and Mackem, S. (1998). Continuing organizer function during chick tail development. *Development* **125,** 1791–1801.
Kondo, T., and Duboule, D. (1999). Breaking colinearity in the mouse HoxD complex. *Cell* **97,** 407–417.
Krumlauf, R. (1994). Hox genes in vertebrate development. *Cell* **78,** 191–201.
Kuratani, S., Matsuo, I., and Aizawa, S. (1997). Developmental patterning and evolution of the mammalian viscerocranium: Genetic insights into comparative morphology. *Dev. Dyn.* **209,** 139–155.
Lander, E. S., Linton, L. M., Birren, B., Nusbaum, C., Zody, M. C., Baldwin, J., Devon, K., Dewar, K., Doyle, M., FitzHugh, W., Funke, R., Gage, D., *et al.* (2001). Initial sequencing and analysis of the human genome. *Nature* **409,** 860–921.
Lawson, K. A., Meneses, J. J., and Pedersen, R. A. (1991). Clonal analysis of epiblast fate during germ layer formation in the mouse embryo. *Development* **113,** 891–911.
Lohnes, D. (2003). The Cdx1 homeodomain protein: An integrator of posterior signaling in the mouse. *Bioessays* **25,** 971–980.
Lohnes, D., Mark, M., Mendelsohn, C., Dolle, P., Dierich, A., Gorry, P., Gansmuller, A., and Chambon, P. (1994). Function of the retinoic acid receptors (RARs) during development (I). Craniofacial and skeletal abnormalities in RAR double mutants. *Development* **120,** 2723–2748.
Lufkin, T., Mark, M., Hart, C. P., Dolle, P., LeMeur, M., and Chambon, P. (1992). Homeotic transformation of the occipital bones of the skull by ectopic expression of a homeobox gene. *Nature* **359,** 835–841.
Marshall, H., Nonchev, S., Sham, M. H., Muchamore, I., Lumsden, A., and Krumlauf, R. (1992). Retinoic acid alters hindbrain Hox code and induces transformation of rhombomeres 2/3 into a 4/5 identity. *Nature* **360,** 737–741.
McGregor, A. P., Pechmann, M., Schwager, E. E., Feitosa, N. M., Kruck, S., Aranda, M., and Damen, W. G. (2008). Wnt8 is required for growth-zone establishment and development of opisthosomal segments in a spider. *Curr. Biol.* **18,** 1619–1623.
McGrew, M. J., Sherman, A., Lillico, S. G., Ellard, F. M., Radcliffe, P. A., Gilhooley, H. J., Mitrophanous, K. A., Cambray, N., Wilson, V., and Sang, H. (2008). Localised axial progenitor cell populations in the avian tail bud are not committed to a posterior Hox identity. *Development* **135,** 2289–2299.

McIntyre, D. C., Rakshit, S., Yallowitz, A. R., Loken, L., Jeannotte, L., Capecchi, M. R., and Wellik, D. M. (2007). Hox patterning of the vertebrate rib cage. *Development* **134,** 2981–2989.

McPherron, A. C., Lawler, A. M., and Lee, S. J. (1999). Regulation of anterior/posterior patterning of the axial skeleton by growth/differentiation factor 11. *Nat. Genet.* **22,** 260–264.

Moreno, T. A., and Kintner, C. (2004). Regulation of segmental patterning by retinoic acid signaling during Xenopus somitogenesis. *Dev. Cell* **6,** 205–218.

Morey, C., Da Silva, N. R., Perry, P., and Bickmore, W. A. (2007). Nuclear reorganisation and chromatin decondensation are conserved, but distinct, mechanisms linked to Hox gene activation. *Development* **134,** 909–919.

Morey, C., Da Silva, N. R., Kmita, M., Duboule, D., and Bickmore, W. A. (2008). Ectopic nuclear reorganisation driven by a Hoxb1 transgene transposed into Hoxd. *J. Cell Sci.* **121,** 571–577.

Morimoto, M., Takahashi, Y., Endo, M., and Saga, Y. (2005). The Mesp2 transcription factor establishes segmental borders by suppressing Notch activity. *Nature* **435,** 354–359.

Nakashima, M., Toyono, T., Akamine, A., and Joyner, A. (1999). Expression of growth/differentiation factor 11, a new member of the BMP/TGFbeta superfamily during mouse embryogenesis. *Mech. Dev.* **80,** 185–189.

Nicolas, J. F., Mathis, L., Bonnerot, C., and Saurin, W. (1996). Evidence in the mouse for self-renewing stem cells in the formation of a segmented longitudinal structure, the myotome. *Development* **122,** 2933–2946.

Nicolet, G. (1970). Analyse autoradiographique de la localisation des differentes ebauches presomptives dans la ligne primitive de l'embryon de poulet. *J. Embryol. Exp. Morphol.* **23,** 79–108.

Nicolet, G. (1971). Avian gastrulation. *Adv. Morphog.* **9,** 231–262.

Niederreither, K., Subbarayan, V., Dolle, P., and Chambon, P. (1999). Embryonic retinoic acid synthesis is essential for early mouse post-implantation development. *Nat. Genet.* **21,** 444–448.

Niehrs, C., and De Robertis, E. M. (1991). Ectopic expression of a homeobox gene changes cell fate in Xenopus embryos in a position-specific manner. *EMBO J.* **10,** 3621–3629.

Noden, D. M. (1986). Patterning of avian craniofacial muscles. *Dev. Biol.* **116,** 347–356.

Nowicki, J. L., and Burke, A. C. (2000). Hox genes and morphological identity: Axial versus lateral patterning in the vertebrate mesoderm. *Development* **127,** 4265–4275.

Oh, S. P., and Li, E. (1997). The signaling pathway mediated by the type IIB activin receptor controls axial patterning and lateral asymmetry in the mouse. *Genes Dev.* **11,** 1812–1826.

Oh, S. P., Yeo, C. Y., Lee, Y., Schrewe, H., Whitman, M., and Li, E. (2002). Activin type IIA and IIB receptors mediate Gdf11 signaling in axial vertebral patterning. *Genes Dev.* **16,** 2749–2754.

Olivera-Martinez, I., Coltey, M., Dhouailly, D., and Pourquié, O. (2000). Mediolateral somitic origin of ribs and dermis determined by quail-chick chimeras. *Development* **127,** 4611–4617.

Ordahl, C. P., and Le Douarin, N. M. (1992). Two myogenic lineages within the developing somite. *Development* **114,** 339–353.

Partanen, J., Schwartz, L., and Rossant, J. (1998). Opposite phenotypes of hypomorphic and Y766 phosphorylation site mutations reveal a function for Fgfr1 in anteroposterior patterning of mouse embryos. *Genes Dev.* **12,** 2332–2344.

Pasteels, J. (1937). Etudes sur la gastrulation des Vertébrés méroblastiques. III. Oiseaux. IV. Conclusions générales. *Arch. Biol.* **48,** 381–488.

Pilon, N., Oh, K., Sylvestre, J. R., Bouchard, N., Savory, J., and Lohnes, D. (2006). Cdx4 is a direct target of the canonical Wnt pathway. *Dev. Biol.* **289,** 55–63.

Pilon, N., Oh, K., Sylvestre, J. R., Savory, J. G., and Lohnes, D. (2007). Wnt signaling is a key mediator of Cdx1 expression *in vivo*. *Development* **134,** 2315–2323.

Pourquié, O., and Tam, P. P. (2001). A nomenclature for prospective somites and phases of cyclic gene expression in the presomitic mesoderm. *Dev. Cell* **1,** 619–620.

Pownall, M. E., Tucker, A. S., Slack, J. M., and Isaacs, H. V. (1996). eFGF, Xcad3 and Hox genes form a molecular pathway that establishes the anteroposterior axis in Xenopus. *Development* **122,** 3881–3892.

Prince, V., and Lumsden, A. (1994). Hoxa-2 expression in normal and transposed rhombomeres: Independent regulation in the neural tube and neural crest. *Development* **120,** 911–923.

Prinos, P., Joseph, S., Oh, K., Meyer, B. I., Gruss, P., and Lohnes, D. (2001). Multiple pathways governing Cdx1 expression during murine development. *Dev. Biol.* **239,** 257–269.

Psychoyos, D., and Stern, C. D. (1996). Fates and migratory routes of primitive streak cells in the chick embryo. *Development* **122,** 1523–1534.

Putnam, N. H., Butts, T., Ferrier, D. E., Furlong, R. F., Hellsten, U., Kawashima, T., Robinson-Rechavi, M., Shoguchi, E., Terry, A., Yu, J. K., Benito-Gutierrez, E. L., Dubchak, I., *et al.* (2008). The amphioxus genome and the evolution of the chordate karyotype. *Nature* **453,** 1064–1071.

Rancourt, S. L., and Rancourt, D. E. (1997). Murine subtilisin-like proteinase SPC6 is expressed during embryonic implantation, somitogenesis, and skeletal formation. *Dev. Genet.* **21,** 75–81.

Richards, S., Gibbs, R. A., Weinstock, G. M., Brown, S. J., Denell, R., Beeman, R. W., Gibbs, R., Beeman, R. W., Brown, S. J., Bucher, G., Friedrich, M., Grimmelikhuijzen, C. J., *et al.* (2008). The genome of the model beetle and pest *Tribolium castaneum*. *Nature* **452,** 949–955.

Richardson, M. K., Allen, S. P., Wright, G. M., Raynaud, A., and Hanken, J. (1998). Somite number and vertebrate evolution. *Development* **125,** 151–160.

Richardson, M. K., Crooijmans, R. P., and Groenen, M. A. (2007). Sequencing and genomic annotation of the chicken (*Gallus gallus*) Hox clusters, and mapping of evolutionarily conserved regions. *Cytogenet. Genome Res.* **117,** 110–119.

Roelen, B. A., de Graaff, W., Forlani, S., and Deschamps, J. (2002). Hox cluster polarity in early transcriptional availability: A high order regulatory level of clustered Hox genes in the mouse. *Mech. Dev.* **119,** 81–90.

Ronshaugen, M., Biemar, F., Piel, J., Levine, M., and Lai, E. C. (2005). The Drosophila microRNA iab-4 causes a dominant homeotic transformation of halteres to wings. *Genes Dev.* **19,** 2947–2952.

Rosenquist, G. C. (1966). A radioautographic study of labeled grafts in the chick blastoderm. Development from primitive-streak stages to stage 12. *Contrib. Embryol. Carnegie Inst. Wash.* **38,** 71–110.

Salsi, V., and Zappavigna, V. (2006). Hoxd13 and Hoxa13 directly control the expression of the EphA7 Ephrin tyrosine kinase receptor in developing limbs. *J. Biol. Chem.* **281,** 1992–1999.

Schoenwolf, G. C. (1992). Morphological and mapping studies of the paranodal and postnodal levels of the neural plate during chick neurulation. *Anat. Rec.* **233,** 281–290.

Schoenwolf, G. C., Garcia-Martinez, V., and Dias, M. S. (1992). Mesoderm movement and fate during avian gastrulation and neurulation. *Dev. Dyn.* **193,** 235–248.

Selleck, M. A., and Stern, C. D. (1991). Fate mapping and cell lineage analysis of Hensen's node in the chick embryo. *Development* **112,** 615–626.

Seo, H. C., Edvardsen, R. B., Maeland, A. D., Bjordal, M., Jensen, M. F., Hansen, A., Flaat, M., Weissenbach, J., Lehrach, H., Wincker, P., Reinhardt, R., and Chourrout, D. (2004). Hox cluster disintegration with persistent anteroposterior order of expression in *Oikopleura dioica*. *Nature* **431,** 67–71.

Solnica-Krezel, L. (2005). Conserved patterns of cell movements during vertebrate gastrulation. *Curr. Biol.* **15,** R213–R228.

Spratt, N. T. Jr. (1947). Regression and shortening of the primitive streak in the explanted chick blastoderm. *J. Exp. Zool.* **104,** 69–100.

Stadler, H. S., Higgins, K. M., and Capecchi, M. R. (2001). Loss of Eph-receptor expression correlates with loss of cell adhesion and chondrogenic capacity in Hoxa13 mutant limbs. *Development* **128,** 4177–4188.

Stern, C. D. (2006). Evolution of the mechanisms that establish the embryonic axes. *Curr. Opin. Genet. Dev.* **16,** 413–418.

Stollewerk, A., Schoppmeier, M., and Damen, W. G. (2003). Involvement of Notch and Delta genes in spider segmentation. *Nature* **423,** 863–865.

Struhl, G. (1983). Role of the esc+ gene product in ensuring the selective expression of segment-specific homeotic genes in Drosophila. *J. Embryol. Exp. Morphol.* **76,** 297–331.

Szumska, D., Pieles, G., Essalmani, R., Bilski, M., Mesnard, D., Kaur, K., Franklyn, A., El Omari, K., Jefferis, J., Bentham, J., Taylor, J. M., Schneider, J. E., *et al.* (2008). VACTERL/caudal regression/Currarino syndrome-like malformations in mice with mutation in the proprotein convertase Pcsk5. *Genes Dev.* **22,** 1465–1477.

Takahashi, Y., Koizumi, K., Takagi, A., Kitajima, S., Inoue, T., Koseki, H., and Saga, Y. (2000). Mesp2 initiates somite segmentation through the Notch signalling pathway. *Nat. Genet.* **25,** 390–396.

Tam, P. P., and Beddington, R. S. (1987). The formation of mesodermal tissues in the mouse embryo during gastrulation and early organogenesis. *Development* **99,** 109–126.

Tam, P. P., and Trainor, P. A. (1994). Specification and segmentation of the paraxial mesoderm. *Anat. Embryol. (Berl.)* **189,** 275–305.

Tarchini, B., and Duboule, D. (2006). Control of Hoxd genes' collinearity during early limb development. *Dev. Cell* **10,** 93–103.

Trainor, P., and Krumlauf, R. (2000). Plasticity in mouse neural crest cells reveals a new patterning role for cranial mesoderm. *Nat. Cell Biol.* **2,** 96–102.

Tvrdik, P., and Capecchi, M. R. (2006). Reversal of Hox1 gene subfunctionalization in the mouse. *Dev. Cell* **11,** 239–250.

van den Akker, E., Fromental-Ramain, C., de Graaff, W., Le Mouellic, H., Brulet, P., Chambon, P., and Deschamps, J. (2001). Axial skeletal patterning in mice lacking all paralogous group 8 Hox genes. *Development* **128,** 1911–1921.

van den Akker, E., Forlani, S., Chawengsaksophak, K., de Graaff, W., Beck, F., Meyer, B. I., and Deschamps, J. (2002). Cdx1 and Cdx2 have overlapping functions in anteroposterior patterning and posterior axis elongation. *Development* **129,** 2181–2193.

van der Hoeven, F., Zakany, J., and Duboule, D. (1996). Gene transpositions in the HoxD complex reveal a hierarchy of regulatory controls. *Cell* **85,** 1025–1035.

Vermot, J., and Pourquié, O. (2005). Retinoic acid coordinates somitogenesis and left-right patterning in vertebrate embryos. *Nature* **435,** 215–220.

Wacker, S. A., Jansen, H. J., McNulty, C. L., Houtzager, E., and Durston, A. J. (2004a). Timed interactions between the Hox expressing non-organiser mesoderm and the Spemann organiser generate positional information during vertebrate gastrulation. *Dev. Biol.* **268,** 207–219.

Wacker, S. A., McNulty, C. L., and Durston, A. J. (2004b). The initiation of Hox gene expression in *Xenopus laevis* is controlled by Brachyury and BMP-4. *Dev. Biol.* **266,** 123–137.

Waddington, C. H. (1952). "The Epigenetics of Birds." Cambridge University Press, Cambridge.

Wellik, D. M. (2007). Hox patterning of the vertebrate axial skeleton. *Dev. Dyn.* **236,** 2454–2463.

Wellik, D. M., and Capecchi, M. R. (2003). Hox10 and Hox11 genes are required to globally pattern the mammalian skeleton. *Science* **301,** 363–367.
Woltering, J. M., and Durston, A. J. (2008). MiR-10 represses HoxB1a and HoxB3a in zebrafish. *PLoS ONE* **3,** e1396.
Yekta, S., Shih, I. H., and Bartel, D. P. (2004). MicroRNA-directed cleavage of HOXB8 mRNA. *Science* **304,** 594–596.
Yekta, S., Tabin, C. J., and Bartel, D. P. (2008). MicroRNAs in the Hox network: An apparent link to posterior prevalence. *Nat. Rev. Genet.* **9,** 789–796.
Yokouchi, Y., Nakazato, S., Yamamoto, M., Goto, Y., Kameda, T., Iba, H., and Kuroiwa, A. (1995). Misexpression of Hoxa-13 induces cartilage homeotic transformation and changes cell adhesiveness in chick limb buds. *Genes Dev.* **9,** 2509–2522.
Zakany, J., Kmita, M., Alarcon, P., de la Pompa, J. L., and Duboule, D. (2001). Localized and transient transcription of Hox genes suggests a link between patterning and the segmentation clock. *Cell* **106,** 207–217.
Zakany, J., Kmita, M., and Duboule, D. (2004). A dual role for Hox genes in limb anterior-posterior asymmetry. *Science* **304,** 1669–1672.

CHAPTER EIGHT

Hox, *Cdx*, and Anteroposterior Patterning in the Mouse Embryo

Teddy Young *and* Jacqueline Deschamps

Contents

Abstract

Cdx and *Hox* gene families descend from the same *ProtoHox* cluster, already present in the common ancestors of bilaterians and cnidarians, and thought to act by providing anteroposterior (A-P) positional identity to axial tissues in all bilaterians. Mouse *Cdx* and *Hox* genes still exhibit common features in their early expression and function. The initiation and early shaping of *Hox* and *Cdx* transcriptional domains in mouse embryos are very similar, in keeping with their common involvement in conveying A-P information to the nascent tissues during embryonic axial elongation. Considerations of the impact on axial patterning of the early expression phase of these genes that correlates with the temporally collinear expression of 3′-5′ *Hox* genes suggest that it is concerned with the acquisition of A-P information by the three germ layers as the axis extends. This early A-P information acquired by all cells emerging from the primitive streak or tailbud and their neighbors in the caudal neural plate gets further modulated by the second phase of gene expression occurring later as the tissues mature and differentiate along the growing axis. We discuss the possibility that regulatory phase 1, common to all *Cdx* and *Hox* genes, is inherent to the concerted mechanism sequentially turning on 3′-5′ *Hox*

Hubrecht Institute, Developmental Biology and Stem Cell Research, Uppsalalaan, Utrecht, The Netherlands

Current Topics in Developmental Biology, Volume 88
ISSN 0070-2153, DOI: 10.1016/S0070-2153(09)88008-3

genes at early stages, and keeping expression of the initiated genes subsequently in the new materials added posteriorly at the axis extends. The posterior *Hox* gene expression domain would be subsequently complemented by *Hox* regulatory phase 2, consisting in a variety of gene-specific, region-specific, and/or tissue-specific gene expression controls. We also touch on the unanswered question whether vertebrate *Cdx* gene expression delivers A-P positional information in its own right, as *Caudal* does in *Drosophila*, or whether it does so exclusively by upregulating *Hox* genes.

1. The *Hox* and *Cdx* Gene Family

A considerable extent of literature has recently solidly anchored the notion that *Hox* and *ParaHox* genes are close relatives that arose from common ancestors during the evolution of a unique *ProtoHox* cluster. This *ProtoHox* cluster would have comprised two or three genes, including one or two ancestors of anterior *Hox* genes and a precursor of posterior genes (the orthologs of Drosophila *AbdB*) (Brooke *et al.*, 1998; Gauchat *et al.*, 2000; Ryan *et al.*, 2007; Quiquand *et al.*, 2009), and might or might not have been concerned with anteroposterior (A-P) axial patterning. The most widely supported hypothetical scenario assumes that a *cis*-transposition of an initial set of genes would have caused their tandem duplication generating the ancestor megacluster from which the *Hox*, *ParaHox*, and *Nkx* genes would have derived after dissociation (Garcia-Fernàndez, 2005; Ryan *et al.*, 2007; Quiquand *et al.*, 2009). After separation of the *Hox* and *ParaHox* clusters, the *ParaHox* genes would have kept their gene content, whereas additional duplications in *cis* would have led, in bilaterians to the larger *Hox* family known in protostomes and deuterostomes. According to a number of studies (Deutsch and Lopez, 2008; Garcia-Fernàndez, 2005) a burst of *cis*-duplications of the anterior genes would have serially generated the five "central" *Hox* genes. Similarly, tandem duplication of the primitive (ancestral) posterior *Hox* gene would have generated the series of six *AbdB*-like *Hox9-14* genes in invertebrate chordates (Amphioxus) and in the ancestors of jawed vertebrates. The lineage leading to teleost fishes and the lineage leading to mammals have lost *Hox14* (Ferrier, 2004; Garcia-Fernàndez, 2005). An interesting hypothesis was proposed to explain the intensive transposition activity in the *Hox* clusters in bilaterians: homology between the DNA-binding domain of the homeodomain of the *Hox* transcription factors and that of DNA type II transposase suggested that an *ANTP* super transposon invaded the common ancestor of metazoans, generating the *Hox*-extended family that comprises the *Hox*, *ParaHox*, *Mox*, and *Evx/Eve* genes (Deutsch and Lopez, 2008). The transposase activity would have later been reduced and lost, whereas DNA-binding capability and a pre-existing transcriptional modulation activity would have been maintained in the Hox homeodomain.

The regulation of the initial *ProtoHox* genes might have included some concerted transcriptional control. Be that as it may, the original 3′-5′ polarity of the linear arrangement of the *ProtoHox* genes has become associated with an early (3′) to late (5′) sequential gene expression of the *Hox* genes, a phenomenon called temporal colinearity (Duboule, 1992, 1994). This phenomenon characterizes *Hox* gene expression in bilaterian species that develop according to a sequential, head-to-tail mode. Temporal colinearity is found exclusively for *Hox* genes organized in intact clusters, and thus in vertebrates where the *Hox* clusters have been consolidated (Duboule, 2007). Coordinated expression in space and in time of at least some of the *Hox* genes has been documented in evolutionary widely separated species such as the polychaete *Chaetopterus*, the beetle *Tribolium*, the cephalochordate *amphioxus*, and the mouse (reviewed by Deschamps, 2007). Some features of a spatial transcriptional colinearity are still visible in some of today's *ParaHox* genes (Brooke *et al.*, 1998; Garcia-Fernàndez, 2005), since vertebrates *Gsx*, *Pdx*, and *Cdx* genes are expressed in domains with an anterior to posterior axial disposition. However, these genes are not expressed collinearly sequentially in time, since *Pdx* is initially expressed later than *Cdx*, and the most anteriorly expressed *Gsx* is initiated last (Ahlgren *et al.*, 1996; Ohlsson *et al.*, 1993; Szucsik *et al.*, 1997; Valerius *et al.*, 1995). Whether this means that temporally collinear expression is an exclusive acquisition of *Hox* genes, or whether extensive alterations of the *ParaHox* cluster have perturbed the temporal collinear expression of these genes at some time of their evolution, is not known. From a regulatory point of view, *ParaHox* genes form a less homogeneous gene family than the *Hox* genes. For instance, the transcriptional direction of *Cdx* is opposite to that of *Gsx* and *Pdx* (Ferrier *et al.*, 2005). Even the tissue specificity of expression differs between *ParaHox* cluster members: in mice, *Cdx* genes are expressed in posterior nascent tissues of the three germ layers similarly to their *Hox* counterparts (reviewed in Deschamps and van Nes, 2005; Deschamps *et al.*, 1999), before being restricted to the endoderm of the digestive tract. *Pdx* is first expressed in the embryonic endoderm and later in the brain as well (Perez-Villamil *et al.*, 1999), and *Gsx* is expressed exclusively in the (neur)ectoderm (Corbin *et al.*, 2000; Li *et al.*, 1994; Szucsik *et al.*, 1997; Valerius *et al.*, 1995).

The objective in this review is to take a closer look at the relationship between mouse *Hox* and *ParaHox* genes (Fig. 8.1) during patterning of the trunk and tail. We will mainly—but not exclusively—focus on the transcriptional regulation of "central" (i.e., PG8), and posterior (PG9-13) paralogy groups of *Hox* genes, and on the *ParaHox* genes *Cdx1*, *2*, and *4*, distant paralogs of *Hox* PG9 (Ferrier *et al.*, 2005; Quiquand *et al.*, 2009). All these genes are initially expressed during gastrulation and later, in the

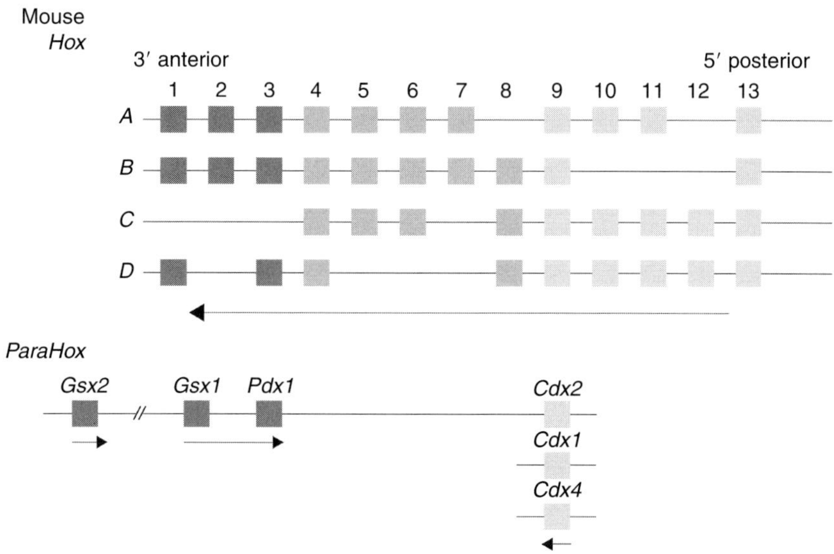

Figure 8.1 Schematic representation (not to scale) of the *Hox* and *ParaHox* genes in the mouse. The four *Hox* clusters are depicted (A-D), with the paralogy group (PG) numbers above the genes. The color blue stands for *Hox* and *ParaHox* genes derived from Anterior *ProtoHox* genes; green stands for the derived "central" *Hox* genes; and yellow stands for genes derived from the "posterior" *ProtoHox* gene. The transcriptional direction of the genes is indicated by an arrow (see text for references). (See Color Insert.)

embryonic area where the progenitors of the posteriorly extending axial and paraxial structures of the trunk and tail are located. The expression and function of *Hox* and *Cdx* genes in the emerging tissues will also be compared.

2. Similarities and Differences in the Two Expression Phases of *Hox* and *Cdx* Genes in the Mouse Embryo

The spatiotemporal dynamics of the early expression of *Hox* genes and *Cdx* genes are very similar, presumably reflecting their close evolutionary relationship. In the mouse embryo, the earliest *Hox* genes are initially expressed at the boundary between extraembryonic and embryonic tissues, in the epiblast and overlying mesoderm at the posterior extremity of the primitive streak (Deschamps and Wijgerde, 1993; Forlani *et al.*, 2003). Their expression domains then spread anteriorly, progressively invading the progenitor region for the lateral plate, intermediate and paraxial mesoderm generated at successively more anterior levels of the primitive streak, and the epiblast containing the neural plate progenitors. The early *Cdx* expression follows the same dynamics (reviewed by Deschamps and van Nes, 2005; Deschamps *et al.*, 1999) (Fig. 8.2A and B, phase I, depicted in red). All mesoderm and all neurectoderm of the trunk and tail are generated from epiblast progenitors that have experienced *Hox* and *Cdx* expression during their early history. Because anterior *Hox* genes and the *Cdx* genes remain expressed in the posterior embryonic region from early to later stages, anterior trunk tissues have experienced expression of the anterior and central *Hox* genes and of *Cdx* genes when they were in the posterior growth zone, whereas posterior trunk and tail express in addition more and more posterior *Hox* genes initiated later during axial elongation. It is this sequential expression of more and more posterior *Hox* genes that is translated into more posterior axial identity of the newly added tissue. The particularity of *Cdx2* expression in the early (E3.5) trophectoderm where it plays a unique, non-*Hox*-related function (Strumpf *et al.*, 2005) will not be discussed here.

After the initially similar expression of *Hox* and *Cdx* genes in the primitive streak, the expansion of the expression domains of these genes rostrally to the node region differs significantly from gene to gene. For example, the expression of *Cdx2* and *Cdx4* in the mesoderm and neurectoderm never expands into the trunk region, but gets downregulated soon after cells are carried away from the node region anteriorwards, as a result of continuing tissue addition during axial extension. *Cdx2* and *Cdx4* transcription in mesoderm and neurectoderm thus remain confined to the immature posterior of the embryo. Their transcripts decay in the presomitic mesoderm (PSM) and neural tube, creating a gradient decreasing rostrally (Gaunt *et al.*, 2003, 2004, 2005) (Fig. 8.2A). In contrast to this, expression of other genes including the *Hox* genes and *Cdx1* is maintained in mesoderm and neurectoderm cells carried away from the node region as the axis extends (although the rostral expression boundaries still will change slightly in the neurectoderm and in the mesoderm in some cases, see below). The final rostral expression boundaries differ in the neurectoderm and the mesoderm, and are more anterior for 3′ than for 5′ genes (Fig. 8.2B and C shows the situation for *Hoxb2* and *Hoxb9*) (reviewed by Deschamps and van Nes, 2005). In describing the dynamics of the expanding *Hox* and *Cdx* expression patterns in the terms of the earlier model of the two *Hox* expression phases (Deschamps and Wijgerde, 1993), we note that all *Hox* and *Cdx* genes behave similarly during phase 1 of expression, when transcripts first appear in the posterior primitive streak and spread anteriorly along the streak in a way clearly at odds with the movements of cell descendants, thus relying on inductive processes (Deschamps and Wijgerde, 1993; Forlani *et al.*, 2003). Posterior *Hox* genes lag behind 3′ *Hox* genes in this initial phase I (Fig. 8.2B and C). *Cdx2* and *Cdx4* in the mesoderm and neurectoderm remain in phase I during

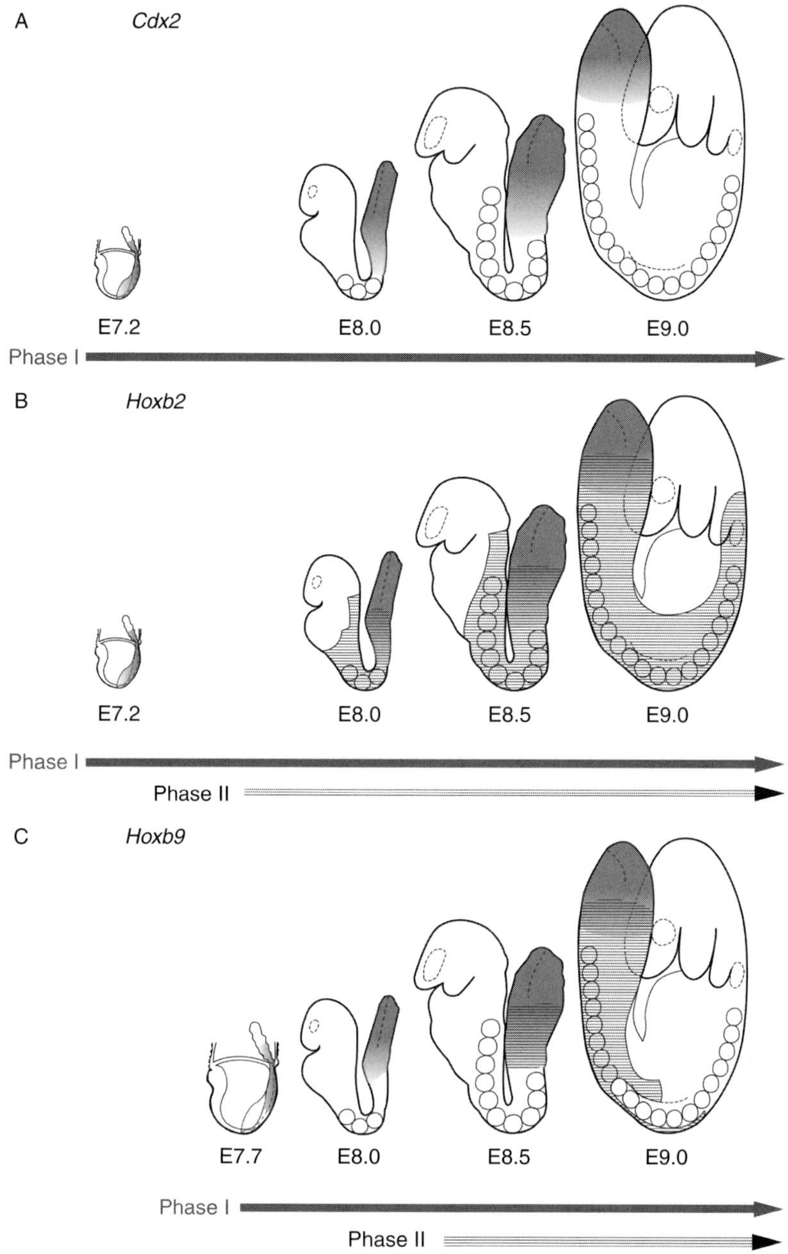

Figure 8.2 (A, B) Representation of phases I and II of expression of a typical *Cdx* gene (A, *Cdx2*) and one of the *Hox* genes (B, *Hoxb2*). The expression of the two genes is shown at four developmental stages. Expression starts for both genes in the posterior part of the primitive streak at the late primitive streak stage, and the transcript domains expand anteriorwards (early head-fold stage represented here). At the early somite stage,

the rest of their expression time while axial extension progresses. *Hox* gene transcription subsequently enters phase II, during which expression in the maturing mesoderm is maintained for most genes and regresses for some (Chen and Capecchi, 1997; Forlani *et al.*, 2003; Iimura and Pourquié, 2006),

---

expression has passed the node and anterior streak, and reached the posterior presomitic mesoderm (PSM) and overlying caudal neural plate. This expression, shown in red in all schemes, corresponds to expression phase I. At E8.5 and E9.0, the posterior part of the embryos is still in *Hox* expression phase I, since the genes are turned on in the new materials continuously generated in the streak and its continuation in the tailbud (at later stages not shown here). Phase I is therefore represented underneath the schemes by a continuous red line, which would stop at the end of posterior tissue addition. For *Hoxb2* (B), at a stage between E7.2 and E8.0, expression continues rostrally to the node region under regulation phase II, shown here with horizontal shading. Expression in the maturing mesoderm is maintained for this gene, while new mesoderm is continuously produced from the streak with time, resulting in an expanding expression domain with a rostral expression boundary farther and farther from the posterior end (B). *Hoxb2* transcript domain in the neurectoderm is not only maintained, but expands further anteriorly in a way that is not clonally supported (Forlani *et al.*, 2003), until reaching the rostral-most expression boundaries (rhombomere R2/R3 boundary), before E8.0 already. Phase II is represented by a time line underneath the scheme for *Hoxb2*. Concerning the mesoderm and neurectoderm, all *Hox* genes and *Cdx1* would undergo a phase I and phase II type of regulation, whereas *Cdx2* and *Cdx4* would only undergo phase I. The situation in the endoderm (not shown here, since data on *Hox* genes in the endoderm throughout development have not been exhaustively reported) is different, since *Cdx* genes are expressed in overlapping patterns in the gut endoderm until late embryonic stages, suggesting that they are all submitted to a type phase I and II regulation in that tissue. It should be stressed that the categories of genes illustrated here with *Hoxb2* (phase I + phase II) and *Cdx2* (only phase I) do not match with the *Hox* versus *Cdx* gene classes since one of the *Cdx* genes, *Cdx1*, behaves like the *Hox* genes and has a phase II regulation in neurectoderm and mesoderm. *Cdx1* expression is then downregulated rapidly from anterior tissues. The downregulation of *Cdx1* is similar to the anterior downregulation of the most 3′ gene of *HoxB*, *Hoxb1* (with the exception that *Hoxb1* expression in R4 is specifically reinforced by a molecular interaction at the level of one of the enhancers). These similarities between *Cdx1* and the *Hox* genes emphasize even more the similarities between *Hox* and *Cdx* families of genes. Expression of the *Hox* genes other than *Hoxb1* generally remains longer active in their entire expression domain in neurectoderm and mesoderm along the axis as shown here for *Hoxb2* phase II (see text for references). (B, C) Comparison of the phases I and II of expression of a 3′ *Hox* gene (B, *Hoxb2*) and a more 5′ gene (C, *Hoxb9*) during development. The evolution of the expression of *Hoxb2* has been described in the legend of (B). Expression of *Hoxb9* in phase I starts in the posterior part of the streak at a later stage than *Hoxb2* (temporal colinearity). Therefore, the transcription domain would not have past the streak/node region before E8.0, after which phase II would start. Transcription of *Hoxb9* under phase II would be maintained in the maturing mesoderm, although its expression boundary regresses somewhat caudally subsequently (Chen and Capecchi, 1997), whereas expression in the neurectoderm is maintained (see text for more references). Expression in the endoderm is not indicated in the schemes, since it has not been documented systematically for these *Hox* genes at the stages represented here. Dark red dashed lines in all schemes represent the primitive streak. (See Color Insert.)

whereas expression in the neurectoderm expands further anteriorly in many cases, in a way that is not clonally supported (Forlani *et al.*, 2003). The second phase of regulation (phase II) thus also critically depends on inductive processes. The cells have not acquired their definitive Hox code in the node region (Forlani *et al.*, 2003), and both the temporally collinear phase I and the later phase II contribute to the final Hox code of axial and paraxial tissues.

The pioneer model of Ed Lewis in *Drosophila* proposed a mechanism of action of the homeotic gene family, whereby embryonic cells on both sides of a *Hox* expression boundary assume a different positional identity as a result of expressing or not expressing a given gene. Given the complex elaboration of the *Hox* expression domains during mouse embryogenesis described above, one can wonder which of these boundaries (phase I, phase II, or intermediate states) is functionally relevant for the acquisition of tissue identity during mouse embryonic development. We will address that issue for the mesoderm, neurectoderm, and endoderm in the following sections.

3. *Hox* and *Cdx* Gene Expression and A-P Patterning

3.1. Patterning of the mesoderm

Hox and *Cdx* genes are expressed all along the primitive streak and its remnant in the tailbud. This comprises the epiblast, which contains progenitors of all types of mesoderm (extraembryonic, lateral plate, intermediate, and paraxial) except the axial mesoderm of the notochord. The extraembryonic mesoderm is not considered here.

3.1.1. Paraxial mesoderm (presomitic and somitic)

Functional inactivation of *Hox* and *Cdx* genes has proven that these genes are required for correct patterning of embryonic structures along the A-P axis. This function has been most extensively documented for paraxial mesoderm which gives rise to the somites and later to the prevertebrae (pvs). Because it is possible to distinguish individual rostral and caudal pvs in mice, the patterning defects resulting from inactivation of single, paralogous and contiguous *Hox* genes could be determined. This aspect of the *Hox* function is more exhaustively addressed by D. Wellik in this issue. Work in many laboratories over 2.5 decades has led to the notion that *Hox* genes pattern the vertebral column in a roughly collinear way, 3′ genes playing a role at relatively anterior (occipital and cervical) levels, whereas more 5′ members affect the identity of trunk vertebrae (thoracic, sacral, and caudal) (reviewed by Wellik, 2007).

The stages of vertebral development at which *Hox* and *Cdx* expression exerts an instructive function seem to vary, depending on the genes considered, and possibly the axial level. The work of Iimura and Pourquié in the chicken embryo showed that overexpression of *Hox* genes in the epiblast before its ingression in the primitive streak can alter the level at which the cells contribute to the mesoderm (Iimura and Pourquié, 2006; see chapter in this issue). In the mouse, it was shown that the level of expression of *Hoxa10* in the PSM exclusively had a deep impact on vertebral morphogenesis (Carapuço *et al.*, 2005) whereas it had no impact when the gene was overexpressed in formed somites. On the other hand, other experiments have shown that the expression of particular *Hox* genes, such as *Hoxa11*, in the already formed somites is important for correct morphogenesis of the caudal part of the axis. Similarly to the stage of action of some *Hox* genes in the PSM, a decrease in *Cdx2* expression in the PSM causes pattern abnormalities of vertebrae at their ultimate upper thoracic level (van den Akker *et al.*, 2002), although the gene is not expressed at all in formed somites. The PSM is therefore a phase in vertebral morphogenesis when *Hox* and *Cdx* genes impose positional information on the paraxial mesoderm.

Because *Cdx* gene products are able to upregulate some at least of the *Hox* genes after binding sites in their regulatory regions (Charité *et al.*, 1998; Subramanian *et al.*, 1995; Tabariès *et al.*, 2005; Taylor *et al.*, 1997), and since *Cdx* and many *Hox* genes are coexpressed in the posterior region of the embryo including the PSM, *Cdx* genes may activate *Hox* genes in the PSM, and thereby modulate axial identity. In spite of the fact that all three *Cdx* genes are paralogs of each other (and distant paralogs of *Hox9-13*), the impact on vertebral patterning of functionally impairing each of the *Cdx* genes is unequal. *Cdx1*-null mice mainly exhibit abnormalities in the cervical and thoracic region, some of which obey the rules defined for anterior transformations (van den Akker *et al.*, 2002 and references therein). *Cdx2* and *Cdx4*, both exclusively expressed in the PSM, give rise to transformations of vertebrae located in the thoracic region exclusively, with a severity and a penetrance that are much lower for *Cdx4* than for *Cdx2* (van Nes *et al.*, 2006). Analysis of compound *Cdx* mutants has revealed that *Cdx* genes act on vertebral patterning in a redundant way (T. Young, J. van Nes, W. de Graaff, and J. Deschamps, unpublished results).

Altogether these data suggest that the *Hox* gene family might affect the positional identity of paraxial mesoderm during their expression phase I (*Cdx* genes and some *Hox* genes) as well as in phase II (certain *Hox* genes).

3.1.2. Lateral plate mesoderm

Studies of vertebral phenotypes of mutants in "Central *Hox* genes" (PG6, 7, and 8) have illustrated the patterning function of these genes in the lateral plate mesoderm contributing to the thoracic region of the axial skeleton (Nowicki and Burke, 2000), and shown that it is regulated independently from that in the paraxial mesoderm (McIntyre *et al.*, 2007). *Cdx* genes are likely also to affect the contribution of the lateral plate mesoderm since their loss-of-function phenotypes include sternal rib defects (van den Akker *et al.*, 2002).

Patterning of the limbs is another manifestation of the role of the *Hox* genes in lateral plate mesoderm-derived structures. Positioning of the limb field along the axis is thought to be relatively independent from patterning of the axial skeleton, since limbs developed at the normal axial position in *Hox* mutants with highly deregionalized axial structure and a transformed sacrum (Wellik and Capecchi, 2003). But the early limb buds themselves are patterned by the *Hox* genes along their anteroposterior axis concomitantly with A-P patterning along the main axis (Tarchini and Duboule, 2006; Zákány *et al.*, 2004). This early A-P polarization of the limb buds results from a restriction of 5′ *HoxD* gene expression to the posterior side of the bud, causing subsequent activation of *Shh* on the posterior limb margin. *Shh* activation is essential for the second phase of *HoxD* gene expression in the distal limb area leading to the development and patterning of the digits (Kmita *et al.*, 2005; Zakany and Duboule, 2007). The 5′ members of *Hox* clusters A and D (*AbdB*-like) are the only *Hox* genes that activate *Shh* on the posterior side of the limb bud (Tarchini *et al.*, 2006). Non-*AbdB Hox* genes, such as *Hoxb8*, had also been found to contribute posterior information to limb bud cells when overexpressed in the anterior limb bud tissue (Charité *et al.*, 1994). However, subsequent studies indicated that only particularly strong transgenic *Hoxb8* overexpression in anterior limb tissue, causing upregulation of 5′ *HoxD* genes, could evoke anterior *Shh* expression and elicit development of posterior digits at the limb anterior side (J. van Nes and J. Deschamps, unpublished results).

In addition to playing a role in A-P patterning of the limbs, *Hox* genes also instruct proximodistal morphogenesis of the growing limbs. This was evidenced by the absence of distal limb segments in mice lacking *Hox* clusters A and D (Kmita *et al.*, 2005). Previous work by the Chambon, Capecchi, and Duboule laboratories had shown that *AbdB*-related *Hox* genes of paralogy groups 9-13 pattern the limbs along the proximodistal axis in a gene-dependent, colinear way.

Cdx genes, as distant paralogs of *Hox9-13* genes, might theoretically be expected to exhibit some patterning activity in the early limb buds. However, no abnormal limb phenotype (except in one incidental instance, see van den Akker *et al.*, 2002) was observed so far in mice with decreased *Cdx* function. This situation may result from the fact that the expression of *Cdx* genes is absent (*Cdx2* and *Cdx4*) or in any case not maintained long enough (*Cdx1*) in the lateral plate mesoderm to play a role in limb bud development. The observation that *Hox* genes are involved in limb development while *Cdx* are not makes it likely that *Hox* expression phase II is essential for early and later limb bud patterning.

Another lateral mesoderm tissue that requires *Cdx* and *Hox* genes is the hematopoietic progenitor cell population. Work on mouse *Hox* regulatory mutants affected in the *Mll* (*trithorax*-like) gene (Ernst *et al.*, 2004a,b) and alteration of *Hox* and *Cdx* expression in mouse ES cells (Lengerke *et al.*, 2008) suggest that *Hox* and *Cdx* genes play a role in early hematopoietic development. Investigations in zebrafish *Cdx* loss-of-function mutants has proven that *Cdx* and *Hox* genes are required to specify blood (Davidson and Zon, 2006), and analysis of primitive hematopoiesis in ES cell-derived embryoid bodies and early mouse *Cdx* mutant embryos (Wang *et al.*, 2008) indicate that this function is at work as well in the mouse. The action of *Hox* and *Cdx* genes on primitive hematopoietic progenitors would take place during expression phase I, common for *Cdx* and *Hox* genes.

3.1.3. Intermediate mesoderm

Some at least of the mouse *Hox* genes are and remain expressed in the derivatives of the mesonephric duct and in the metanephric kidney mesenchyme. Posterior *Hox* genes have been shown to be involved in the ontogenesis of the metanephros (Mugford *et al.*, 2008; Wellik *et al.*, 2002). Recent work in the Duboule laboratory revealed that two overlapping sets of *HoxD* genes are expressed in the epithelial (anterior genes *Hoxd1-9*) versus mesenchymal (posterior genes *Hoxd9-12*) cells of the metanephros, and function, respectively, in maintaining the integrity of the renal tubular epithelia (*Hoxd8* and *Hoxd9*) and in regulating metanephric mesenchyme-ureteric bud interactions (*Hoxd11-Hoxd13*) (Di-Poï *et al.*, 2007). *Cdx* genes are involved in kidney patterning in the zebrafish, as shown by Wingert *et al.* (2007), who reported that *Cdx* loss of function causes a posterior shift in the axial position of the pronephros. The situation in the mouse has not been examined in the most severe *Cdx* loss-of-function allelic combination possible, *Cdx*-null embryos.

3.2. *Hox* and *Cdx* genes in the neurectoderm

The regulation of *Hox* gene expression takes place at both ends of the neural territory patterned by the *Hox* genes, the forming hindbrain at the rostral end, and the posteriorly extending neural tube on the caudal side.

At the caudal end of the axis, *Hox* genes are regulated within the posteriorly elongating neural plate, encompassing the area where neural progenitors reside around the node region (Delfino-Machín *et al.*, 2005; Diez del Corral and Storey, 2004; Mathis *et al.*, 2001). The sequential expression of 3′-5′ *Hox* genes and of *Cdx* genes during phase I of their regulation confers positional identity to the nascent neural tissue at the caudal end of the extending axis. The regulation of *Hox* gene expression by Fgf and RA (Bel-Vialar *et al.*, 2002; Olivera-Martinez and Storey, 2007) modulates neuronal cell fate specification in the ventral spinal cord (Dasen *et al.*, 2003; Liu *et al.*, 2001). Nordström *et al.* (2006) showed that differential *Cdx* and *Hox* expression profiles exert a rostrocaudal patterning role on the hindbrain and spinal cord progenitors in response to early Wnt signals. Experiments in zebrafish *Cdx* mutants revealed that *Cdx* expression contributes to posterior neural identity. Loss of *Cdx* function causes posterior expansion of the hindbrain at the expense of trunk and tail (Shimizu *et al.*, 2006; Skromne *et al.*, 2007). The early role of *Hox* and *Cdx* expression during the acquisition of early axial identity of motor neurons occurring in regulation phase I must be followed by a later function of these genes during motor neuron pool diversification and topographic organization of target muscle connectivity (Dasen *et al.*, 2005) presumed to happen during expression phase II. The evaluation of the functional impact of this regulation of the *Hox* genes on the neural tube in mouse *Hox* mutants has mainly been assessed at later stages, after morphological and functional maturation of the neural tube occurring around and after birth. Gait and walking defects of some of the *Hox* mutant mice have indicated central and posterior *Hox* genes of the four clusters as playing patterning roles in the trunk motoneurons (Tarchini *et al.*, 2005; Tiret *et al.*, 1998; Wu *et al.*, 2008). The role of *Hox* genes in the sensory nervous system has also been studied in the mouse. Recently, a function of "central" *Hox* genes in the elaboration of the sensory nervous system in the mouse spinal cord was described. *Hoxb8* was shown to be required for correct projection of sensory afferents from the lower lumbar area to the dorsal horn (Holstege *et al.*, 2008). *Hoxb8*-null mice exhibit an altered response to temperature and nociceptive stimuli at the level of the hindlimbs, and have abnormal secondary neuron organization in the dorsal-most laminae of the lumbar neural tube (Holstege *et al.*, 2008). The data show an A-P specificity in the axial location of the defects, compatible with a role of the *Hox* gene during early regulation phase 1. However, the defect in *Hoxb8*-null mice can also be seen as affecting the specific dorsoventral confinement of the *Hox*-expressing cells in the dorsal horn, possibly linked to expression phase II. Motor and sensory neural defects have not been described yet in *Cdx* mouse mutants, and await the availability of a *Cdx*-null genotypic situation.

On the anterior side of the early neural tube, the expression of the *Hox* genes 3′ to PG8 is subjected to a specific regulation establishing the definitive *Hox* rostral expression boundaries in the hindbrain. The extreme rostral expression boundary of 3′ (*Hoxb1-Hoxb4*) and "central" *Hox* genes (*Hoxb5-Hoxb8*) in the hindbrain has been shown to be achieved in a temporally sequential manner, and to depend on endogenous retinoic acid (RA) signaling from the flanking paraxial mesoderm. A collinear window of RA response of these latter genes in the hindbrain was shown to exist between E8.5 (for *Hoxb1* and *b2*) and E10.5 (for *Hoxb8*). Retinoic acid-responsive elements (RAREs) have been identified and functionally characterized for the 3′ most genes (*Hoxb1-Hoxb2*) and for *Hoxb4*. An additional RARE was discovered between *Hoxb4* and *Hoxb5*, and shown to function on *Hoxb5* and *Hoxb8* reporter transgenes *in vivo* (Oosterveen *et al.*, 2003). However, inactivation of this RARE element by gene targeting did not affect the expression boundaries of any 3′ or central *Hoxb* gene, suggesting functional redundancy of this RARE with the 3′ RAREs of the *HoxB* cluster (W. de Graaff and J. Deschamps, unpublished results). This temporally collinear regulation of 3′ and "central" *Hox* genes by RA in the hindbrain, superimposed to the early *Hox* expression pattern, would take place during expression phase 2, and reflect the important patterning role of the *Hox* genes in this part of the central nervous system (reviewed by Trainor and Krumlauf, 2000). A specific function of 3′ *Hox* genes was documented by the study of Goddard *et al.* (1996), describing the absence of the somatic motor component of the facial nerve and subsequent facial paralysis of *Hoxb1* mutant mice. Gaufo *et al.* (2004) studied the contribution of *Hox* genes to the diversity of the hindbrain sensory system. An essential function was documented for *Hoxa2* in the innervation map of the sensory nervous system corresponding to the whiskers (Oury *et al.*, 2006) (see Chapter 3). Mouse *Cdx2* and *Cdx4* are never expressed in the hindbrain and upper neural tube, and *Cdx1* is only expressed there very early and transiently. These genes do not play a patterning role in the differentiating hindbrain.

3.3. *Hox* and *Cdx* and the endoderm of the digestive system

Trunk endoderm is generated differently from the mesoderm. While trunk mesoderm is produced continuously from epiblast through the primitive streak and its continuation in the tailbud, only fore- and anterior midgut endoderm arises from epiblast ingressing through the anterior-most part of the streak. Caudal midgut and hindgut endoderm are formed from expansion of the endoderm produced earlier (Franklin *et al.*, 2008; Tam *et al.*, 2007). The posterior endoderm of the early-somite stage embryo harbors the progenitors of the full length of the gut from the level of the forelimb bud to the end of the embryonic gut (Franklin *et al.*, 2008).

The early endoderm is patterned from early stages on along the A-P axis by diffusible growth factors, among which Fgf (Wells and Melton, 2000). Since Fgf signaling is known to regulate *Cdx* expression in vertebrates (Keenan *et al.*, 2006; Lohnes, 2003), it is likely that the Fgf/Cdx pathway plays an early patterning role in the endoderm (Stringer *et al.*, 2008). Work on zebrafish documents early functions of *Cdx* genes in endoderm patterning. *Cdx1b* plays a very early role in endoderm formation (Cheng *et al.*, 2008), and loss of function of *Cdx4* and *Cdx1a* leads to a caudal shift of the foregut-derived pancreas and liver, and patterning defects in these foregut derivatives (Kinkel *et al.*, 2008). It has not been reported so far whether any *Cdx* loss-of-function mice affects foregut organs.

The expression and function of mouse *Cdx* genes has mostly been studied at midgestation. *Cdx* genes are expressed in gene-specific patterns in the endoderm (Beck *et al.*, 2000). Expression of *Cdx1* is detected after E12 in the gut endoderm, at levels decreasing from the rostral midgut till the end of the hindgut (Duprey *et al.*, 1988). *Cdx2* is the most anteriorly expressed *Cdx* gene in mouse endoderm. It is expressed in the posterior foregut down to the hindgut (Beck *et al.*, 1995). *Cdx4* is exclusively expressed in the hindgut endoderm (Gamer and Wright, 1993; T. Young and J. Deschamps, unpublished results). Homozygote-null mutants for both *Cdx1* and *Cdx4* are healthy, reflecting the lack of functional gut dysmorphogenesis. A-P patterning defects in the endoderm of *Cdx* loss-of-function mutants was exclusively observed in the case of *Cdx2*. Chimerism of mice with *Cdx2*-null cells, or haploinsufficiency of *Cdx2* in *Cdx2* heterozygote mice lead to areas of heteroplasia around the ceacum where the epithelium has adopted a stomach-like identity (Beck *et al.*, 1999; Chawengsaksophak *et al.*, 1997). Ectopic expression of *Cdx2* in the stomach on the other hand causes the appearance of areas of intestine epithelium (Mutoh *et al.*, 2002; Silberg *et al.*, 2002). These anterior and posterior transformations, respectively, of epithelium identity in *Cdx2* loss or gain of function fit into the category of homeotic-like transformations. Recent work reveals that endodermal expression of *Cdx2* is the initiating step for gut histodifferentiation, and that subsequent endoderm/mesoderm crosstalk involving *Barx1* in the mesenchyme lead to the postnatal gut phenotype (Stringer *et al.*, 2008). Absence of *Cdx2* expression in the endoderm of the gut is associated with the expression of *Barx1* in the underlying mesoderm and the subsequent histodifferentiation of stomach mucosa (Stringer *et al.*, 2008).

The endoderm of the digestive tract around day 12 expresses some at least of the *Hox* genes in a spatially collinear way along the axis (reviewed by Beck *et al.*, 2000). *Hox* mutations have been shown to give rise, in several cases, to specific defects at morphological boundaries in the digestive tract, such as the ceacum (Zacchetti *et al.*, 2007) and the anal sphincters (Kondo *et al.*, 1996).

Since initial emergence of definitive endoderm from the anterior primitive streak is relatively early and short-lasting, phase I of *Hox* and *Cdx* expression is probably short in the endoderm, and subsequent expression of the genes during expansion of the posterior endoderm to form the caudal midgut to the hindgut, and during histodifferentiation and patterning of the digestive tract would correspond to gene expression phase II (endoderm expression is not shown in Fig. 8.2).

4. Conclusion

The close evolutionary relationship between *Hox* and *Cdx* genes, the similarities between the regulation of their early expression domains in nascent tissue, and the functional involvement of both gene families in patterning mesoderm (lateral, intermediate, and paraxial), neurectoderm, and endoderm in vertebrate embryos are consistent with the view that *Cdx* and *Hox* genes act as closely related homeotic-like family members. They both participate in patterning the PSM of the extending trunk, the nascent neural tissue and the early endoderm. This collaboration would include a *Cdx-Hox* crossregulatory link during phase I, at a time when the genes are coexpressed in the primitive streak, PSM, and nascent neural and endoderm tissue.

Cdx loss-of-function mutations affect central *Hox* gene expression (van den Akker *et al.*, 2002; T. Young and J. Deschamps, unpublished results). *Cdx* genes may therefore contribute some additional posteriorizing information to that of these *Hox* genes. Whether *Cdx* genes exert homeotic patterning functions in their own right, in addition to upregulating *Hox* genes is difficult to test in the absence of a *Hox*-null mutation, but the truly homeotic function of *Caudal* in *Drosophila* (Moreno and Morata, 1999) proves the potential of a homeotic role for vertebrate Cdx proteins.

Hox and *Cdx* genes appear to share the early phase (phase I, Fig. 8.2) of their transcriptional regulation in the embryo proper. *Cdx* and 3′ *Hox* genes may respond to the same initiating signals during gastrulation, in a way inherently linked to the inductive events in the early embryo. *Wnt* signals seem to be the earliest modulators of rostrocaudal identity in the neural progenitors during gastrulation (Nordström *et al.*, 2006) and *Wnt* is a candidate to initially induce *Hox* genes in the primitive streak (Forlani *et al.*, 2003).

The beginning of phase I of *Hox* expression, common to all *Cdx* and *Hox* genes, would be inherent to the concerted mechanism sequentially turning on more and more 5′ *Hox* genes in nascent tissues from the three germ layers as the axis extends. It may thus underlie temporal colinearity of sequentially induced 3′-5′ *Hox* genes, and it is likely to be ancestral and to have been evolutionary conserved in bilaterians that extend their body by posterior additions. Alterations in phase I *Hox* and *Cdx* expression would cause early modifications in axial identity of mesoderm and neuron progenitors.

Phase II of *Hox*/*Cdx* expression constitutes a second and later regulatory phase. For each gene, expression phase II would be engaged after its domain of early expression has encompassed the whole primitive streak. It would concern events occurring later than the initial setting of A-P instructions to the emerging caudal tissues. *Hox* expression phase II would account for gene-specific, region-specific, and sometimes tissue-specific expression and function of the different genes, and therefore consist in a sum of gene transcription modulation events, including auto- and crossregulations in some cases. It would contribute to morphogenesis of the axial skeleton, and control processes such as late segmental differentiation of hindbrain neurons, proper connectivity of spinal motor neurons, dorsoventral organization of the dorsal horn, proximodistal development of the limbs, and rostrocaudal histogenesis of gut endoderm. This phase II, which follows the temporally collinear phase I, would end with the setting of the *Hox* expression boundaries in a spatially collinear way along the axis in the different germ layers. The relative independence of phase I and II has recently got support from work in the Duboule laboratory, showing that temporal and spatial regulation of mouse HoxD genes can be uncoupled (Tschopp *et al.*, 2009).

Phase I expression of *Hox* and *Cdx* genes has an impact on A-P identity of cellular precursors of trunk tissues, but does not impose a definitive Hox code on these precursors (Forlani *et al.*, 2003) because this Hox code will still be modified in most cases subsequently during phase II, in a tissue-specific and axial position-dependent way. In the end, a likely answer to the question asked above regarding the most relevant *Hox* and *Cdx* expression phase for the acquisition of tissue identity during mouse embryonic development is that every stage of expression of these genes during the history of the tissues must be important. *Hox* and *Cdx* expression are likely to impact on successive, sometimes transient states of cell populations interacting with their environment, more than representing discrete definitive "Hox identity codes" for individual cells at a given moment.

ACKNOWLEDGMENTS

We warmly thank Indayani Young for help with the artwork, and Felix Beck, Jean Deutsch, Denis Duboule, David Ferrier, Brigitte Galliot, and Olivier Pourquié for critically reading the manuscript. We apologize for not being able to cite all the work of colleagues relevant to the issues discussed here. J.D. is supported by the Dutch Research Organization NWO ALW, the Dutch Bsik Program "Stem Cells in Development and Disease" and the EU Network of Excellence (framework 6) "Cells into Organs." T.Y. is financed by the EU Network of Excellence (framework 6) "Cells into Organs."

REFERENCES

Ahlgren, A., Jonsson, J., and Edlund, H. (1996). The morphogenesis of the pancreatic mesenchyme is uncoupled from that of the pancreatic epithelium in IPF1/PDX1-deficient mice. *Development* **122,** 1409–1416.

Beck, F., Erler, T., Russell, A., and James, R. (1995). Expression of Cdx-2 in the mouse embryo and placenta: Possible role in patterning of the extra-embryonic membranes. *Dev. Dyn.* **204,** 219–227.

Beck, F., Chawengsaksophak, K., Waring, P., Playford, R. J., and Furness, J. B. (1999). Reprogramming of intestinal differentiation and intercalary regeneration in Cdx2 mutant mice. *Proc. Natl. Acad. Sci. USA* **96,** 7318–7323.

Beck, F., Tata, F., and Chawengsaksophak, K. (2000). Homeobox genes and gut development. *Bioessays* **22,** 431–441.

Bel-Vialar, S., Itasaki, N., and Krumlauf, R. (2002). Initiating Hox gene expression: In the early chick neural tube differential sensitivity to FGF and RA signaling subdivides the HoxB genes in two distinct groups. *Development* **129,** 5103–5115.

Brooke, N. M., Garcia-Fernàndez, J., and Holland, P. W. (1998). The ParaHox gene cluster is an evolutionary sister of the Hox gene cluster. *Nature* **392,** 920–922.

Carapuço, M., Nóvoa, A., Bobola, N., and Mallo, M. (2005). Hox genes specify vertebral types in the presomitic mesoderm. *Genes Dev.* **19,** 2116–2121.

Charité, J., de Graaff, W., Shen, S., and Deschamps, J. (1994). Ectopic expression of Hoxb-8 causes duplication of the ZPA in the forelimb and homeotic transformation of axial structures. *Cell* **78,** 589–601.

Charité, J., de Graaff, W., Consten, D., Reijnen, M. J., Korving, J., and Deschamps, J. (1998). Transducing positional information to the Hox genes: Critical interaction of cdx gene products with position-sensitive regulatory elements. *Development* **125,** 4349–4358.

Chawengsaksophak, K., James, R., Hammond, V. E., Köntgen, F., and Beck, F. (1997). Homeosis and intestinal tumours in Cdx2 mutant mice. *Nature* **386,** 84–87.

Chen, F., and Capecchi, M. R. (1997). Targeted mutations in hoxa-9 and hoxb-9 reveal synergistic interactions. *Dev. Biol.* **181,** 186–196.

Cheng, P. Y., Lin, C. C., Wu, C. S., Lu, Y. F., Lin, C. Y., Chung, C. C., Chu, C. Y., Huang, C. J., Tsai, C. Y., Korzh, S., Wu, J. L., and Hwang, S. P. (2008). Zebrafish cdx1b regulates expression of downstream factors of Nodal signaling during early endoderm formation. *Development* **135,** 941–952.

Corbin, J. G., Gaiano, N., Machold, R. P., Langston, A., and Fishell, G. (2000). The gsh2 homeodomain gene controls multiple aspects of telencephalic development. *Development* **127,** 5007–5020.

Dasen, J. S., Liu, J. P., and Jessell, T. M. (2003). Motor neuron columnar fate imposed by sequential phases of Hox-c activity. *Nature* **425,** 926–933.

Dasen, J. S., Tice, B. C., Brenner-Morton, S., and Jessell, T. M. (2005). A Hox regulatory network establishes motor neuron pool identity and target-muscle connectivity. *Cell* **123,** 477–491.

Davidson, A. J., and Zon, L. I. (2006). The caudal-related homeobox genes cdx1a and cdx4 act redundantly to regulate hox gene expression and the formation of putative hematopoietic stem cells during zebrafish embryogenesis. *Dev. Biol.* **292,** 506–518.

Delfino-Machín, M., Lunn, J. S., Breitkreuz, D. N., Akai, J., and Storey, K. G. (2005). Specification and maintenance of the spinal cord stem zone. *Development* **132,** 4273–4283.

Deschamps, J. (2007). Ancestral and recently recruited global control of the Hox genes in development. *Curr. Opin. Genet. Dev.* **17,** 422–427.

Deschamps, J., and van Nes, J. (2005). Developmental regulation of the Hox genes during axial morphogenesis in the mouse. *Development* **132,** 2931–2942.

Deschamps, J., and Wijgerde, M. (1993). Two phases in the establishment of HOX expression domains. *Dev. Biol.* **156,** 473–480.

Deschamps, J., van den Akker, E., Forlani, S., de Graaff, W., Oosterveen, T., Roelen, B., and Roelfsema, J. (1999). Initiation, establishment and maintenance of Hox gene expression patterns in the mouse. *Int. J. Dev. Biol.* **43,** 635–650.

Deutsch, J. S., and Lopez, P. (2008). Are transposition events at the origin of the bilaterian Hox complexes? *In* "Evolving Pathways: Key Themes in Evolutionary Developmental Biology" (A. Minelli, and G. Fusco, Eds.), pp. 236–258. Cambridge University Press, Cambridge.

Diez del Corral, R., and Storey, K. G. (2004). Opposing FGF and retinoid pathways: A signalling switch that controls differentiation and patterning onset in the extending vertebrate body axis. *Bioessays* **26,** 857–869.

Di-Poï, N., Zákány, J., and Duboule, D. (2007). Distinct roles and regulations for HoxD genes in metanephric kidney development. *PLoS Genet.* **3,** e232.

Duboule, D. (1992). The vertebrate limb, temporal colinearity and the Hox/HOM gene network. *Bioessays* **14,** 375–384.

Duboule, D. (1994). Temporal colinearity and the phylotypic progression: A basis for the stability of a vertebrate Bauplan and the evolution of morphologies through heterochrony. *Dev. Suppl.* 135–142.

Duboule, D. (2007). The rise and fall of Hox gene clusters. *Development* **134,** 2549–2560.

Duprey, P., Chowdhury, K., Dressler, G. R., Balling, R., Simon, D., Guenet, J. L., and Gruss, P. (1988). A mouse gene homologous to the Drosophila gene caudal is expressed in epithelial cells from the embryonic intestine. *Genes Dev.* **2,** 1647–1654.

Ernst, P., Fisher, J. K., Avery, W., Wade, S., Foy, D., and Korsmeyer, S. J. (2004a). Definitive hematopoiesis requires the mixed-lineage leukemia gene. *Dev. Cell* **6,** 437–443.

Ernst, P., Mabon, M., Davidson, A. J., Zon, L. I., and Korsmeyer, S. J. (2004b). An Mll-dependent Hox program drives hematopoietic progenitor expansion. *Curr. Biol.* **14,** 2063–2069.

Ferrier, D. E. (2004). Hox genes: Did the vertebrate ancestor have a Hox14? *Curr. Biol.* **14,** R210–R211.

Ferrier, D. E., Dewar, K., Cook, A., Chang, J. L., Hill-Force, A., and Amemiya, C. (2005). The chordate ParaHox cluster. *Curr. Biol.* **15,** R820–R822.

Forlani, S., Lawson, K. A., and Deschamps, J. (2003). Acquisition of Hox codes during gastrulation and axial elongation in the mouse embryo. *Development* **130,** 3807–3819.

Franklin, V., Khoo, P. L., Bildsoe, H., Wong, N., Lewis, S., and Tam, P. L. (2008). Regionalisation of the endoderm progenitors and morphogenesis of the gut portals of the mouse embryo. *Mech. Dev.* **125,** 587–600.

Gamer, L. W., and Wright, C. V. (1993). Murine Cdx-4 bears striking similarities to the Drosophila caudal gene in its homeodomain sequence and early expression pattern. *Mech. Dev.* **43,** 71–81.

Garcia-Fernàndez, J. (2005). The genesis and evolution of homeobox gene clusters. *Nat. Rev. Genet.* **6,** 881–892.

Gauchat, D., Mazet, F., Berney, C., Schummer, M., Kreger, S., Pawlowski, J., and Galliot, B. (2000). Evolution of Antp-class genes and differential expression of Hydra Hox/ParaHox genes in anterior patterning. *Proc. Natl. Acad. Sci. USA* **97,** 4493–4498.

Gaufo, G. O., Wu, S., and Capecchi, M. R. (2004). Contribution of Hox genes to the diversity of the hindbrain sensory system. *Development* **131,** 1259–1266.

Gaunt, S. J., Drage, D., and Cockley, A. (2003). Vertebrate caudal gene expression gradients investigated by use of chick cdx-A/lacZ and mouse cdx-1/lacZ reporters in transgenic mouse embryos: Evidence for an intron enhancer. *Mech. Dev.* **120,** 573–586.

Gaunt, S. J., Cockley, A., and Drage, D. (2004). Additional enhancer copies, with intact cdx binding sites, anteriorize Hoxa-7/lacZ expression in mouse embryos: Evidence in keeping with an instructional cdx gradient. *Int. J. Dev. Biol.* **48,** 613–622.

Gaunt, S. J., Drage, D., and Trubshaw, R. C. (2005). cdx4/lacZ and cdx2/lacZ protein gradients formed by decay during gastrulation in the mouse. *Int. J. Dev. Biol.* **49,** 901–908.

Goddard, J. M., Rossel, M., Manley, N. R., and Capecchi, M. R. (1996). Mice with targeted disruption of Hoxb-1 fail to form the motor nucleus of the VIIth nerve. *Development* **122,** 3217–3228.

Holstege, J. C., de Graaff, W., Hossaini, M., Cano, S. C., Jaarsma, D., van den Akker, E., and Deschamps, J. (2008). Loss of Hoxb8 alters spinal dorsal laminae and sensory responses in mice. *Proc. Natl. Acad. Sci. USA* **105,** 6338–6343.

Iimura, T., and Pourquié, O. (2006). Collinear activation of Hoxb genes during gastrulation is linked to mesoderm cell ingression. *Nature* **442,** 568–571.

Keenan, I. D., Sharrard, R. M., and Isaacs, H. V. (2006). FGF signal transduction and the regulation of Cdx gene expression. *Dev. Biol.* **299,** 478–488.

Kinkel, M. D., Eames, S. C., Alonzo, M. R., and Prince, V. E. (2008). Cdx4 is required in the endoderm to localize the pancreas and limit beta-cell number. *Development* **135,** 919–929.

Kmita, M., Tarchini, B., Zàkàny, J., Logan, M., Tabin, C. J., and Duboule, D. (2005). Early developmental arrest of mammalian limbs lacking HoxA/HoxD gene function. *Nature* **435,** 1113–1116.

Kondo, T., Dollé, P., Zákány, J., and Duboule, D. (1996). Function of posterior HoxD genes in the morphogenesis of the anal sphincter. *Development* **122,** 2651–2659.

Lengerke, C., Schmitt, S., Bowman, T. V., Jang, I. H., Maouche-Chretien, L., McKinney-Freeman, S., Davidson, A. J., Hammerschmidt, M., Rentzsch, F., Green, J. B., Zon, L. I., and Daley, G. Q. (2008). BMP and Wnt specify hematopoietic fate by activation of the Cdx-Hox pathway. *Cell Stem Cell* **2,** 72–82.

Li, H., Witte, D. P., Branford, W. W., Aronow, B. J., Weinstein, M., Kaur, S., Wert, S., Singh, G., Schreiner, C. M., Whitsett, J. A., Scott, W. J., and Potter, S. S. (1994). Gsh-4 encodes a LIM-type homeodomain, is expressed in the developing central nervous system and is required for early postnatal survival. *EMBO J.* **13,** 2876–2885.

Liu, J. P., Laufer, E., and Jessell, T. M. (2001). Assigning the positional identity of spinal motor neurons: Rostrocaudal patterning of Hox-c expression by FGFs, Gdf11, and retinoids. *Neuron* **32,** 997–1012.

Lohnes, D. (2003). The Cdx1 homeodomain protein: An integrator of posterior signaling in the mouse. *Bioessays* **25,** 971–980.

Mathis, L., Kulesa, P. M., and Fraser, S. E. (2001). FGF receptor signalling is required to maintain neural progenitors during Hensen's node progression. *Nat. Cell Biol.* **3,** 559–566.

McIntyre, D. C., Rakshit, S., Yallowitz, A. R., Loken, L., Jeannotte, L., Capecchi, M. R., and Wellik, D. M. (2007). Hox patterning of the vertebrate rib cage. *Development* **134,** 2981–2989.

Moreno, E., and Morata, G. (1999). *Caudal* is the Hox gene that specifies the most posterior Drosophila segment. *Nature* **400,** 873–877.

Mugford, J. W., Sipilä, P., Kobayashi, A., Behringer, R. R., and McMahon, A. P. (2008). Hoxd11 specifies a program of metanephric kidney development within the intermediate mesoderm of the mouse embryo. *Dev. Biol.* (doi:10.1016/j.ydbio.2008.03.044).

Mutoh, H., Hakamata, Y., Sato, K., Eda, A., Yanaka, I., Honda, S., Osawa, H., Kaneko, Y., and Sugano, K. (2002). Conversion of gastric mucosa to intestinal metaplasia in Cdx2-expressing transgenic mice. *Biochem. Biophys. Res. Commun.* **294,** 470–479.

Nordström, U., Maier, E., Jessell, T. M., and Edlund, T. (2006). An early role for WNT signaling in specifying neural patterns of Cdx and Hox gene expression and motor neuron subtype identity. *PLoS Biol.* **4,** e252.

Nowicki, J. L., and Burke, A. C. (2000). Hox genes and morphological identity: Axial versus lateral patterning in the vertebrate mesoderm. *Development* **127,** 4265–4275.

Ohlsson, H., Karlsson, K., and Edlund, T. (1993). IPF1, a homeodomain-containing transactivator of the insulin gene. *EMBO J.* **12,** 4251–4259.

Olivera-Martinez, I., and Storey, K. G. (2007). Wnt signals provide a timing mechanism for the FGF-retinoid differentiation switch during vertebrate body axis extension. *Development* **134,** 2125–2135.

Oosterveen, T., Niederreither, K., Dollé, P., Chambon, P., Meijlink, F., and Deschamps, J. (2003). Retinoids regulate the anterior expression boundaries of 5′ Hoxb genes in posterior hindbrain. *EMBO J.* **22,** 262–269.

Oury, F., Murakami, Y., Renaud, J. S., Pasqualetti, M., Charnay, P., Ren, S. Y., and Rijli, F. M. (2006). Hoxa2- and rhombomere-dependent development of the mouse facial somatosensory map. *Science* **313,** 1408–1413.

Perez-Villamil, B., Schwartz, P. T., and Vallejo, M. (1999). The pancreatic homeodomain transcription factor IDX1/IPF1 is expressed in neural cells during brain development. *Endocrinology* **140,** 3857–3860.

Quiquand, M., Yanze, N., Schmich, J., Schmid., V., Galliot, B., and Piraino, S. (2009). More constraint on ParaHox than Hox gene families in early metazoan evolution. *Dev. Biol.* **328,** 173–187.

Ryan, J. F., Mazza, M. E., Pang, K., Matus, D. Q., Baxevanis, A. D., Martindale, M. Q., and Finnerty, J. R. (2007). Pre-bilaterian origins of the Hox cluster and the Hox code: Evidence from the sea anemone, *Nematostella vectensis. PLoS ONE* **2,** e153.

Shimizu, T., Bae, Y. K., and Hibi, M. (2006). Cdx-Hox code controls competence for responding to Fgfs and retinoic acid in zebrafish neural tissue. *Development* **133,** 4709–4719.

Silberg, D. G., Sullivan, J., Kang, E., Swain, G. P., Moffett, J., Sund, N. J., Sackett, S. D., and Kaestner, K. H. (2002). Cdx2 ectopic expression induces gastric intestinal metaplasia in transgenic mice. *Gastroenterology* **122,** 689–696.

Skromne, I., Thorsen, D., Hale, M., Prince, V. E., and Ho, R. K. (2007). Repression of the hindbrain developmental program by Cdx factors is required for the specification of the vertebrate spinal cord. *Development* **134,** 2147–2158.

Stringer, E., Pritchard, C. A., and Beck, F. (2008). Cdx2 initiates histodifferentiation of the midgut endoderm. *FEBS Lett.* **582,** 2555–2560.

Strumpf, D., Mao, C. A., Yamanaka, Y., Ralston, A., Chawengsaksophak, K., Beck, F., and Rossant, J. (2005). Cdx2 is required for correct cell fate specification and differentiation of trophectoderm in the mouse blastocyst. *Development* **132,** 2093–2102.

Subramanian, V., Meyer, B. I., and Gruss, P. (1995). Disruption of the murine homeobox gene Cdx1 affects axial skeletal identities by altering the mesodermal expression domains of Hox genes. *Cell* **83,** 641–653.

Szucsik, J. C., Witte, D. P., Li, H., Pixley, S. K., Small, K. M., and Potter, S. S. (1997). Altered forebrain and hindbrain development in mice mutant for the Gsh-2 homeobox gene. *Dev. Biol.* **191,** 230–242.

Tabariès, S., Lapointe, J., Besch, T., Carter, M., Woollard, J., Tuggle, C. K., and Jeannotte, L. (2005). Cdx protein interaction with Hoxa5 regulatory sequences contributes to Hoxa5 regional expression along the axial skeleton. *Mol. Cell. Biol.* **25,** 1389–1401.

Tam, P. P., Khoo, P. L., Lewis, S. L., Bildsoe, H., Wong, N., Tsang, T. E., Gad, J. M., and Robb, L. (2007). Sequential allocation and global pattern of movement of the definitive endoderm in the mouse embryo during gastrulation. *Development* **134,** 251–260.

Tarchini, B., and Duboule, D. (2006). Control of Hoxd genes' collinearity during early limb development. *Dev. Cell* **10,** 93–103.

Tarchini, B., Huynh, T. H., Cox, G. A., and Duboule, D. (2005). HoxD cluster scanning deletions identify multiple defects leading to paralysis in the mouse mutant Ironside. *Genes Dev.* **19,** 2862–2876.

Tarchini, B., Duboule, D., and Kmita, M. (2006). Regulatory constraints in the evolution of the tetrapod limb anterior-posterior polarity. *Nature* **443,** 985–988.

Taylor, J. K., Levy, T., Suh, E. R., and Traber, P. G. (1997). Activation of enhancer elements by the homeobox gene Cdx2 is cell line specific. *Nucleic Acids Res.* **25,** 2293–2300.

Tiret, L., Le Mouellic, H., Maury, M., and Brûlet, P. (1998). Increased apoptosis of motoneurons and altered somatotopic maps in the brachial spinal cord of Hoxc-8-deficient mice. *Development* **125,** 279–291.

Trainor, P. A., and Krumlauf, R. (2000). Patterning the cranial neural crest: Hindbrain segmentation and Hox gene plasticity. *Nat. Rev. Neurosci.* **1,** 116–124.

Tschopp, P., Tarchini, B., Spitz, F., Zakany, J., and Duboule, D. (2009). Uncoupling time and space in the collinear regulation of Hox genes. PLOS Genetics, **5,** Issue 3 e1000398.

Valerius, M. T., Li, H., Stock, J. L., Weinstein, M., Kaur, S., Singh, G., and Potter, S. S. (1995). Gsh-1: A novel murine homeobox gene expressed in the central nervous system. *Dev. Dyn.* **203,** 337–351.

van den Akker, E., Forlani, S., Chawengsaksophak, K., de Graaff, W., Beck, F., Meyer, B. I., and Deschamps, J. (2002). Cdx1 and Cdx2 have overlapping functions in anteroposterior patterning and posterior axis elongation. *Development* **129,** 2181–2193.

van Nes, J., de Graaff, W., Lebrin, F., Gerhard, M., Beck, F., and Deschamps, J. (2006). The Cdx4 mutation affects axial development and reveals an essential role of Cdx genes in the ontogenesis of the placental labyrinth in mice. *Development* **133,** 419–428.

Wang, Y., Yabuuchi, A., McKinney-Freeman, S., Ducharme, D. M., Ray, M. K., Chawengsaksophak, K., Archer, T. K., and Daley, G. Q. (2008). Cdx gene deficiency compromises embryonic hematopoiesis in the mouse. *Proc. Natl. Acad. Sci. USA* **105,** 7756–7761.

Wellik, D. M. (2007). Hox patterning of the vertebrate axial skeleton. *Dev. Dyn.* **236,** 2454–2463.

Wellik, D. M., and Capecchi, M. R. (2003). Hox10 and Hox11 genes are required to globally pattern the mammalian skeleton. *Science* **301,** 363–367.

Wellik, D. M., Hawkes, P. J., and Capecchi, M. R. (2002). Hox11 paralogous genes are essential for metanephric kidney induction. *Genes Dev.* **16,** 1423–1432.

Wells, J. M., and Melton, D. A. (2000). Early mouse endoderm is patterned by soluble factors from adjacent germ layers. *Development* **127,** 1563–1572.

Wingert, R. A., Selleck, R., Yu, J., Song, H. D., Chen, Z., Song, A., Zhou, Y., Thisse, B., Thisse, C., McMahon, A. P., and Davidson, A. J. (2007). The cdx genes and retinoic acid control the positioning and segmentation of the zebrafish pronephros. *PLoS Genet.* **3,** 1922–1938.

Wu, Y., Wang, G., Scott, S. A., and Capecchi, M. R. (2008). Hoxc10 and Hoxd10 regulate mouse columnar, divisional and motor pool identity of lumbar motoneurons. *Development* **135,** 171–182.

Zacchetti, G., Duboule, D., and Zakany, J. (2007). Hox gene function in vertebrate gut morphogenesis: The case of the caecum. *Development* **134,** 3967–3973.

Zakany, J., and Duboule, D. (2007). The role of Hox genes during vertebrate limb development. *Curr. Opin. Genet. Dev.* **17,** 359–366.

Zákány, J., Kmita, M., and Duboule, D. (2004). A dual role for Hox genes in limb anterior-posterior asymmetry. *Science* **304,** 1669–1672.

CHAPTER NINE

Hox Genes and Vertebrate Axial Pattern

Deneen M. Wellik

Contents

Abstract

The axial skeleton in all vertebrates is comprised of similar structures that extend from anterior to posterior along the body axis: the occipital skull bones, cervical, thoracic, lumbar, sacral and caudal vertebrae. Despite significant changes in the number and size of these elements during vertebrate evolution, the basic character of these anatomical elements, as well as the order in which they appear, has remained strikingly similar. Extensive expression analysis, classic embryology experiments in chick and targeted loss-of-function mutant analyses in mice have clearly demonstrated that *Hox* genes are key regulators of morphology along the axial skeleton. The cumulative data from this work provides an emerging understanding of *Hox* gene function in patterning the vertebrate axial skeleton. This chapter summarizes genetic, molecular and embryologic findings on role of *Hox* genes in establishing axial morphology and how these combined results impact our current understanding of the '*Hox* code'.

Department of Internal Medicine, Division of Molecular Medicine & Genetics, and Department of Cell and Developmental Biology, University of Michigan Medical Center, Ann Arbor, Michigan, USA

Current Topics in Developmental Biology, Volume 88
ISSN 0070-2153, DOI: 10.1016/S0070-2153(09)88009-5

1. Introduction

During development, animals specify a variety of different cell types. In addition to the assignment of appropriate cell fates, each cell must develop in the context appropriate for its surroundings. "Positional information" is a term used by Lewis Wolpert to describe translation of genetic information into reproducible spatial patterns of cellular differentiation (Wolpert, 1969). The acquisition, or assignment, of positional information to an initially similar set of cells allows these cells to develop into final states that exhibit distinct spatial patterns, the development of form.

Along the anteroposterior (AP) axis of the fruit fly *Drosophila*, *Hox/HomC* genes are a critical component of the system that provides this positional identity (Lewis, 1963, 1978). *Drosophila Hox/HomC* genes possess several unique characteristics as key regulators of AP pattern. All members of this complex are linked on a single cluster. Further, the *Drosophila* homeotic genes map in the same order within the cluster as their expression and functional domains lie along the AP axis, with the most 3′ members of the complex affecting the most anterior structures and more 5′ genes affecting increasingly posterior structures. Recessive and dominant mutations in these genes result in homeotic transformations of the body plan (reviewed in McGinnis and Krumlauf, 1992).

Since the discovery of vertebrate *Hox* genes more than two decades ago, many striking similarities between these genes and the *Drosophila Hox* complex have been found, and much attention has been devoted to their expression and function during vertebrate AP patterning. In mammals, duplication of the ancestral cluster has given rise to four *Hox* complexes and, similar to their pattern in *Drosophila*, their expression from anterior to posterior along the body axis is correlated with their position from 3′ to 5′ in each cluster. The conserved features of closely linked, colinear arrangement and expression of *Hox* genes in most bilaterian organisms suggest critically important roles for these genes in AP patterning that are still being unraveled.

The vertebrate axial skeleton provides an example of how patterning information must be conveyed and integrated during animal development. The first level of information involves the differentiation of a subset of cells along the AP axis to become prechondrogenic, migrate to surround the neural tube, condense to form prevertebral structures, and form cartilage. This appears to happen similarly from the cervical region through the tail of the animal, although these signals alone are not sufficient to generate a functional axial skeleton. In addition to this information, a second layer of information is required to direct the growth and patterning of each individual vertebra to achieve a final morphology that distinguishes a cervical vertebra from a thoracic, lumbar, or sacral vertebra. Vertebrate *Hox* genes are indispensable for this second layer of patterning. It is the combined functions of *Hox* genes that control the establishment of vertebral identity along the axial skeleton.

2. *Hox* Genes and the Axial Skeleton

2.1. Derivation and morphology of the axial skeleton

During gastrulation, cells are directed into distinct cell layers: ectoderm, endoderm, and mesoderm. The ectoderm gives rise to the skin, the nervous system and is the source of placode formation that induces eye and ear development. The endoderm forms the primitive gut tube, which contributes to many organ systems including the liver, pancreas, and lung. The mesoderm also contributes to organ formation in conjunction with the endoderm, as well as distinctly giving rise to the urogenital tract as well as the musculature, dermis, bones, and tendons of the body.

Shortly after the mesodermal layer is specified, it gives rise to the notochord immediately ventral to the neural tube, and the remainder is divided into three compartments based on its mediolateral position with respect to the midline: the lateral plate mesoderm, the intermediate mesoderm, and the somitic mesoderm (Fig. 9.1). The lateral plate mesoderm lies furthest from the midline. The most lateral edges later grow and fuse at the midline to close the body cavity. In most vertebrates, four bulges in the lateral plate tissue grow and give rise to the skeleton and connective tissue of the four limbs. The intermediate mesoderm is positioned medial to the lateral plate mesoderm and forms the urogenital tract of the developing animal. The somitic mesoderm is the most medial and lies adjacent to the neural tube, and shortly after ingression, this tissue becomes segmented into somites.

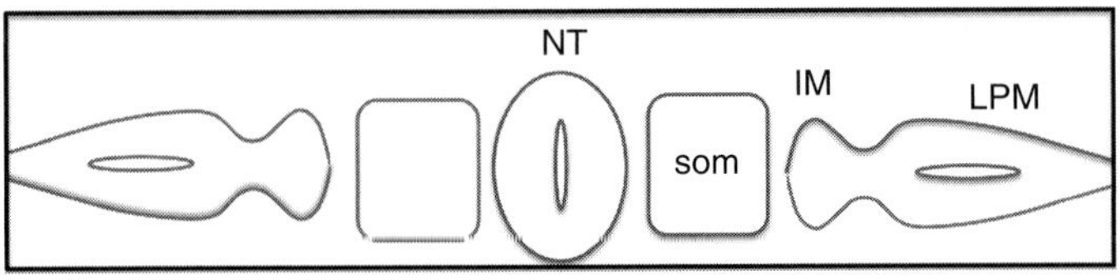

Figure 9.1 Diagram of three mesodermal compartments in early development. On each side of the neural tube (NT, an ectoderm-derived structure), lies the somites (som), which give rise to the axial skeleton, muscle and dermis, the intermediate mesoderm (IM), which gives rise to the urogenital system, and the lateral plate mesoderm (LPM), which gives rise to the sternum of the axial skeleton as well as the tendons and skeletal components of the limb.

Somites provide the metameric pattern that prefigures the axial skeleton. In addition to most of the axial skeleton, the somites also give rise to the dermal layer of the skin and the body musculature. The sclerotomal region of each somite differentiates and gives rise to most of the axial skeleton. The first five somites lose their segmented appearance and contribute collectively to the base of the skull, but from the fifth somite through the posterior end of the embryo, each somite contributes to the formation of half of two vertebral elements. After differentiating into the sclerotome, the cells migrate from their position at the ventromedial part of the somite to surround the neural tube. During this migration, the somites undergo resegmentation, in which the posterior half of one somite fuses with the anterior half of the next caudal somite to form a single vertebral element (Goodrich, 1958). During this process, these cells begin expressing pre-chondrogenic markers, such as Sox9 (Wright *et al.*, 1995). This process appears to be uniform throughout the axial skeleton.

However, along the axial skeleton, vertebrae develop unique morphologies. The axial formula for all tetrapod vertebrates follows a common order: the occipital region (nonsegmental skull), followed by the cervical, thoracic, lumbar, sacral then caudal vertebrae (Fig. 9.2). Over the 350 million years of tetrapod vertebrate history, the number of segments contributing to each region has diverged, but the order and some common morphological traits for each region has been preserved. The first cervical vertebra, the atlas, forms an articulation with the skull and this vertebra, along with the rest of the cervical vertebrae, form the skeleton of the neck. While mammals are constrained to seven cervical vertebrae, other tetrapods have highly variable numbers of cervical vertebrae, with some avian species having as many as 25 (Procter and Lynch, 1993).

Cervical vertebrae are followed by ribbed, thoracic vertebrae, which form the rib cage. The number of thoracic vertebrae varies widely in vertebrates. Birds have only five to seven ribbed, thoracic vertebrae, for instance, while some snakes have several hundred (Richardson *et al.*, 1998). The lumbar vertebrae in land vertebrates are the load-bearing vertebrae posterior to the thoracic rib cage and are generally the largest and densest of the vertebrae. In tetrapods, the sacral vertebrae grow lateral protrusions that fuse and are the site of pelvic and hindlimb attachment. Finally, the numbers of caudal vertebrae vary widely in vertebrates from the long, prehensile tail of some monkeys to the three to five fused vertebrae of the human coccyx.

Despite the obvious differences between a mouse and a human, their vertebral formulas are similar. Like most mammals, mouse and humans have seven cervical vertebrae. Mice have 13 thoracic vertebrae, the first seven grow around the body wall and join at the sternum. Humans have 12 ribbed vertebrae and also have the first seven attached to the sternum. Mice have six lumbar vertebrae compared to five in human, and both species have four sacral vertebrae. Only the caudal region differs markedly, with humans having three

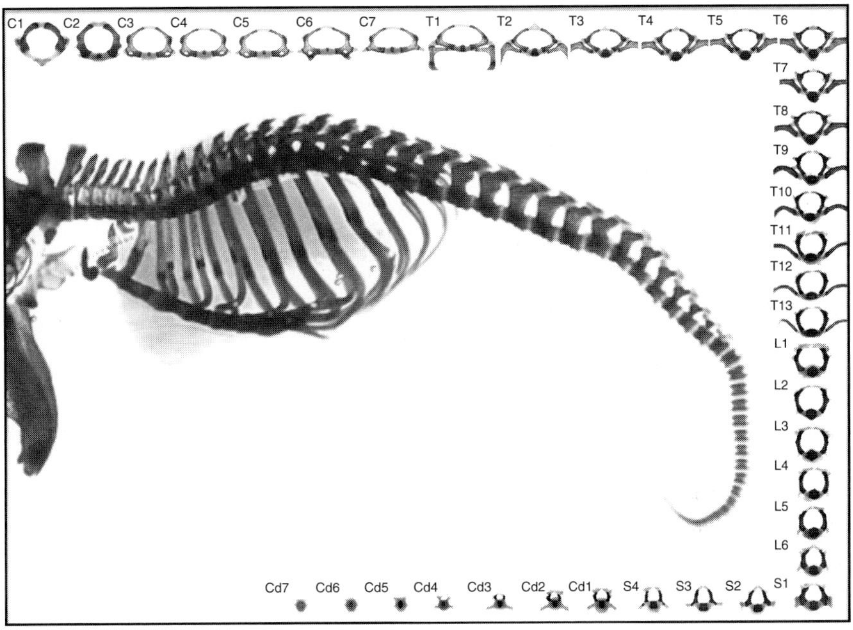

Figure 9.2 A lateral view of an E18.5 mouse skeleton stained with Alcian blue and alizarin red; anterior to the left, posterior to the right. Circumferentially around the edges of the panel, the individual vertebral elements are pictures, beginning with the first cervical vertebra, the atlas, in the top, left side with the elements in order, clockwise. C, cervical, T, thoracic, L, lumbar, S, sacral and Cd, caudal; numbers reflect their position in the skeleton. (Only 7 of the approximately 30 caudal vertebrae are shown). (See Color Insert.)

to five fused caudal vertebrae, which make up the specialized coccyx compared to approximately 30 caudal vertebrae that make up the tail in mice.

2.2. *Hox* expression

The earliest indication that vertebrate *Hox* genes might play a role in vertebrate axial patterning was obtained by in situ hybridization analyses. *Hox* genes are expressed from 3′ to 5′ in the clusters, with the earliest genes expressed in the posterior primitive streak at late streak stages, and more 5′ genes expressed at progressively later stages (Dressler and Gruss, 1989; Duboule and Dolle, 1989; Gaunt, 1991; Gaunt and Strachan, 1996; Gaunt *et al.*, 1986, 1990; Graham *et al.*, 1989; Izpisua-Belmonte *et al.*, 1991). This temporal control of *Hox* expression onset, coupled with growth and elongation of the embryo, results in spatially graded anterior boundaries of expression where 3′ genes (*Hox1* and *Hox2*) display anterior expression limits in the hindbrain region of the embryo and increasingly 5′ genes demonstrate increasingly posterior limits of expression, a unique characteristic of *Hox* gene expression termed colinearity. While the initial spatiotemporal expression domains are colinear along the cluster, the definitive domains of expression along the AP axis are not static and final boundaries are adjusted during later developmental stages (reviewed in Deschamps *et al.*, 1999; McGinnis and Krumlauf, 1992). The collective evidence, including locus-specific position effects on expression as well as regulatory elements within and outside the cluster (Herault *et al.*, 1999; Kmita *et al.*, 2000, 2002; Kondo and Duboule, 1999; Lehoczky and Innis, 2008; Spitz *et al.*, 2001, 2005), indicates that additional control mechanisms are layered onto the initial regulatory mechanism.

The duplication of the genome during vertebrate evolution has resulted in the existence of four clusters, *HoxA*, *B*, *C*, and *D*, in the mammalian genome and, thus there are several members of each of 13 paralog groups. In mammals, each group has two to four members. These paralogous genes are defined based on their sequence similarity and position within the clusters. Much of the regulation of expression from the clusters is likely to be conserved as well. In situ hybridization analyses demonstrate that paralogs share highly similar expression domains (Burke, 1995, 2000; Gaunt *et al.*, 1989; Kessel and Gruss, 1990; McIntyre *et al.*, 2007). Thus, the idea that redundant genetic information is likely to be shared between paralogous genes was postulated very early (Gruss and Kessel, 1991; Kessel and Gruss, 1990; McGinnis and Krumlauf, 1992). How this conserved expression translates into morphological pattern, however, has been more difficult to understand.

2.3. Fate determination

An essential prerequisite to understanding the control of axial patterning events is to establish the stage at which cells become committed to their fate. Classic embryologic manipulation experiments in chick have shown that commitment to axial fate occurs early in development, prior to overt segmentation. In a landmark study, heterotypic transplantation of segmental plate from chick embryos (the tissue analogous to presomitic mesoderm in mice) producing thoracic level mesoderm to host segmental plate regions at earlier stages and at more anterior regions resulted in the ectopic presence of ribs in the neck. These experiments demonstrate that cells from the presomitic mesoderm have already established the intrinsic patterning information for proper AP-specific axial patterning prior to overt somite formation (Kieny *et al.*, 1972).

Examination of *Hox* expression in these transplantation experiments demonstrates that transplanted segmental plate tissue that maintains its original identity also retains its original *Hox* expression (Nowicki and Burke, 2000). Transplantation to a heterotopic location does not reprogram *Hox* expression in these tissues. While not a direct test of function, these experiments strengthen the correlation of *Hox* gene expression with axial specification.

2.4. Comparative data

Comparison of *Hox* expression boundaries between species provides further correlative evidence for a role for these genes in global patterning of the vertebrate axial skeleton. The most detailed comparisons have been performed between mice and chicks (Burke *et al.*, 1995). Since the fifth somite contributes partially to the occipital skull and partially to the first cervical vertebra in both species, the axial skeleton of both mice and chicks begin their vertebral columns with the same somite. However, mice have seven cervical vertebrae while chickens have 14, therefore, the cervical to thoracic transition differs by seven somites between these two species. Analysis of *Hoxa4* expression in both species demonstrates its anterior expression boundary at the level of the second cervical vertebra, or somite 7 in both species (Burke *et al.*, 1995; Gaunt *et al.*, 1989, 1994). In contrast, analysis of *Hoxc6* expression in both species reveals that its anterior boundary expression is at the cervical to thoracic transition (Burke *et al.*, 1995; Gaunt, 1994; Gaunt *et al.*, 1986; Jegalian and De Robertis, 1992). In mice, this is at the level of somite 12, while in chick the anterior boundary of *Hoxc6* is at somite 19. Thus, the anterior level of *Hoxc6* expression marks the change in morphology, not absolute somite number in different species.

This analysis was extended by Burke *et al.* to 23 chicken *Hox* genes and 16 mouse *Hox* genes. This extensive comparison establishes that *Hox* expression boundaries reflect morphological boundaries, not the number of segments (Burke *et al.*, 1995). These data support the view that *Hox* genes regulate the establishment of distinct morphological structures along the AP axis in vertebrates and suggest that evolutionary changes in the level for transition between vertebral types might occur by changes in expression boundaries of *Hox* genes.

3. *Hox* Function in Axial Patterning

While early studies provided strong, correlative evidence that vertebrate *Hox* genes play an important role in AP patterning of the axial skeleton, direct functional evidence for this role has been provided by both gain- and loss-of-function experiments.

3.1. Ectopic expression experiments

Ectopic expression of *HomC* genes in *Drosophila* consistently produces transformations of the identity of segments toward the identity specified by the ectopically expressed gene (Gibson *et al.*, 1990; Kuziora and McGinnis, 1988; Mann and Hogness, 1990). Expression of the ectopic gene in regions anterior to its normal expression boundary leads to posterior homeotic transformations, but ectopic expression in regions posterior to the normal expression pattern leads to promotion of more anterior identities in posterior regions (Gibson *et al.*, 1990; Kuziora and McGinnis, 1988). Overall, these experiments suggest that each *Hox* gene promotes its own identity programs, provided its expression is in a region permissive for this activity (McGinnis and Krumlauf, 1992).

In the mouse, gain-of-function experiments support similar roles for vertebrate *Hox* genes in axial patterning. Forced expression of *Hoxd4* in regions anterior to its normal expression domain results in transformation of the occipital region to cervical identity, a posterior homeotic transformation (Lufkin *et al.*, 1992). Gain-of-function of expression of *Hoxa7* under the control of the chick β-actin promoter results in mice with eight cervical vertebrae instead of seven, an anterior homeotic transformation (Kessel *et al.*, 1990). Overexpression of *Hoxc6* in regions posterior to its normal expression domain also led to anterior transformations of the posterior vertebra, exhibiting a dominant role for the overexpressed *Hox* gene in posterior regions (Jegalian and De Robertis, 1992). Anteriorizations are also observed upon ectopic, anteriorized expression of *Hoxc8* (Pollock *et al.*, 1992). Global alteration of *Hox* axial expression and transformations of the axial skeleton are also caused by teratogenic does of retinoic acid in developing mouse embryo (Kessel and Gruss, 1991). In these experiments, high doses of retinoic acid were given at early stages (E7.0) lead to global anterior shifts in *Hox* expression and posterior homeotic transformations along the vertebral column.

Additional evidence that gain-of-function of *Hox* expression leads to dominant transformation of axial identity is provided by Carapuco *et al.* (2005). In these experiments, ectopic expression of *Hoxa10* under the control of the *Dll1* enhancer leads to expression of *Hoxa10* in all somites, beginning at presomitic mesoderm stages. The remarkable phenotype in these mutant mice is the complete absence of rib formation in the thoracic region, a function consistent with the endogenous *Hox10* function of repressing rib formation in the lumbosacral region (Wellik and Capecchi, 2003, discussed below). Further examination of posterior regions by the authors revealed a repression of sacral wing outgrowth as well, demonstrating that *Hoxa10* function, when ectopically overexpressed, is dominant in both posterior and anterior regions.

3.2. Genetic loss-of-function mutations

Since the discovery of technology that permits designed genetic manipulation of the mouse genome (Thomas and Capecchi, 1987), targeted mutations have been generated in each of the 39 mouse *Hox* genes (Boulet and Capecchi, 1996; Carpenter *et al.*, 1997; Chen and Capecchi, 1997, 1999;

Chen *et al.*, 1998; Chisaka and Capecchi, 1991; Chisaka *et al.*, 1992; Condie and Capecchi, 1993; Davis and Capecchi, 1994; Fromental-Ramain *et al.*, 1996a,b; Garcia-Gasca and Spyropoulos, 2000; Horan *et al.*, 1994, 1995a,b; Jeannotte *et al.*, 1993; Kostic and Capecchi, 1994; Le Mouellic *et al.*, 1992; Lufkin *et al.*, 1991; Manley and Capecchi, 1995, 1998; McIntyre *et al.*, 2007; Rancourt *et al.*, 1995; Rijli *et al.*, 1993, 1994, 1995; Small and Potter, 1993; Suemori *et al.*, 1995; van den Akker *et al.*, 1999, 2001; Wahba *et al.*, 2001; Wellik and Capecchi, 2003). Mutations in paralog groups from the *Hox3* to the *Hox11* group lead to defects in the axial skeleton. As expected from their expression patterns and previous genetic work in *Drosophila*, 3′ anterior *Hox* mutants exhibit axial phenotypes in anterior regions of the axial skeleton and increasingly 5′ *Hox* mutants display phenotypes in more posterior regions. For instance, loss of *Hoxd3* leads to defects in the first and second cervical vertebrae (Condie and Capecchi, 1993), while loss of *Hoxd11* results in changes in sacral patterning (Davis and Capecchi, 1994).

In *Drosophila*, loss-of-function mutants for many of the genes (*Scr*, *Antp*, *Ubx*, *abdA*, *AbdB*; Lewis, 1978; Sanchez-Herrero *et al.*, 1985; Wakimoto and Kaufman, 1981) result in anterior homeotic transformations of the *Drosophila* body plan. However, transformations occur only in regions where there is overlapping expression with other *Hox* genes, suggesting that the phenotype from loss of one *Hox* gene is a result of the function of the remaining *Hox* gene expressed in that region. In cases where there is no overlapping *Hox* expression (*lab*, *Dfd*; Merrill *et al.*, 1987; Regulski *et al.*, 1987), loss-of-function mutations result in structural deficiencies, but no detectable transformations toward other segment identities (McGinnis and Krumlauf, 1992).

Unlike in *Drosophila*, single loss-of-function mutations in mice have not led to consistent phenotypic changes in the axial skeleton. Some loss-of-function mutants result in anterior homeotic transformations. In *Hoxa4* mutants, for instance, anterior homeotic transformations of the third cervical vertebra (C3) to C2 occur (Horan *et al.*, 1994). Loss of *Hoxc8* leads to conversion of the first thoracic vertebra (T1) to an eighth cervical vertebra (Chen *et al.*, 1998), and loss of *Hoxb9* results in the conversion of the first lumbar vertebra (L1) to a 14th ribbed, thoracic vertebra (Le Mouellic *et al.*, 1992). These results are consistent with genetic results from *Drosophila*, but only very small numbers of the 30 precaudal mouse vertebrae are affected by loss of function of any single *Hox* gene.

Other single loss-of-function *Hox* mouse mutants result in posterior homeotic transformations. Examples of these phenotypes include ectopic rib formation on the seventh cervical vertebra (a C7-T1 conversion) in *Hoxa5* and *Hoxa6* single mutants (Jeannotte *et al.*, 1993; Kostic and Capecchi, 1994), and conversion of the 13th thoracic vertebra into a lumbar phenotype (T13-L1) with loss of *Hoxa11* (Small and Potter, 1993). These phenotypes are not consistent with results expected from previous

Drosophila genetics. Further, as observed in single gene loss-of-function *Hox* mutations that lead to anterior transformations, only one or a small number of vertebrae are posteriorly transformed. Of note, changes in the axial skeleton occur almost entirely at the regions of morphological transition, that is, at the cervicothoracic junction, thoracolumbar or lumbosacral transitions. Taken together, single loss-of-function analyses suggest a role for *Hox* genes in patterning the vertebrate skeleton, but coherent genetic rules for their function are not forthcoming from these studies.

3.3. *Hox* cluster deletions

Duplication of the clusters during vertebrate radiation has resulted in four *Hox* clusters in mice. It had been previously appreciated that some cross-regulation exists among *Hox* genes within a cluster (Gould *et al.*, 1997; Rancourt *et al.*, 1995), and some evidence suggested that functional redundancy might exist between adjacent genes within a cluster, at least in the limb (Davis and Capecchi, 1996; Zakany and Duboule, 1996). Elegant genetic tests have examined the effects of deletion of whole clusters from the mouse genome, and these results do not support a synergistic role for genes within the clusters in axial patterning (Kmita *et al.*, 2005; Kondo *et al.*, 1996; Medina-Martinez *et al.*, 2000; Suemori *et al.*, 1995).

Deletion of the *HoxC* cluster was reported to cause only mild transformations of the axial skeleton (Suemori *et al.*, 1995). Deletions in the *HoxA* and *HoxD* cluster have also been generated (Kmita *et al.*, 2005; Kondo *et al.*, 1996), and while the axial phenotypes have not been described in detail, the authors state that the phenotypes caused by the *HoxD* deletions are largely a combination of single mutant phenotypes (Spitz *et al.*, 2001). The most severe phenotypes have been reported for mice carrying a deletion of the *HoxB* cluster (Medina-Martinez *et al.*, 2000). In these mutants, however, the somite-derived portion of the axial skeleton is only mildly perturbed. The sternum, which is derived from lateral plate mesoderm (not paraxial mesoderm like the rest of the axial skeleton), is severely mispatterned. Recent work has shown that somite-derived mesoderm and lateral plate-derived mesoderm are patterned with a high degree of independence (Burke, 2000; McIntyre *et al.*, 2007; Nowicki and Burke, 2000; Nowicki *et al.*, 2003; reviewed in Burke, 2000; McIntyre *et al.*, 2007; Nowicki and Burke, 2000; Nowicki *et al.*, 2003; Wellik, 2007), thus, it is likely that defects in these mice are mainly in lateral plate-derived structures.

3.4. Redundancy of paralogous genes in axial patterning

That duplication of the clusters during vertebrate radiation provided redundant genetic material has been noted by several groups (Kessel and Gruss, 1990; Maconochie *et al.*, 1996; McGinnis and Krumlauf, 1992). The four mammalian clusters are made up of 13 paralogous groups of genes. Each group is made up of two to four members. The genes in each paralog group are much more related to one another with respect to sequence and global expression patterns than they are to members outside the group, raising the likelihood that patterning information might be shared among these genes. Exploring the possibility that each paralog group controls AP regional identity, however, requires the generation of and interbreeding of many combinations of unlinked mutations; a task that has only been accomplished in recent years, after the generation of the single mutant animals (Chen *et al.*, 1998; McIntyre *et al.*, 2007; van den Akker *et al.*, 2001; Wellik and Capecchi, 2003).

In every case where combinations of more than one member of a *Hox* paralog group have been mutated, synergistic phenotypes have resulted (Chen and Capecchi, 1997, 1999; Chen *et al.*, 1998; Condie and Capecchi, 1994; Davis *et al.*, 1995; Fromental-Ramain *et al.*, 1996a,b; Gavalas *et al.*, 1998; Horan *et al.*, 1995a; Manley and Capecchi, 1998; McIntyre *et al.*, 2007; Studer *et al.*, 1998; van den Akker *et al.*, 2001; Wahba *et al.*, 2001; Wellik and Capecchi, 2003; Wellik *et al.*, 2002). This is especially true in axial patterning. Condie and Capecchi (1994) originally showed that loss of function of *Hoxd3* resulted in a remodeling of the craniocervical joint and a partial fusion of C1, the atlas, to the occipital bone. Single mutants for *Hoxa3* do not display defects in the craniocervical joint. However, when *Hoxa3/Hoxd3* double mutant embryos were generated, the atlas is no longer present at newborn stages (Condie and Capecchi, 1994). This is likely to represent an anterior transformation whereby the first prevertebral skeletal element that normally gives rise to the atlas becomes part of the base of the skull, following the fate of the first four and a half somites (Goodrich, 1930). Horan *et al.* demonstrated a similar redundant function for the *Hox4* paralogous genes. While single mutants from this group result in incompletely penetrant defects only in C2 or C3, removal of three of the four *Hox4* genes (*Hoxa4*, *Hoxb4*, and *Hoxd4*) result in fully penetrant anterior homeotic transformations of C2 through C5 (Horan *et al.*, 1995). These early studies demonstrate the existence of functional redundancy among *Hox* paralogous genes and the importance of removing this redundancy in order to understand their genetic function in vertebrates.

Complete paralogous mutants have since been generated for the *Hox5*, *Hox6*, *Hox9*, *Hox10*, and *Hox11* genes and the axial skeletons have been analyzed in detail (McIntyre *et al.*, 2007; Wellik and Capecchi, 2003). Extended, regional axial transformations are produced in all sets of paralogous mutants compared to effects in only one or two vertebrae in the majority of single or double mutant combinations. *Hox5* triple mutants demonstrate anterior homeotic transformations of C3 through the first thoracic vertebra (T1) to a C2, or axis-like phenotype, while *Hox6* paralog mutants show anteriorizing transformations in C6 through T6 (McIntyre *et al.*, 2007). *Hox9* quadruple mutants show dramatic transformations of the posterior thoracic vertebrae, with 13 or 14 ribbed vertebrae attached to the sternum instead of the usual complement of seven (McIntyre *et al.*, 2007). Mutation of the *Hox10* paralog group results in a transformation of lumbar and sacral vertebrae to posterior thoracic fate, with transformed vertebrae showing small rib projections. In the *Hox11* paralog mutants, the sacral vertebrae and early caudal vertebrae are anteriorly transformed to a lumbar-like fate (Wellik and Capecchi, 2003). Together, mutations in paralog groups from the *Hox3* through *Hox11* families consistently result in regional anterior transformations and affect each vertebra along the somite-derived axial skeleton (Fig. 9.3).

Perhaps the most unexpected feature of this patterning comes from comparing the phenotypes from adjacent paralog groups. In each case where adjacent paralog group mutants have been carefully examined, significant overlap in the affected AP regions is observed. Further, within these overlapping regions, the phenotypes caused by loss of each paralog group are distinct. For instance, in both *Hox5* and *Hox6* paralogous mutants, the first thoracic vertebra displays anterior homeotic transformations. In the case of the *Hox5* mutants, however, the first thoracic vertebra reveals transformations of the dorsal neural arch towards a C2 fate (Fig. 9.4, compare *Hox5* to wild-type T1). Ribs initiate, but do not extend from this vertebrae. In *Hox6* mutants, T1 is converted to the phenotype of the next most anterior vertebra, C7, with no neural arch morphology and no

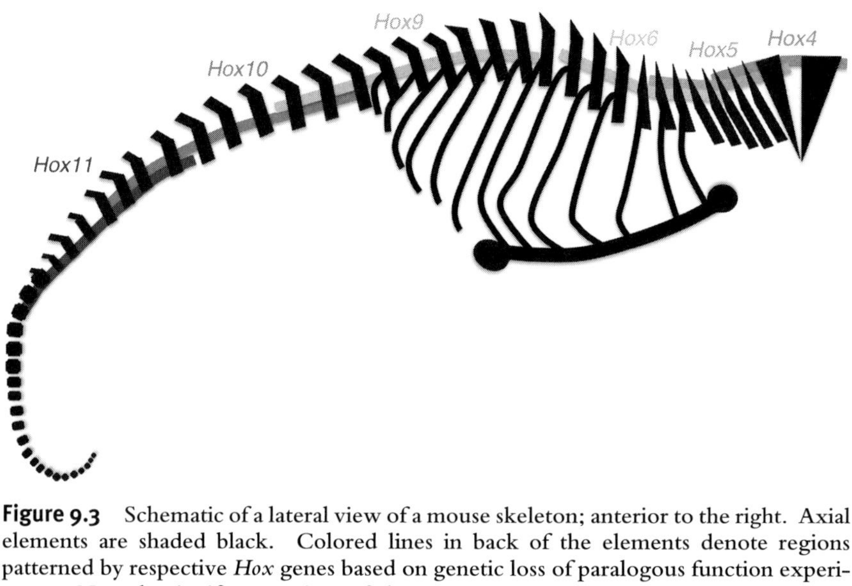

Figure 9.3 Schematic of a lateral view of a mouse skeleton; anterior to the right. Axial elements are shaded black. Colored lines in back of the elements denote regions patterned by respective *Hox* genes based on genetic loss of paralogous function experiments. Note the significant regions of phenotypic overlap between adjacent paralogous groups. (See Color Insert.)

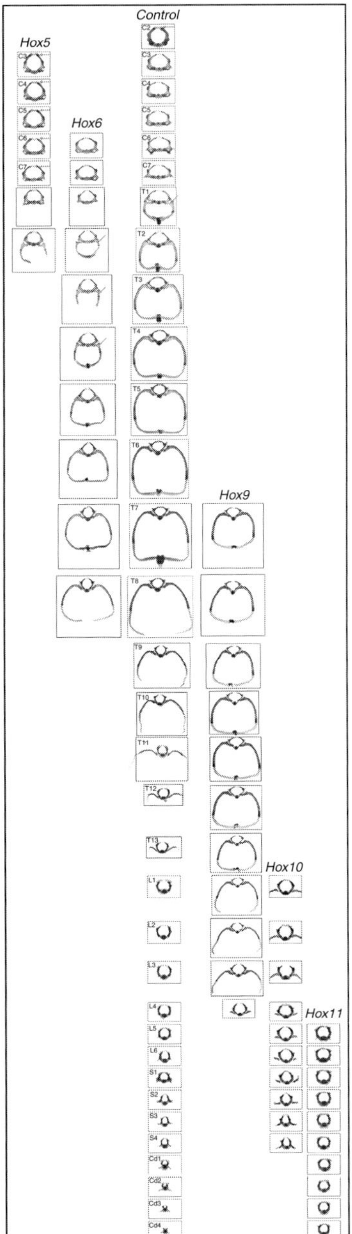

Figure 9.4 Dissected vertebral elements from a control E18.5 embryo, and paralog mutants of *Hox5*, *Hox6*, *Hox9*, *Hox10* and *Hox11* stained with Alcian blue and alizarin red. The third column of elements shows the entire vertebral column from the second cervical vertebra (C2) until the fourth caudal vertebra (Cd4). The affected vertebral elements are shown for each set of mutants. Each paralogous mutant group displays anterior homeotic transformations along AP-restricted regions of the axial skeleton. Note the differences in phenotypes in overlapping regions in adjacent mutant groups. (See Color Insert.)

rib formation (Fig. 9.4). Similar examples are shown in the lumbar region of the *Hox9* and *Hox10* paralogous mutants and in the sacral region of the *Hox10* and *Hox11* paralogous mutants (Fig. 9.4). Thus, the combined data support a model wherein each set of paralog *Hox* genes directs specific morphologies along AP-restricted regions of the somite-derived skeleton. While these functions are redundant between paralogs, the function of each group is distinct, with each paralog group imparting specific morphologies to regions of the axial skeleton.

These studies also highlight one further distinction in considering patterning of the axial skeleton, particularly the rib cage. It has recently been demonstrated that while the rest of the axial skeleton is comprised of only somitic mesoderm, the sternum of the rib cage is derived from lateral plate mesoderm (Burke and Nowicki, 2003; Nowicki and Burke, 2000; Nowicki *et al.*, 2003). Consistent with this, *Hox* patterning of the sternum demonstrates a high degree of independence from somitic axial patterning (McIntyre *et al.*, 2007; Wellik, 2007). In several of the *Hox* mutants from paralog groups 5 through 9 (Chen and Capecchi, 1997; Chen *et al.*, 1998; McIntyre *et al.*, 2007; Medina-Martinez *et al.*, 2000; van den Akker *et al.*, 2001), the patterning of the entire sternum is affected. *Hox* mutant phenotypes do not exhibit the colinearity that is characteristic in patterning the rest of the axial skeleton. Therefore, with respect to patterning the rib cage, it is important to separate lateral plate-derived sternal patterning from the patterning of the rest of the somite-derived skeleton (reviewed in Wellik, 2007).

4. Conclusions—The Nature of the Mammalian "*Hox* Code"

Recent studies examining the loss of paralogous *Hox* function in mice confirms the critical role for *Hox* genes in global patterning of the vertebrate axial skeleton. Perhaps the most important conclusion from these studies is that the genetic functions of the mammalian *Hox* genes are indeed similar to the functions originally described for the *Drosophila HomC* complex. Loss of *Hox* paralog function consistently results in anterior homeotic transformations along colinear regions of the somite-derived skeleton. As predicted shortly after the discovery of duplicated clusters in vertebrates, these analyses have been grueling and have required complicated combinations of mutations that remove the functional redundancy that is highly conserved among paralogs (Kessel and Gruss, 1990).

The extent of the functional redundancy retained among paralogs is quite remarkable. In all paralog groups that contribute to axial morphology, loss of multiple paralogs has resulted in more severe phenotypes without exception (Chen *et al.*, 1998; Condie and Capecchi, 1994; Horan *et al.*, 1995a; McIntyre *et al.*, 2007; van den Akker *et al.*, 2001; Wellik and Capecchi, 2003). The reasons for this redundancy are likely to be directly linked to the conserved mechanism of colinearity of regulation. It is their conserved colinear expression that endows the *Hox* genes with the ability to globally pattern (Duboule, 1994). However, evolutionary conservation of this colinearity also serves as a constraint to divergence of gene sequence (and thus function) within the complex. The maintenance of the clustered structure over time, one of the keys to its colinear character, is a profound functional burden (Akam, 1989). Because of this burden, the *Hox* complexes (and, thus, the genes within the complex) are more protected from evolutionary divergence than other regions of the genome. One of the results of this protection is that duplicated genomic material (including paralogous genes within the clusters as well as many of their regulatory elements) is more likely to remain functionally redundant.

The current combined genetic analyses have also led to less anticipated findings regarding *Hox* function in vertebrate axial patterning. The overlap in phenotypes between adjacent paralog group mutants was not predicted based on single mutant analyses. While it had been previously appreciated that homeotic transformations in *Drosophila* are only observed for *HomC* loss-of-function mutations that have expression domains that overlap with other *HomC* genes (McGinnis and Krumlauf, 1992), single gene loss-of-function *Hox* mutations in mouse generally result in transformations only at the most anterior limit of gene expression. Loss of paralog function, however, demonstrates that each group patterns extended regions of the axial skeleton (generally 6–10 consecutive vertebrae), and these regions overlap significantly between adjacent groups (McIntyre *et al.*, 2007; Wellik and Capecchi, 2003). These analyses are incomplete, because not all paralog group mutants have been examined at the level of individual vertebrae, but current work suggests at least two paralog groups contribute significantly to the patterning of each vertebral element.

These findings directly impact our current understanding of the "*Hox* code," which holds that the identity of a vertebral segment is specified by the combination of functionally active *Hox* genes (Kessel and Gruss, 1991). Current data support a combinatorial "*Hox* code" wherein the activity of at least two paralog groups contributes to the patterning of the axial skeleton. Combinations of different levels of each of these proteins would allow for the continuum of morphologies that exist in the vertebrate axial skeleton. At the anterior limits of some paralog groups, the new *Hox* activities provide dramatic changes to phenotype, such as cessation of rib formation at L1 by the *Hox10* genes or the generation of sacral wings at the lumbosacral border by *Hox11* genes (Wellik and Capecchi, 2003). In other regions, more gradual changes in morphology occur, such as broadening of the rib angles by *Hox6* genes (McIntyre *et al.*, 2007). Together, reported *Hox* patterning activities account for most of the modifications of vertebral morphology along the AP axis.

Strictly speaking, the concept of a combinatorial "*Hox* code" and "posterior prevalence," which posits that posteriorly expressed genes are dominant in imposing their posterior regional identity on overlapping anterior genes (Duboule, 1991), are mutually exclusive. Certainly, the combined genetic results demonstrate that, in adjacent paralog groups, the next most posterior set of *Hox* genes do not prevent the function of the anterior group, and no posterior prevalence exists between adjacent paralog groups examined to date (McIntyre *et al.*, 2007; Wellik and Capecchi, 2003). Further, overexpression studies demonstrate that the ectopic expression of a *Hox* gene affects regions both anterior to its normal expression boundaries and posterior to it (Carapuco *et al.*, 2005; Gibson *et al.*, 1990; Kuziora and McGinnis, 1988). Until positively or negatively regulated downstream genes of the *Hox* genes in axial patterning are identified, it will be difficult to resolve these models at a molecular level, but the current data do not provide genetic support for posterior prevalence in axial patterning.

Only one characteristic of axial morphology is not dramatically altered with loss of *Hox* function—the ability to generate ribs. The generation of ectopic ribs along the lumbar region (with loss of *Hox10* function; Wellik and Capecchi, 2003), dramatic changes in rib angles (with loss of *Hox5* and *Hox6*; McIntyre *et al.*, 2007), and number of ribs attaching to the sternum (with loss of *Hox9*; McIntyre *et al.*, 2007) are modified substantially with loss of Hox function, but no mutant combinations have resulted in a significant loss of the number of ribbed vertebrae (in *Hox6* mutants, there is one less than normal; McIntyre *et al.*, 2007). Thus, there is no evidence that induction of rib formation is controlled by *Hox* genes. The combined genetic results suggest that the more anterior *Hox* genes (*Hox3* through *Hox5*) may have been co-opted to allow anterior somites to form vertebrae instead of becoming part of the skull. *Hox6* through *Hox8* genes modify the morphology of the thoracic rib cage, and *Hox9* genes repress the ability of posterior ribs to extend and attach to the sternum. *Hox10* genes repress rib formation posterior to the thoracic vertebrae, and *Hox11* genes actively promote sacral morphology. Therefore, ribbed thoracic vertebrae that extend laterally around much of the body wall are the only axial phenotype that has not been clearly shown to be *Hox* dependent. Thus, this may represent a developmental, and perhaps evolutionary, "ground state." However, the argument that more sophisticated combinations of *Hox* mutations are required to see these kinds of changes in of this region of the thoracic vertebrae is at least one alternative explanation for this observation. Loss of function of all four complexes would need to be generated, conditionally and only in the somites, to address this question fully.

The combined molecular, embryological and genetic data accumulated over the past two decades has contributed substantially to our understanding of the important roles of *Hox* genes in axial patterning, but the molecular mechanisms involved in these events remain completely unknown. One of the most critical pieces of information that is lacking is the timing of regulatory events that lead to the morphological changes that define the different axial identities. It is only several days after the initiation of *Hox* expression in the embryo that sclerotomal cells begin to acquire morphologies specific to their AP position. At these later time points, *Hox* expression in the skeletal regions no long extends to the posterior extremities, and is excluded from condensing cartilage and found in perichondrial cells surrounding the shaping cartilage rudiments in at least a few examples (Hostikka and Capecchi, 1998; Nelson *et al.*, 2008). This leads to the possibility that the early, nested expression patterns observed for many of the *Hox* genes are a result of the complex regulatory initiation programs that are shared among the cluster, and that the later expression patterns that correlate with the regions that demonstrate phenotypes in mutants may be the critical developmental window in which *Hox* genes function in axial patterning. Careful analyses of Hox protein expression along with conditional, temporal ablation of these genes, will afford a greater understanding of the function of *Hox* genes in axial patterning.

REFERENCES

Akam, M. (1989). Hox and HOM: Homologous gene clusters in insects and vertebrates. *Cell* **57**(3), 347–349.

Boulet, A. M., and Capecchi, M. R. (1996). Targeted disruption of hoxc-4 causes esophageal defects and vertebral transformations. *Dev. Biol.* **177**(1), 232–249.

Burke, A. C. (2000). *Hox* genes and the global patterning of the somitic mesoderm. *Curr. Top. Dev. Biol.* **47,** 155–181.

Burke, A. C., and Nowicki, J. L. (2003). A new view of patterning domains in the vertebrate mesoderm. *Dev. Cell* **4**(2), 159–165.

Burke, A. C., Nelson, C. E., Morgan, B. A., and Tabin, C. (1995). *Hox* genes and the evolution of vertebrate axial morphology. *Development* **121**(2), 333–346.

Carapuco, M., Novoa, A., Bobola, N., and Mallo, M. (2005). *Hox* genes specify vertebral types in the presomitic mesoderm. *Genes Dev.* **19**(18), 2116–2121.

Carpenter, E. M., Goddard, J. M., Davis, A. P., Nguyen, T. P., and Capecchi, M. R. (1997). Targeted disruption of Hoxd-10 affects mouse hindlimb development. *Development* **124** (22), 4505–4514.

Chen, F., and Capecchi, M. R. (1997). Targeted mutations in hoxa-9 and hoxb-9 reveal synergistic interactions. *Dev. Biol.* **181**(2), 186–196.

Chen, F., and Capecchi, M. R. (1999). Paralogous mouse *Hox* genes, *Hoxa9*, *Hoxb9*, and *Hoxd9*, function together to control development of the mammary gland in response to pregnancy. *Proc. Natl. Acad. Sci. USA* **96**(2), 541–546.

Chen, F., Greer, J., and Capecchi, M. R. (1998). Analysis of Hoxa7/Hoxb7 mutants suggests periodicity in the generation of the different sets of vertebrae. *Mech. Dev.* **77**(1), 49–57.

Chisaka, O., and Capecchi, M. R. (1991). Regionally restricted developmental defects resulting from targeted disruption of the mouse homeobox gene *hox-1.5*. *Nature* **350**(6318), 473–479.

Chisaka, O., Musci, T. S., and Capecchi, M. R. (1992). Developmental defects of the ear, cranial nerves and hindbrain resulting from targeted disruption of the mouse homeobox gene *Hox-1.6*. *Nature* **355**(6360), 516–520.

Condie, B. G., and Capecchi, M. R. (1993). Mice homozygous for a targeted disruption of Hoxd-3 (Hox-4.1) exhibit anterior transformations of the first and second cervical vertebrae, the atlas and the axis. *Development* **119**(3), 579–595.

Condie, B. G., and Capecchi, M. R. (1994). Mice with targeted disruptions in the paralogous genes *hoxa-3* and *hoxd-3* reveal synergistic interactions. *Nature* **370**(6487), 304–307.

Davis, A. P., and Capecchi, M. R. (1994). Axial homeosis and appendicular skeleton defects in mice with a targeted disruption of hoxd-11. *Development* **120**(8), 2187–2198.

Davis, A. P., and Capecchi, M. R. (1996). A mutational analysis of the *5′ HoxD* genes: Dissection of genetic interactions during limb development in the mouse. *Development* **122**(4), 1175–1185.

Davis, A. P., Witte, D. P., Hsieh-Li, H. M., Potter, S. S., and Capecchi, M. R. (1995). Absence of radius and ulna in mice lacking hoxa-11 and hoxd-11. *Nature* **375**(6534), 791–795.

Deschamps, J., van den Akker, E., Forlani, S., De Graaff, W., Oosterveen, T., Roelen, B., and Roelfsema, J. (1999). Initiation, establishment and maintenance of *Hox* gene expression patterns in the mouse. *Int. J. Dev. Biol.* **43**(7), 635–650.

Dressler, G. R., and Gruss, P. (1989). Anterior boundaries of *Hox* gene expression in mesoderm-derived structures correlate with the linear gene order along the chromosome. *Differentiation* **41**(3), 193–201.

Duboule, D. (1991). Patterning in the vertebrate limb. *Curr. Opin. Genet. Dev.* **1**(2), 211–216.

Duboule, D. (1994). Temporal colinearity and the phylotypic progression: A basis for the stability of a vertebrate Bauplan and the evolution of morphologies through heterochrony. *Development (Suppl.)* 135–142.

Duboule, D., and Dolle, P. (1989). The structural and functional organization of the murine *HOX* gene family resembles that of *Drosophila* homeotic genes. *EMBO J.* **8**(5), 1497–1505.

Fromental-Ramain, C., Warot, X., Lakkaraju, S., Favier, B., Haack, H., Birling, C., Dierich, A., Doll e, P., and Chambon, P. (1996a). Hoxa-13 and Hoxd-13 play a crucial role in the patterning of the limb autopod. *Development* **122**(10), 2997–3011.

Fromental-Ramain, C., Warot, X., Messadecq, N., LeMeur, M., Dolle, P., and Chambon, P. (1996b). Specific and redundant functions of the paralogous *Hoxa-9* and *Hoxd-9* genes in forelimb and axial skeleton patterning. *Development* **122**(2), 461–472.

Garcia-Gasca, A., and Spyropoulos, D. D. (2000). Differential mammary morphogenesis along the anteroposterior axis in *Hoxc6* gene targeted mice. *Dev. Dyn.* **219**(2), 261–276.

Gaunt, S. J. (1991). Expression patterns of mouse *Hox* genes: Clues to an understanding of developmental and evolutionary strategies. *Bioessays* **13**(10), 505–513.

Gaunt, S. J. (1994). Conservation in the Hox code during morphological evolution. *Int. J. Dev. Biol.* **38**(3), 549–552.

Gaunt, S. J., and Strachan, L. (1996). Temporal colinearity in expression of anterior *Hox* genes in developing chick embryos. *Dev. Dyn.* **207**(3), 270–280.

Gaunt, S. J., Miller, J. R., Powell, D. J., and Duboule, D. (1986). Homoeobox gene expression in mouse embryos varies with position by the primitive streak stage. *Nature* **324**(6098), 662–664.

Gaunt, S. J., Krumlauf, R., and Duboule, D. (1989). Mouse homeo-genes within a subfamily, Hox-1.4, -2.6 and -5.1, display similar anteroposterior domains of expression in the embryo, but show stage- and tissue-dependent differences in their regulation. *Development* **107**(1), 131–141.

Gaunt, S. J., Coletta, P. L., Pravtcheva, D., and Sharpe, P. T. (1990). Mouse Hox-3.4: Homeobox sequence and embryonic expression patterns compared with other members of the *Hox* gene network. *Development* **109**(2), 329–339.

Gavalas, A., Studer, M., Lumsden, A., Rijli, F. M., Krumlauf, R., and Chambon, P. (1998). Hoxa1 and Hoxb1 synergize in patterning the hindbrain, cranial nerves and second pharyngeal arch. *Development* **125**(6), 1123–1136.

Gibson, G., Schier, A., LeMotte, P., and Gehring, W. J. (1990). The specificities of Sex combs reduced and Antennapedia are defined by a distinct portion of each protein that includes the homeodomain. *Cell* **62**(6), 1087–1103.

Goodrich, E. S. (1930). "Studies on the Structure and Development of Vertebrates." Macmillan, London.

Goodrich, E. S. (1958). "Studieson the Structure and Development of Vertebrates" vol. I, p. 485. Dover Publications, New York, NY.

Gould, A., Morrison, A., Sproat, G., White, R. A., and Krumlauf, R. (1997). Positive cross-regulation and enhancer sharing: Two mechanisms for specifying overlapping Hox expression patterns. *Genes Dev.* **11**(7), 900–913.

Graham, A., Papalopulu, N., and Krumlauf, R. (1989). The murine and *Drosophila* homeobox gene complexes have common features of organization and expression. *Cell* **57**(3), 367–378.

Gruss, P., and Kessel, M. (1991). Axial specification in higher vertebrates. *Curr. Opin. Genet. Dev.* **1**(2), 204–210.

Herault, Y., Beckers, J., Gerard, M., and Duboule, D. (1999). *Hox* gene expression in limbs: Colinearity by opposite regulatory controls. *Dev. Biol.* **208**(1), 157–165.

Horan, G. S., Wu, K., Wolgemuth, D. J., and Behringer, R. R. (1994). Homeotic transformation of cervical vertebrae in Hoxa-4 mutant mice. *Proc. Natl. Acad. Sci. USA* **91**(26), 12644–12648.

Horan, G. S., Kovacs, E. N., Behringer, R. R., and Featherstone, M. S. (1995a). Compound mutants for the paralogous *hoxa-4*, *hoxb-4*, and *hoxd-4* genes show more complete homeotic transformations and a dose-dependent increase in the number of vertebrae transformed. *Genes Dev.* **9**(13), 1667–1677.

Horan, G. S., Ramirez-Solis, R., Featherstone, M. S., Wolgemuth, D. J., Bradley, A., and Behringer, R. R. (1995b). Mutations in paralogous *Hox* genes result in overlapping homeotic transformations of the axial skeleton: Evidence for unique and redundant function. *Dev. Biol.* **169**(1), 359–372.

Hostikka, S. L., and Capecchi, M. R. (1998). The mouse *Hoxc11* gene: Genomic structure and expression pattern. *Mech. Dev.* **70**(1–2), 133–145.

Izpisua-Belmonte, J. C., Falkenstein, H., Dolle, P., Renucci, A., and Duboule, D. (1991). Murine genes related to the *Drosophila AbdB* homeotic genes are sequentially expressed during development of the posterior part of the body. *EMBO J.* **10**(8), 2279–2289.

Jeannotte, L., Lemieux, M., Charron, J., Poirier, F., and Robertson, E. J. (1993). Specification of axial identity in the mouse: Role of the *Hoxa-5 (Hox1.3)* gene. *Genes Dev.* **7**(11), 2085–2096.

Jegalian, B. G., and De Robertis, E. M. (1992). Homeotic transformations in the mouse induced by overexpression of a human *Hox3.3* transgene. *Cell* **71**(6), 901–910.

Kessel, M., and Gruss, P. (1990). Murine developmental control genes. *Science* **249**(4967), 374–379.

Kessel, M., and Gruss, P. (1991). Homeotic transformations of murine vertebrae and concomitant alteration of Hox codes induced by retinoic acid. *Cell* **67**(1), 89–104.

Kessel, M., Balling, R., and Gruss, P. (1990). Variations of cervical vertebrae after expression of a *Hox-1.1* transgene in mice. *Cell* **61**(2), 301–308.

Kieny, M., Mauger, A., and Sengel, P. (1972). Early regionalization of somitic mesoderm as studied by the development of axial skeleton of the chick embryo. *Dev. Biol.* **28**(1), 142–161.

Kmita, M., van Der Hoeven, F., Zakany, J., Krumlauf, R., and Duboule, D. (2000). Mechanisms of *Hox* gene colinearity: Transposition of the anterior *Hoxb1* gene into the posterior HoxD complex. *Genes Dev.* **14**(2), 198–211.

Kmita, M., Fraudeau, N., Herault, Y., and Duboule, D. (2002). Serial deletions and duplications suggest a mechanism for the collinearity of *Hoxd* genes in limbs. *Nature* **420**(6912), 145–150.

Kmita, M., Tarchini, B., Zakany, J., Logan, M., Tabin, C. J., and Duboule, D. (2005). Early developmental arrest of mammalian limbs lacking *HoxA/HoxD* gene function. *Nature* **435**(7045), 1113–1116.

Kondo, T., and Duboule, D. (1999). Breaking colinearity in the mouse HoxD complex. *Cell* **97**(3), 407–417.

Kondo, T., Dolle, P., Zakany, J., and Duboule, D. (1996). Function of posterior *HoxD* genes in the morphogenesis of the anal sphincter. *Development* **122**(9), 2651–2659.

Kostic, D., and Capecchi, M. R. (1994). Targeted disruptions of the murine *Hoxa-4* and *Hoxa-6* genes result in homeotic transformations of components of the vertebral column. *Mech. Dev.* **46**(3), 231–247.

Kuziora, M. A., and McGinnis, W. (1988). Autoregulation of a *Drosophila* homeotic selector gene. *Cell* **55**(3), 477–485.

Lehoczky, J. A., and Innis, J. W. (2008). BAC transgenic analysis reveals enhancers sufficient for *Hoxa13* and neighborhood gene expression in mouse embryonic distal limbs and genital bud. *Evol. Dev.* **10**(4), 421–432.

Le Mouellic, H., Lallemand, Y., and Brulet, P. (1992). Homeosis in the mouse induced by a null mutation in the *Hox-3.1* gene. *Cell* **69**(2), 251–264.

Lewis, E. B. (1963). Genes and developmental pathways. *Am. Zool.* **3,** 33–56.

Lewis, E. B. (1978). A gene complex controlling segmentation in *Drosophila. Nature* **276** (5688), 565–570.

Lufkin, T., Dierich, A., LeMeur, M., Mark, M., and Chambon, P. (1991). Disruption of the *Hox-1.6* homeobox gene results in defects in a region corresponding to its rostral domain of expression. *Cell* **66**(6), 1105–1119.

Lufkin, T., Mark, M., Hart, C. P., Dolle, P., LeMeur, M., and Chambon, P. (1992). Homeotic transformation of the occipital bones of the skull by ectopic expression of a homeobox gene. *Nature* **359**(6398), 835–841.

Maconochie, M., Nonchev, S., Morrison, A., and Krumlauf, R. (1996). Paralogous *Hox* genes: Function and regulation. *Annu. Rev. Genet.* **30,** 529–556.

Manley, N. R., and Capecchi, M. R. (1995). The role of Hoxa-3 in mouse thymus and thyroid development. *Development* **121**(7), 1989–2003.

Manley, N. R., and Capecchi, M. R. (1998). Hox group 3 paralogs regulate the development and migration of the thymus, thyroid, and parathyroid glands. *Dev. Biol.* **195**(1), 1–15.

Mann, R. S., and Hogness, D. S. (1990). Functional dissection of Ultrabithorax proteins in *D. melanogaster. Cell* **60**(4), 597–610.

McGinnis, W., and Krumlauf, R. (1992). Homeobox genes and axial patterning. *Cell* **68**(2), 283–302.

McIntyre, D. C., Rakshit, S., Yallowitz, A. R., Loken, L., Jeannotte, L., Capecchi, M. R., and Wellik, D. M. (2007). Hox patterning of the vertebrate rib cage. *Development* **134**(16), 2981–2989.

Medina-Martinez, O., Bradley, A., and Ramirez-Solis, R. (2000). A large targeted deletion of Hoxb1-Hoxb9 produces a series of single-segment anterior homeotic transformations. *Dev. Biol.* **222**(1), 71–83.

Merrill, V. K., Turner, F. R., and Kaufman, T. C. (1987). A genetic and developmental analysis of mutations in the Deformed locus in *Drosophila melanogaster. Dev. Biol.* **122**(2), 379–395.

Nelson, L. T., Rakshit, S., Sun, H., and Wellik, D. M. (2008). Generation and expression of a targeted *Hoxa11eGFP* allele in mice, *Dev. Dyn.* **237**(11), 3410–3416.

Nowicki, J. L., and Burke, A. C. (2000). *Hox* genes and morphological identity: Axial versus lateral patterning in the vertebrate mesoderm. *Development* **127**(19), 4265–4275.

Nowicki, J. L., Takimoto, R., and Burke, A. C. (2003). The lateral somitic frontier: Dorso-ventral aspects of anterio-posterior regionalization in avian embryos. *Mech. Dev.* **120**(2), 227–240.

Pollock, R. A., Jay, G., and Bieberich, C. J. (1992). Altering the boundaries of Hox3.1 expression: Evidence for antipodal gene regulation. *Cell* **71**(6), 911–923.

Procter, N. S., and Lynch, P. J. (1993). "Manual of Ornithology: Avian Structure and Function." Yale University Press, New Haven, CT.

Rancourt, D. E., Tsuzuki, T., and Capecchi, M. R. (1995). Genetic interaction between *hoxb-5* and *hoxb-6* is revealed by nonallelic noncomplementation. *Genes Dev.* **9**(1), 108–122.

Regulski, M., McGinnis, N., Chadwick, R., and McGinnis, W. (1987). Developmental and molecular analysis of Deformed: A homeotic gene controlling *Drosophila* head development. *EMBO J.* **6**(3), 767–777.

Richardson, M. K., Allen, S. P., Wright, G. M., Raynaud, A., and Hanken, J. (1998). Somite number and vertebrate evolution. *Development* **125**(2), 151–160.

Rijli, F. M., Mark, M., Lakkaraju, S., Dierich, A., Dolle, P., and Chambon, P. (1993). A homeotic transformation is generated in the rostral branchial region of the head by disruption of *Hoxa-2*, which acts as a selector gene. *Cell* **75**(7), 1333–1349.

Rijli, F. M., Dolle, P., Fraulob, V., LeMeur, M., and Chambon, P. (1994). Insertion of a targeting construct in a Hoxd-10 allele can influence the control of Hoxd-9 expression. *Dev. Dyn.* **201**(4), 366–377.

Rijli, F. M., Matyas, R., Pellegrini, M., Dierich, A., Gruss, P., Dolle, P., and Chambon, P. (1995). Cryptorchidism and homeotic transformations of spinal nerves and vertebrae in Hoxa-10 mutant mice. *Proc. Natl. Acad. Sci. USA* **92**(18), 8185–8189.

Sanchez-Herrero, E., Vernos, I., Marco, R., and Morata, G. (1985). Genetic organization of *Drosophila* bithorax complex. *Nature* **313**(5998), 108–113.

Small, K. M., and Potter, S. S. (1993). Homeotic transformations and limb defects in Hox A11 mutant mice. *Genes Dev.* **7**(12A), 2318–2328.

Spitz, F., Gonzalez, F., Peichel, C., Vogt, T. F., Duboule, D., and Zakany, J. (2001). Large scale transgenic and cluster deletion analysis of the HoxD complex separate an ancestral regulatory module from evolutionary innovations. *Genes Dev.* **15**(17), 2209–2214.

Spitz, F., Herkenne, C., Morris, M. A., and Duboule, D. (2005). Inversion-induced disruption of the Hoxd cluster leads to the partition of regulatory landscapes. *Nat. Genet.* **37**(8), 889–893.

Studer, M., Gavalas, A., Marshall, H., Ariza-McNaughton, L., Rijli, F. M., Chambon, P., and Krumlauf, R. (1998). Genetic interactions between *Hoxa1* and *Hoxb1* reveal new roles in regulation of early hindbrain patterning. *Development* **125**(6), 1025–1036.

Suemori, H., Takahashi, N., and Noguchi, S. (1995). Hoxc-9 mutant mice show anterior transformation of the vertebrae and malformation of the sternum and ribs. *Mech. Dev.* **51**(2–3), 265–273.

Thomas, K. R., and Capecchi, M. R. (1987). Site-directed mutagenesis by gene targeting in mouse embryo-derived stem cells. *Cell* **51**(3), 503–512.

van den Akker, E., Reijnen, M., Korving, J., Brouwer, A., Meijlink, F., and Deschamps, J. (1999). Targeted inactivation of Hoxb8 affects survival of a spinal ganglion and causes aberrant limb reflexes. *Mech. Dev.* **89**(1–2), 103–114.

van den Akker, E., Fromental-Ramain, C., de Graaff, W., Le Mouellic, H., Brulet, P., Chambon, P., and Deschamps, J. (2001). Axial skeletal patterning in mice lacking all paralogous group 8 *Hox* genes. *Development* **128**(10), 1911–1921.

Wahba, G. M., Hostikka, S. L., and Carpenter, E. M. (2001). The paralogous *Hox* genes *Hoxa10* and *Hoxd10* interact to pattern the mouse hindlimb peripheral nervous system and skeleton. *Dev. Biol.* **231**(1), 87–102.

Wakimoto, B. T., and Kaufman, T. C. (1981). Analysis of larval segmentation in lethal genotypes associated with the *Antennapedia* gene complex in *Drosophila melanogaster*. *Dev. Biol.* **81**(1), 51–64.

Wellik, D. M. (2007). Hox patterning of the vertebrate axial skeleton. *Dev. Dyn.* **236**(9), 2454–2463.

Wellik, D. M., and Capecchi, M. R. (2003). *Hox10* and *Hox11* genes are required to globally pattern the mammalian skeleton. *Science* **301**(5631), 363–367.

Wellik, D. M., Hawkes, P. J., and Capecchi, M. R. (2002). *Hox11* paralogous genes are essential for metanephric kidney induction. *Genes Dev.* **16**(11), 1423–1432.

Wolpert, L. (1969). Positional information and the spatial pattern of cellular differentiation. *J. Theor. Biol.* **25**(1), 1–47.

Wright, E., Hargrave, M. R., Christiansen, J., Cooper, L., Kun, J., Evans, T., Gangadharan, U., Greenfield, A., and Koopman, P. (1995). The Sry-related gene *Sox9* is expressed during chondrogenesis in mouse embryos. *Nat. Genet.* **9**(1), 15–20.

Zakany, J., and Duboule, D. (1996). Synpolydactyly in mice with a targeted deficiency in the HoxD complex. *Nature* **384**(6604), 69–71.

Subject Index

T

U

V

W

Y

Z

Contents of Previous Volumes

Volume 49

Volume 50

Volume 51

Volume 52

Volume 53

Volume 54

Volume 55

Volume 56

Volume 57

Volume 58

Volume 59

Volume 60

Volume 61

Volume 62

Volume 63

Volume 64

Volume 65

Volume 66

Volume 67

Volume 68

Volume 69

Volume 70

Volume 73

Volume 74

Volume 75

Volume 76

Volume 77

Volume 78

Volume 79

Volume 81

Volume 82

Volume 85

Volume 86

Volume 87

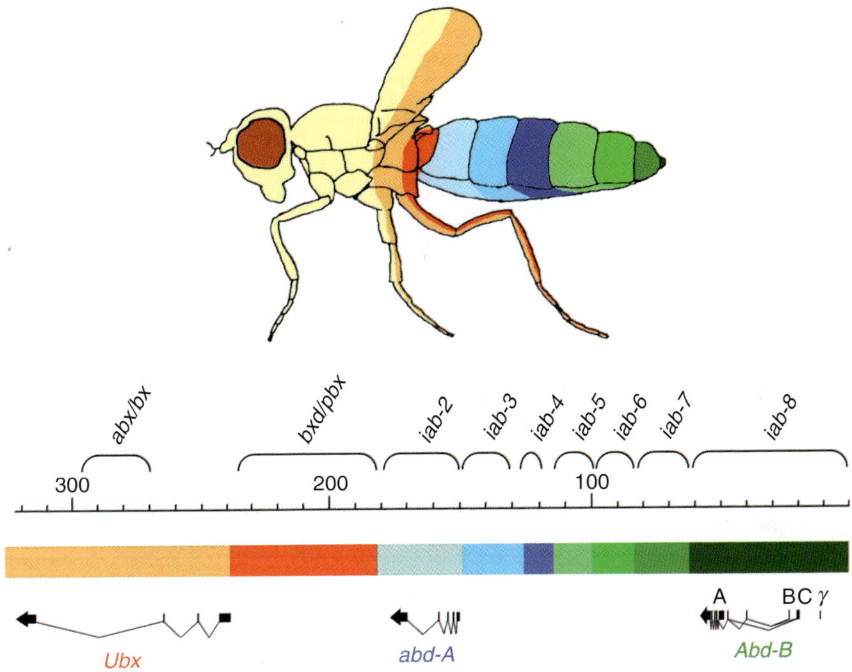

Robert K. Maeda and François Karch, Figure 1.1 Please refer to the legend in the text.

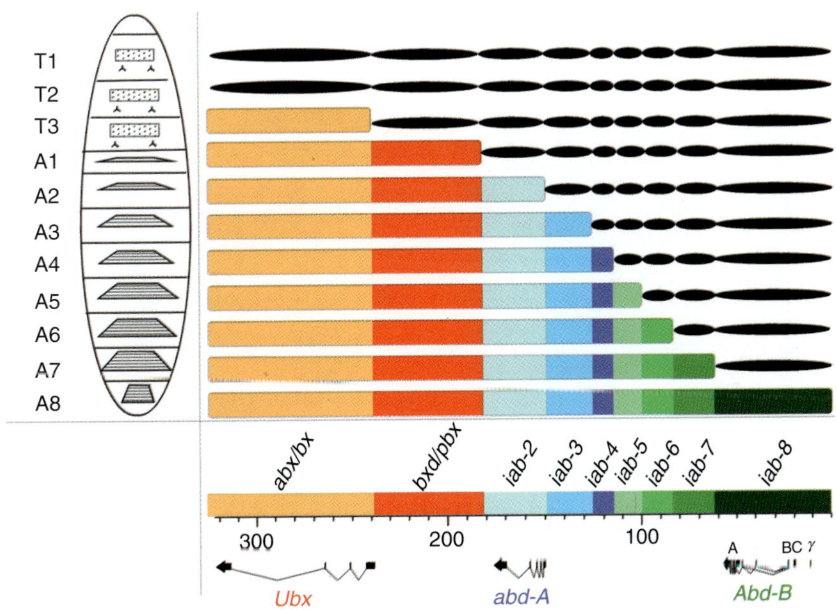

Robert K. Maeda and François Karch, Figure 1.2 Please refer to the legend in the text.

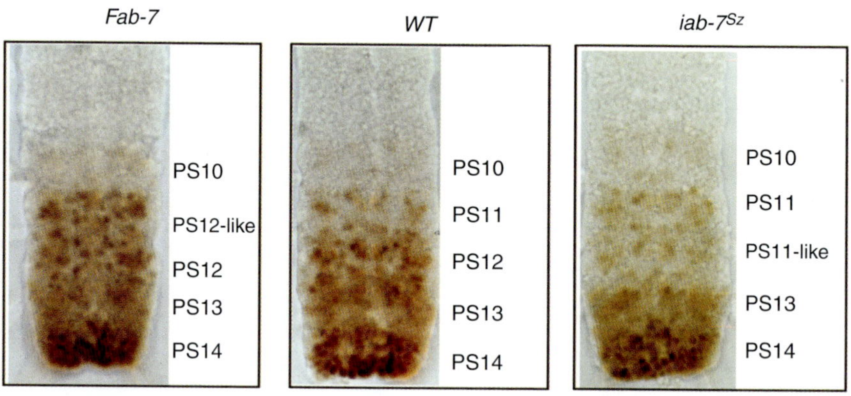

Robert K. Maeda and François Karch, Figure 1.3 Please refer to the legend in the text.

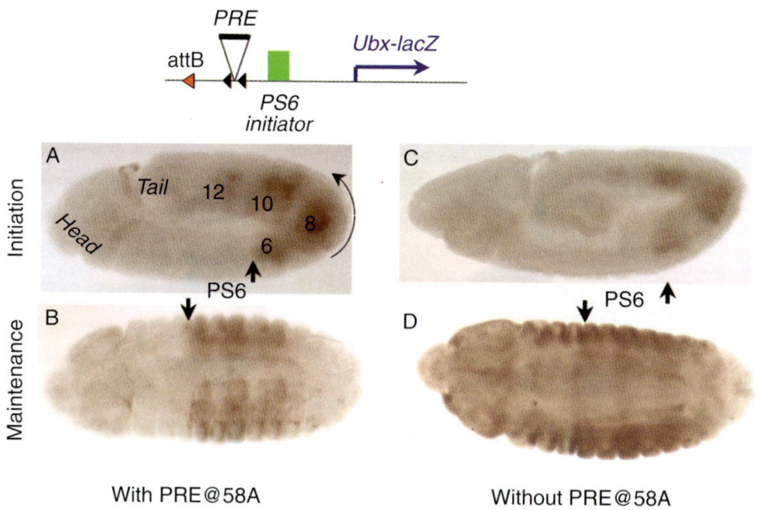

Robert K. Maeda and François Karch, Figure 1.4 Please refer to the legend in the text.

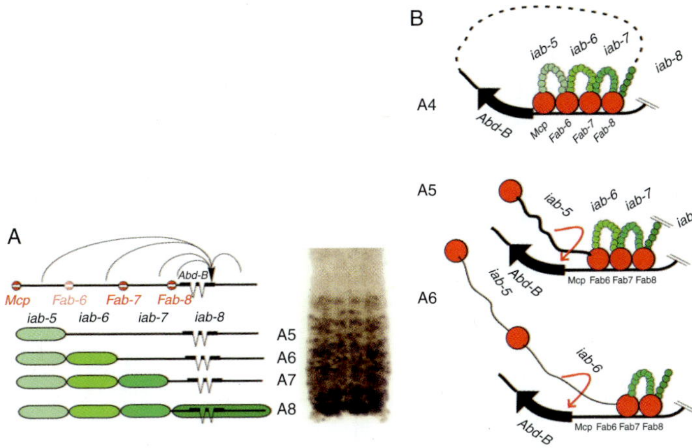

Robert K. Maeda and François Karch, Figure 1.6 Please refer to the legend in the text.

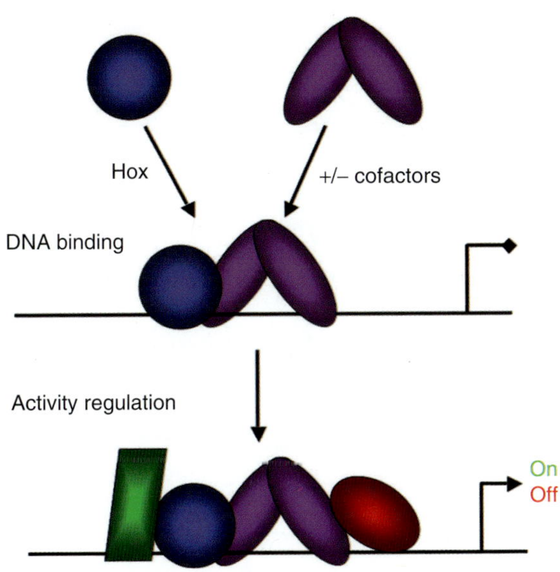

Richard S. Mann *et al.*, Figure 3.1 Please refer to the legend in the text.

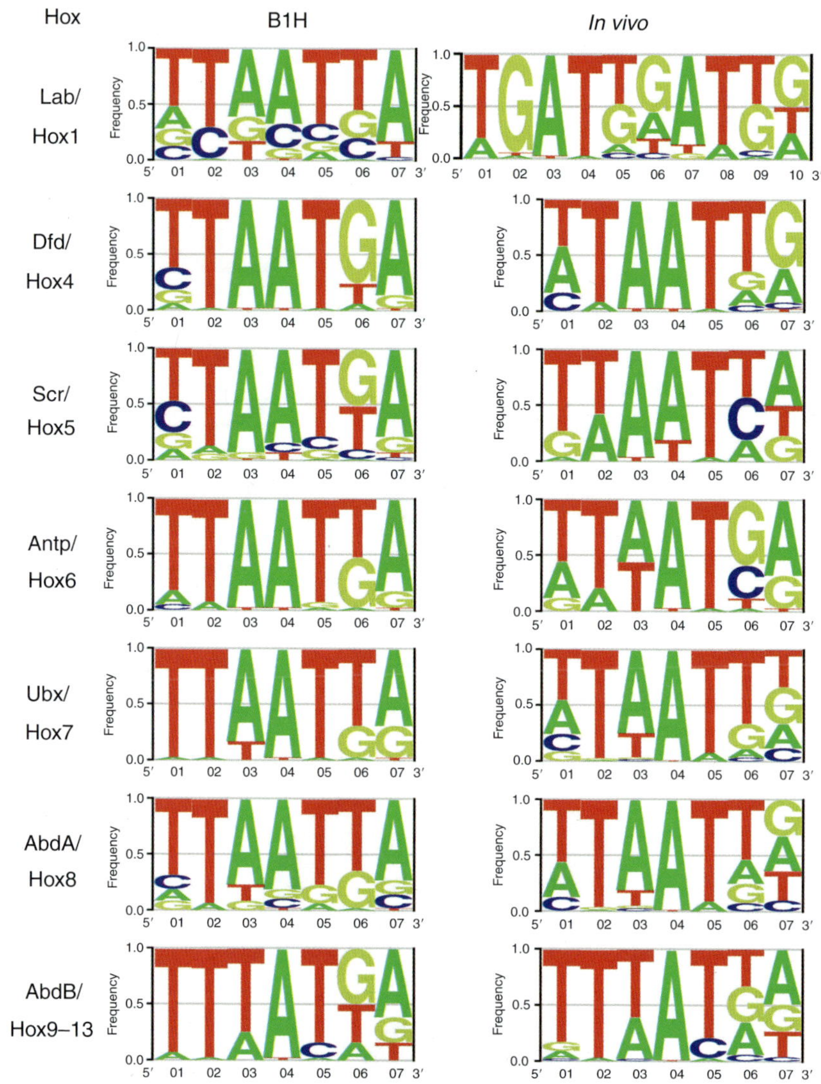

Richard S. Mann *et al*., Figure 3.2 Please refer to the legend in the text.

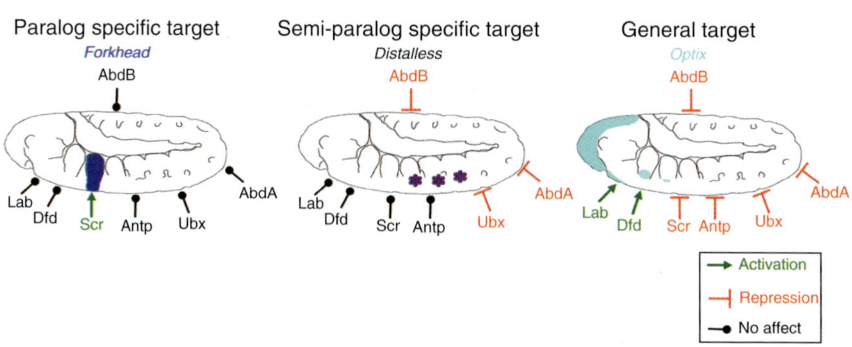

Richard S. Mann *et al*., Figure 3.3 Please refer to the legend in the text.

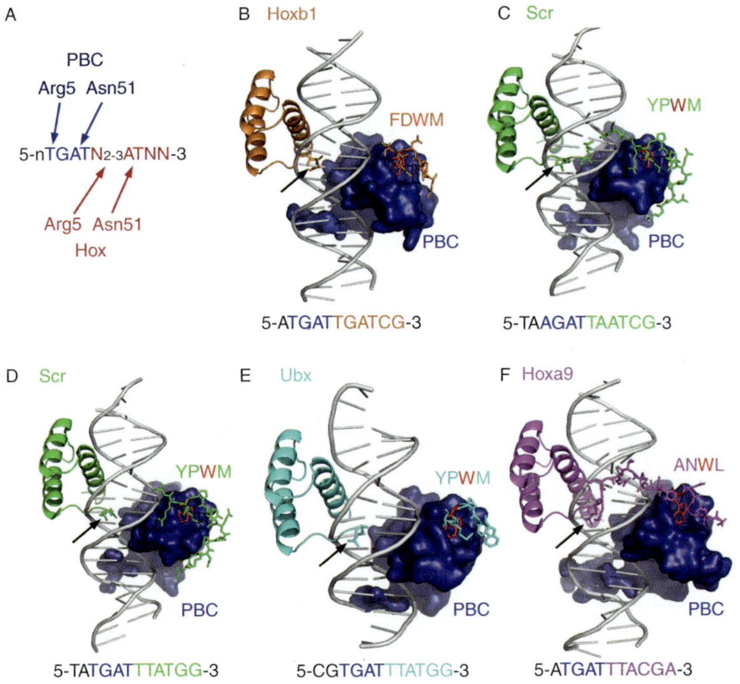

Richard S. Mann *et al.*, Figure 3.4 Please refer to the legend in the text.

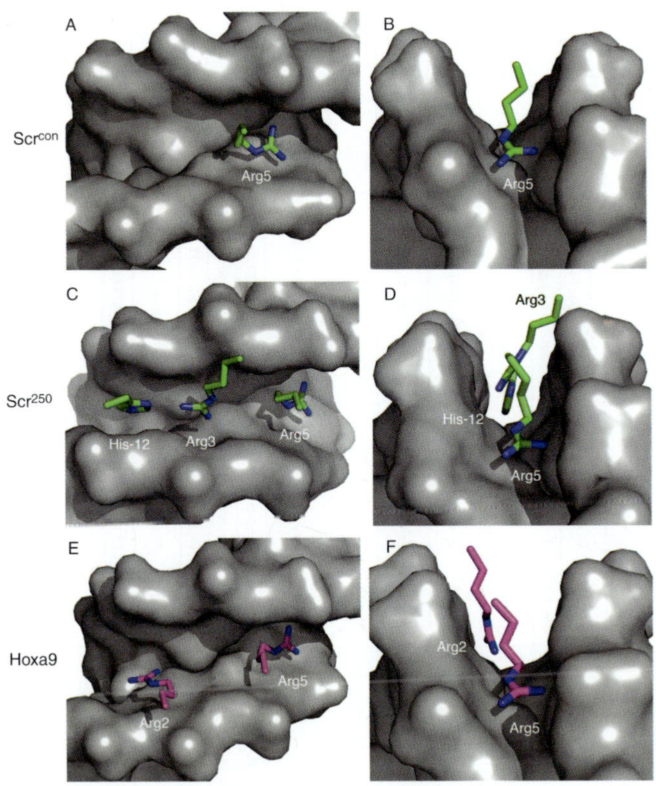

Richard S. Mann *et al.*, Figure 3.5 Please refer to the legend in the text.

Name (gene)	Hox	Reporter Validation	DNA Binding Site (Hox, PBC, Meis/Hth, Other)	Cooperative Cofactors	Other Factors	Target Origin; Notes	References
EphA2-repeat E (*EphA2/eck*)	Hoxa1 Hoxb1	*Ex vivo*	gcaTGATGGATGGgct	Pbx‡	ND	Mouse	(Chen and Ruley, 1998)
b1-ARE (*Hoxb1*)	Hoxb1 Lab	*In vivo*	ctcAGATGGATGGgctgcgggTGATTGAAGTgTCTTTGTCATGCTAAT gcttggggggTGATGGATGGgcgctggggCTGCCAaac	Pbx1, Prep1	Sox2, Oct1	Mouse; Autoregulatory element, R3 site (also known as Repeat3) functions as a Labial target in *Drosophila*	(Chan and Mann, 1996; Di Rocco et al., 2001; Ferretti et al., 2005; Pöpperl et al., 1995)
r4-Hoxa2 (*Hoxa2*)	Hoxb1	*In vivo*	ggaTGATTTATTTgag; attTGACAGtaatgaagagTGATAGATTGctc; ggcTGATGCATTAatt	Pbx1, Prep1	ND	Chicken and Mouse	(Tümpel et al., 2007)
b2-PP2 (*Hoxb2*)	Hoxb1	*In vivo*	gagCTGTCAgggggctaAGATTGATCGccc	Pbx1†, Prep1	ND	Mouse	(Ferretti et al., 2000)
A3-PP2; A3-PhP1 (*Hoxa3*)	Hoxb1	*In vivo*	ggtTGATTATTGacc; gagTCATAAATCTtgcccagccataaaTGACAAaaa	Pbx1a†, Prep1	ND	Mouse and Chicken; Mutation of Prep site did not affect reporter expression in vivo	(Manzanares et al., 2001)
P2b_0.38 (*Phox2b*)	Hoxb1	*In vivo*	gcgTGATTGAATtaa; ttaTTGTCAtgt	Pbx1a†, Prep1	ND	Mouse	(Samad et al., 2004)
FPB (*COL5A2*)	Hoxb1	*Ex vivo*	gtcTGATTGATGGtaa	Pbx1a*	Prep1*	Human; Cotransfection of Prep1 increased expression ex vivo	(Penkov et al., 2000)
125bpHb9 (*Hb9*)	Hoxb1, Hoxb3	*In vivo*	agcTGATGAATTGacaaaaACTAATCA	Pbx1†	ND	Mouse	(Nakano et al., 2005)
EVIII (*CG11339*)	Lab	*In vivo*	tcgTGATCAATTAcagCTGACTggg	Exd, Hth	ND	*Drosophila*; Although the Exd half-site is required in vivo, cooperative binding of Lab/Exd dimers was not observed in vitro	(Ebner et al., 2005)
lab48/95 (*lab*)	Lab	*In vivo*	aatTGATGGATTGcccggcgccgaCTGTCAccg	Exd, Hth	ND	*Drosophila*; Autoregulatory element	(Ryoo et al., 1999)
enh450 (*ceh-13*)	CEH-13	*In vivo*	aaaTGATGGATGGttc	CEH-20	ND	*C. elegans*; Autoregulatory element	(Streit et al., 2002)
E2-A2RE (*Hoxa2*)	Hoxa2	*Ex vivo*	cggTGATTGATGGaag	Pbx1†	ND	Mouse, Autoregulatory element	(Lampe et al., 2004)
HRE1.25 (*Hoxa2*)	Hoxa2	*In vivo*	tctTGATTGATGAact	Pbx1a*	Prep1*	Mouse; Autoregulatory element, cotransfection of Prep1 increased expression ex vivo	(Lampe et al., 2008)
TTF-1-BS1&2 (*TTF-1*)	Hoxb3	*Ex vivo*	cctTAATTGgct; agaTAATTAgct	ND	ND	Rat; Both sites are necessary for expression ex vivo	(Guazzi et al., 1994)
CR3-HS1&2 (*Hoxb3, Hoxb4*)	Hoxb4, Hoxd4, Hoxb5	*In vivo*	agTGATTAATGGttttctgtaTAATTCtc	Exd†	ND	Mouse; Autoregulatory element; also responds to Dfd, Hoxb5, Scr, Antp, and Ubx in *Drosophila*	(Gould et al., 1997)
Rarb (*Rarb*)	Hoxb4, Hoxd4	*In vivo*	gggTGATAAATAAtgg	Pbx#	ND	Mouse; Role of Pbx not tested	(Serpente et al., 2005)
Repeat 3-TA (synthetic)	Dfd	*In vivo*	gggTGATTAATGGgcg	Exd	ND	Synthetic; (GG->TA) mutation of Repeat3 of b1-ARE	(Chan et al., 1997)
DfdEAE-module E (*Dfd*)	Dfd	*In vivo*	cccTAATTGccacgCATTAGctc	ND	Exd‡	*Drosophila*; Autoregulatory element, atypical Exd binding site, *exd* required in vivo	(Pinsonneault et al., 1997; Zeng et al., 1994)

Richard S. Mann *et al.*, Table 3.1 (Continued)

Richard S. Mann *et al.*, Table 3.2 (Continued)

Name (gene)	Hox	Reporter Validation	DNA Binding Site (Hox, PBC, Meis/Hth, Other)	Cooperative Cofactors	Other Factors	Target Origin; Notes	References
rprDfd (*rpr*)	Dfd	*In vivo*	caaTAATTAccc; ctcTAATTGccc; aacTAATTGaca; tcaTAATTGagg	Not Exd‡	ND	*Drosophila*; Cooperative binding with Exd not observed	(Lohmann et al., 2002)
1.28DRE (*1.28*)	Dfd	*In vivo*	gttTAATTGgtt; tctTAATAGccg; ccaCTACATTAATTATgaa; ccgCGATAATAAatc	Not Exd‡	ND	*Drosophila*; Sites not tested individually, cooperative binding with Exd not observed	(Pederson et al., 2000)
jkl-216-site1 (*hlh-8*)	LIN-39	*In vivo*	agtTGAAaaATTACCgcg	CEH-20	ND	*C. elegans*	(Liu and Fire, 2000)
egl-17 (*egl-17*)	LIN-39	*In vivo*	ctcTGATTAATCActg	CEH-20	ND	*C. elegans*	(Cui and Han, 2003)
egl-18/elt-6-site 1 (*elg-18, elt-6*)	LIN-39	*In vivo*	gggTGATATATATgtt	CEH-20†	ND	*C. elegans*	(Koh et al., 2002)
Fkh250 (*fkh*)	Scr	*In vivo*	tcaAGATTAATCGcca	Exd, Hth	ND	*Drosophila*; Hth binding site not found	(Ryoo and Mann, 1999)
Lin-39-site2 (*lin-39*)	LIN-39	*In vivo*	catTGATTTATTTttg	CEH-20†	ND	*C. elegans*; Autoregulatory element	(Wagmaister et al., 2006)
mod-84 (*mod*)	Scr	*In vivo*	acaTAATTTgttagtatgTAATATtccctgGTATTTTTCCAATGactgtcaa gcagtgctgGTAAATGGCGGAACaag	ND	Tsh	*Drosophila*; Although both Scr and Tsh regulate *mod*, a cooperative interaction was not observed	(Taghli-Lamallem et al., 2007; Taghli-Lamallem et al., 2008)
p53 (*p53*)	Hoxa5	*Ex vivo*	tctTAATTCaaa	ND	ND	Human and Mouse	(Raman et al., 2000a)
PRP-62 (*PR*)	Hoxa5	*Ex vivo*	cggTAATTGggg	ND	ND	Human	(Raman et al., 2000b)
PTN-106 (*PTN*)	Hoxa5	*Ex vivo*	tttTAATAAgct	ND	ND	Human	(Chen et al., 2005)
HBE (*flk1*)	Hoxb5	*In vivo*	tgcTGATTAATGAaaa	Pbx#	ND	Mouse; Mutation affecting reporter activity located in putative Pbx site	(Wu et al., 2003)
bFGF (*bFGF*)	Hoxb7	*Ex vivo*	cgcTAATCTgg	ND	ND	Human	(Caré et al., 1996)
egl-1+5995 (*egl-1*)	MAB-5	*In vivo*	cgtTGATTTATTTtta	CEH-20	ND	*C. elegans*	(Liu et al., 2006)
apME680 (*ap*)	Antp	*In vivo*	ccaATTAtttTGATTAATGCcaa; cccATAAATAATATtaa; tttTTATGAgtt; ttgAAATGAact; tgaATTATGATTATTGCcat	Exd†	ND	*Drosophila*; *exd* required in vivo	(Capovilla et al., 2001)
Tsh (*tsh*)	Antp, Ubx	*In vivo*	aacTAATGTAATTACaaa; tgaTAATTGact; cacATAAATctt; tttTAATATttt	ND	ND	*Drosophila*; Cluster of Hox binding sites tested by deletion	(McCormick et al., 1995)
sal1.1 (*sal*)	Ubx	*In vivo*	ttaTAATgTGCCCGTCTTAATATgat	Not Exd or Hth	Mad/ Med	*Drosophila*; *exd* and *hth* not required in vivo	(Galant et al., 2002; Walsh and Carroll, 2007)
knot (*knot*)	Ubx	*In vivo*	gctTAATTTg; gctTAATTCt; cacTAATTAt; gccTAATTGTAATTGTTATTATTAATTAa	Not Exd or Hth	ND	*Drosophila*; Hox sites are additive, *exd* and *hth* not required in vivo	(Hersh and Carroll, 2005)
B3 (*βTub60D*)	Ubx	*In vivo*	ttcATAATTGcagcggccacactcCAATTAAatt	ND	ND	*Drosophila*	(Kremser et al., 1999)
CG13222-site1 (*CG13222*)	Ubx	*In vivo*	gtgTAATTTatc	Not Exd or Hth	ND	*Drosophila*; *exd* and *hth* not required in vivo	(Hersh et al., 2007)
Antp-P2 (*Antp*)	Ubx, AbdA	*In vivo*	30 Hox binding sites	ND	ND	*Drosophila*; Activity lost when all 30 sites are mutated.	(Appel and Sakonju, 1993)

Name (gene)	Hox	Reporter Validation	DNA Binding Site (Hox, PBC, Meis/Hth, Other)	Cooperative Cofactors	Other Factors	Target Origin; Notes	References
DMX-R (*Dll*)	Ubx, AbdA	*In vivo*	gcaCTATAAAaCTGTCCgcggGAATGATTTAATTTcccaAATATTgtc	Exd, Hth, En	Slp	*Drosophila*; Slp and En function as repressors	(Gebelein et al., 2004)
dpp-midgut enhancer (*dpp*)	Ubx, AbdA	*In vivo*	caaaTTTATTACTAATTGGGtgTGAATTGCaggcagtgcaagtgctgca tatgcagcatgcagcATAATCGAAatgggtgCTAATTGATag	Exd	ND	*Drosophila*; Atypical Exd binding site, additional Hox and Exd sites required	(Capovilla and Botas, 1998; Capovilla et al., 1994; Chan et al., 1994; Manak et al., 1994; Stultz et al., 2006; Sun et al., 1995)
RhoA (*rho*)	AbdA	*In vivo*	agtTCATTGATTGACATTTTTATTAtgc	Exd, Hth	Sens	*Drosophila*; Sens competes for binding with AbdA-Exd-Hth complexes	(Li-Kroeger et al., 2008)
XC-Box2 (*wg*)	AbdA	*In vivo*	gcaTAATCTAATTGcgg	Not Exd or Hth‡	Mad, Medea, Creb	*Drosophila*; Exd and Hth do not form complexes with AbdA in vitro	(Grienenberger et al., 2003)
AH5 (*Hoxa4*)	Dfd, Scr, Antp, Ubx & AbdA	*In vivo*	ggCAATTAAATTTATGGgggcTATAATTActg	ND	ND	Mouse; Possible role for Exd, repressed by AbdA; activated by other Hox	(Haerry and Gehring, 1997)
FkhCON (synthetic)	Scr, Antp, Ubx, AbdA	*In vivo*	tcaAGATTTATGGcca	Exd, Hth	ND	Synthetic; Fkh250 binding site mutated to consensus Hox-Exd site, repressed by AbdA, activated by other Hox	(Ryoo and Mann, 1999)
MLC (*MLC1/3*)	Hoxc8, Hoxa10	*Ex vivo*	cttATTAAATTAcCATGTGtga	ND	ND	Rat; Factor binding to CATGTG (E-box) not identified, Hoxc8 activates and Hoxa10 represses	(Ceccarelli et al., 1999; Houghton and Rosenthal, 1999; Rao et al., 1996)
SBE (*OPN*)	Hoxc8, Hoxa9	*Ex vivo*	atgCAGTCtataaatgaaaagggtagtTAATGAcat	ND	Smad1, Smad3, Smad4	Mouse; Smad3 directly binds DNA site while Smad1 and Smad4 antagonize Hox binding	(Shi et al., 2001; Shi et al., 1999)
EphB4-1365 (*EphB4*)	Hoxa9	*Ex vivo*	agcTTATT	ND	ND	Human	(Bruhl et al., 2004)
E-510 (*E-selectin*)	Hoxa9	*Ex vivo*	atgCAATTTTATTAAtat	ND	ND	Human	(Bandyopadhyay et al., 2007)
N-CAM (*N-CAM*)	Hoxc6, Hoxb8, Hoxb9	*Ex vivo*	tTAATAATtac; cctTAATCAg	ND	ND	Mouse; Hoxc6 and Hoxb9 activate while Hoxb8 represses reporter expression ex vivo	(Edelman and Jones, 1995; Jones et al., 1993; Jones et al., 1992)
p21A10RE (*p21*)	Hoxa10	*Ex vivo*	tttTTATAAttt	Pbx1a*, Meis1b*	ND	Human; Cotransfection of Pbx1a and Meis1b increased reporter expression ex vivo, no direct binding sites identified	(Bromleigh and Freedman, 2000)
B3A (*B₃-integrin*)	Hoxa10	*Ex vivo*	aatGTTATTTtta	ND	ND	Human; Hox binding sites not confirmed by mutagenesis	(Daftary et al., 2002)
EMXC (*Emx2*)	Hoxa10	*Ex vivo*	tgtTTATGTgat	Not Pbx‡	ND	Human; Pbx did not form complexes with Hoxa10 in vitro	(Troy et al., 2003)
Runx2-site1 (*Runx2*)	Hoxa10	*Ex vivo*	aggTTATAGctt	ND	ND	Rat	(Hassan et al., 2007)

(Continued)

Name (gene)	Hox	Repcrter Validation	DNA Binding Site (Hox, PBC, Meis/Hth, Other)	Cooperative Cofactors	Other Factors	Target Origin; Notes	References
PPE (*Ren-1c*)	Hoxd10	*In vivo*	tgtTTCCACAct; cgcTTCCggc; cttTGATTTATTAccc	Pbx1b, Prep1*	N1IC, Ets-1	Mouse; Homeodomain of Prep1 not necessary for activation ex vivo, Prep1 binding site not identified	(Glenn et al., 2008; Pan et al., 2005; Pan et al., 2001)
Six2 (*Six2*)	Hox11	*In vivo*	gTTATCTgacccggggcctgcccgcgccaGacaatagTCGaGTCA aattaTTc	ND	Pax2, Eya1	Mouse	(Gong et al., 2007)
Enpp2 (*Enpp2*)	Hoxa13	*Ex vivo*	TTAATTG; TTAACAT; TTTATAT	ND	ND	Mouse	(McCabe and Innis, 2005)
Sostdc1 (*Sostdc1*)	Hoxa13	*Ex vivo*	15 sites tested	ND	ND	Mouse	(Knosp et al., 2007)
EphA7-site 3 (*EphA7*)	Hoxa13 Hoxd13	*Ex vivo*	ataTTATTGgag	ND	ND	Mouse	(Salsi and Zappavigna, 2006)
hHa2-motifs: 1, 3, 6, 11 (*hHa2*)	Hoxc13	*Ex vivo*	aaaTTAATTAgcag; tctTTTATTGgaa; cttTTAATGAaag; catTTAATATgttg	Not Meis/Prep	ND	Human; Meis/Prep localization cytoplasmic	(Jave-Suárez and Schweizer, 2006; Jave-Suárez et al., 2002)
hHa5-motifs: 2, 3, 4, 6, 9, and 13 (*hHa5*)	Hoxc13	*Ex vivo*	actTTAATGAgga; gttTTAATAGaaa; aggTTAATGAggg; tatTTTATGAgact; tctTTTATTGgcct; cagTTAATTGgac	Not Meis/Prep	ND	Human; Meis/Prep localization cytoplasmic	
hHa7-motif 2 (*hHa7*)	Hoxc13	*Ex vivo*	gctTTAATGAgct	ND	ND	Human	
Foxq1 (*Foxq1*)	Hoxc13	*Ex vivo*	accTTCATTAcaa; tccTCCATAAAaca; agcTTAATAGGgac	ND	ND	Mouse; Other Hoxc13 sites may be required	(Potter et al., 2006)
BE3-6 (*yellow*)	AbdB	*In vivo*	aggTCGTAAAAcgtatttttacccatttgcatgtTTATTATGcgt	ND	ND	*Drosophila*	(Jeong et al., 2006)
BabDE (*bab*)	AbdB	*In vivo*	7 TTTAT, 7 TTTAC, 8 TTAT, and 2 ACAATGT binding sites	ND	Dsx	*Drosophila*	(Williams et al., 2008)
i4c-S2 (*psa-3*)	NOB-1	*In vivo*	tttTGATAGTAATttt	CEH-20	ND	*C. elegans*; CEH-20 facilitates NOB-1 DNA binding independent of its own binding site	(Arata et al., 2006)

Richard S. Mann *et al.*, Table 3.1 Please refer to the legend in the text.

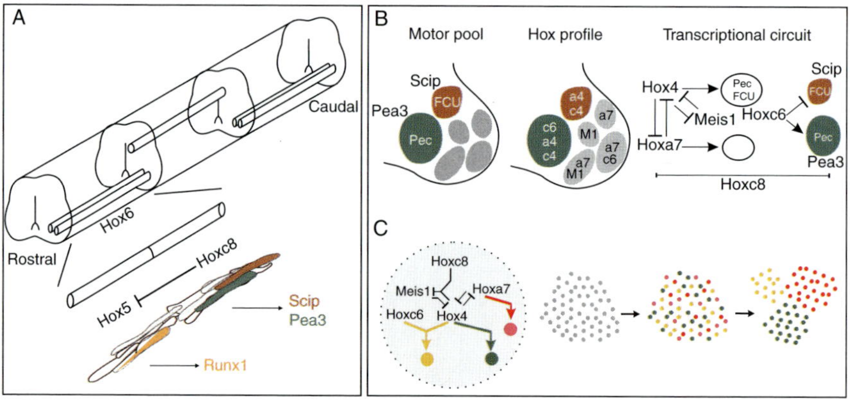

Jeremy S. Dasen and Thomas M. Jessell, Figure 6.4 Please refer to the legend in the text.

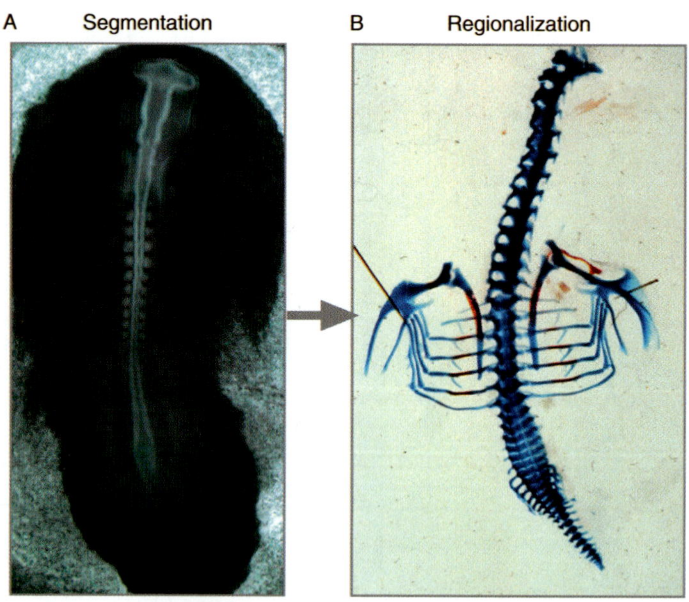

Tadahiro Iimura *et al.*, Figure 7.1 Please refer to the legend in the text.

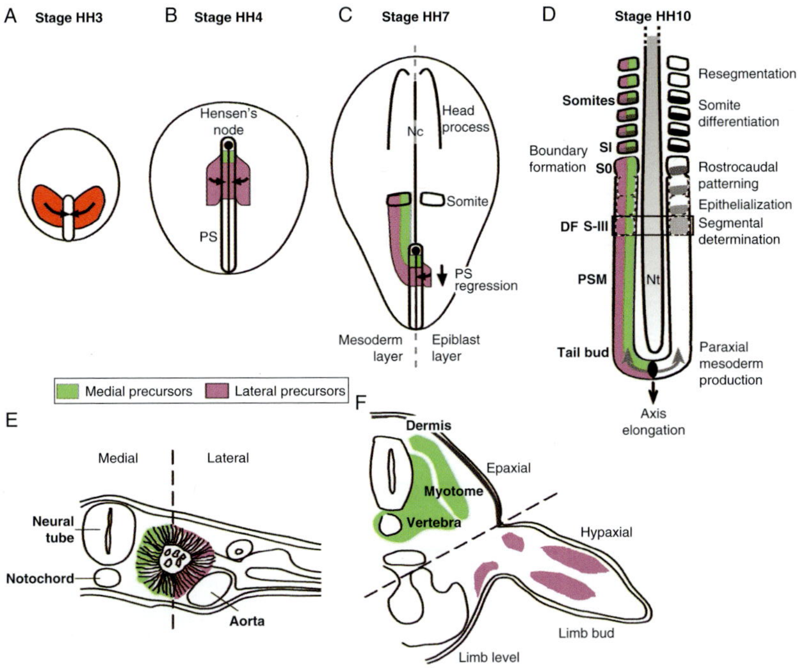

Tadahiro Iimura *et al.*, Figure 7.2 Please refer to the legend in the text.

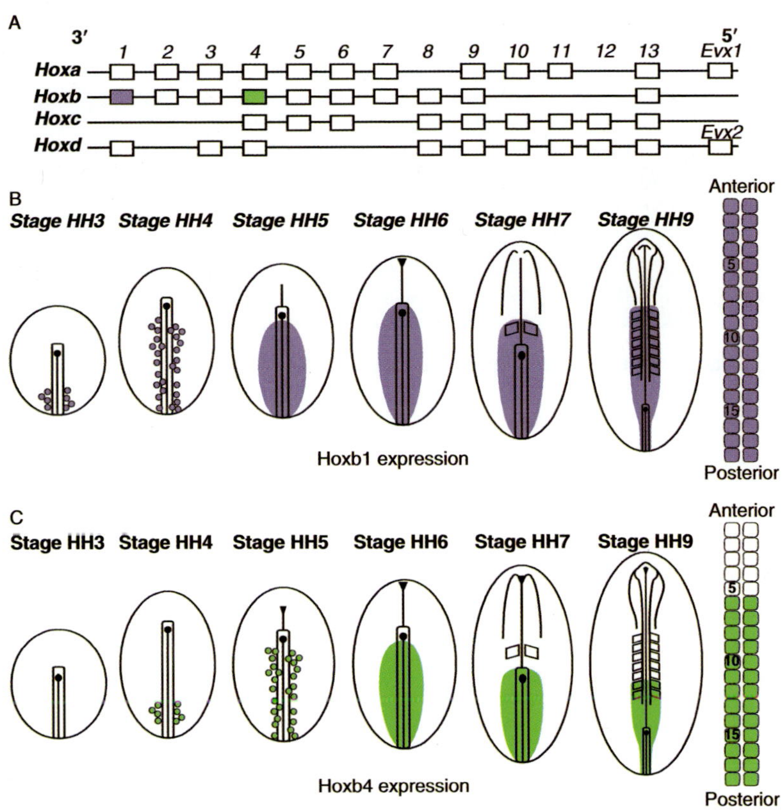

Tadahiro Iimura *et al.*, Figure 7.3 Please refer to the legend in the text.

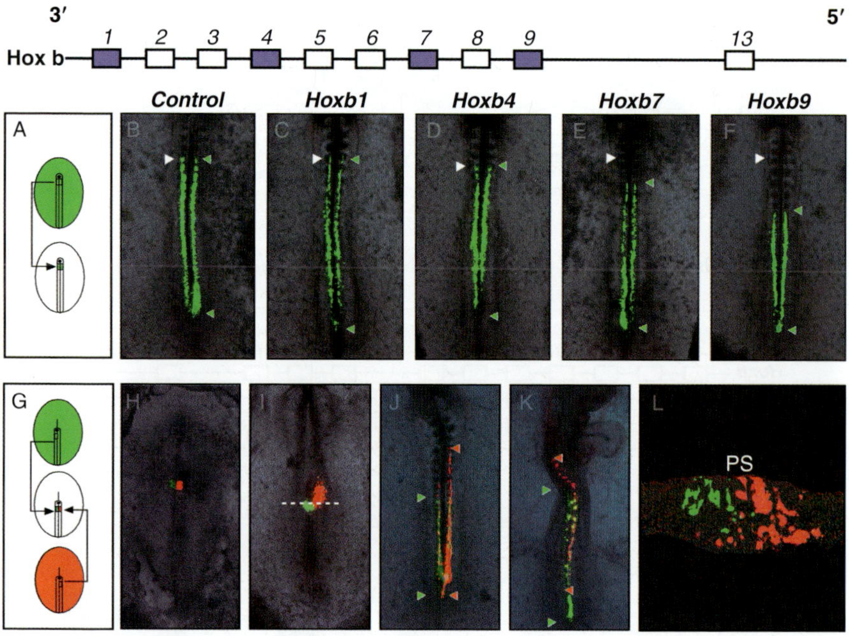

Tadahiro Iimura *et al.*, Figure 7.4 Please refer to the legend in the text.

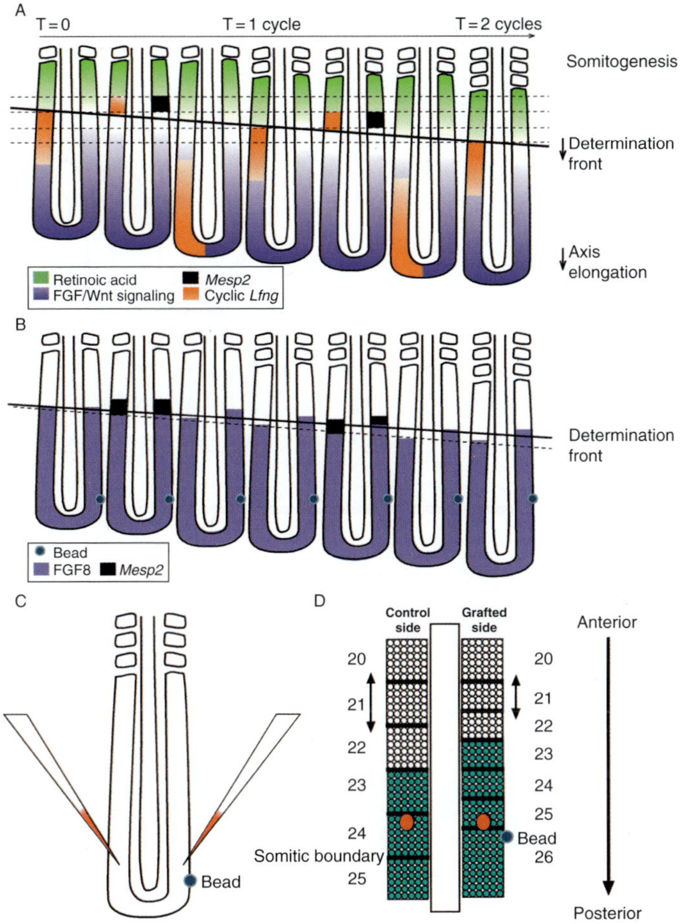

Tadahiro Iimura *et al.*, Figure 7.5 Please refer to the legend in the text.

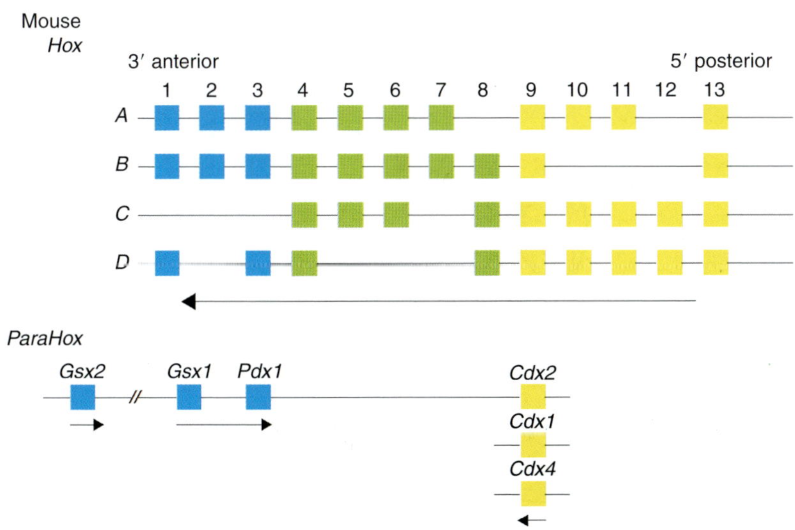

Teddy Young and Jacqueline Deschamps, Figure 8.1 Please refer to the legend in the text.

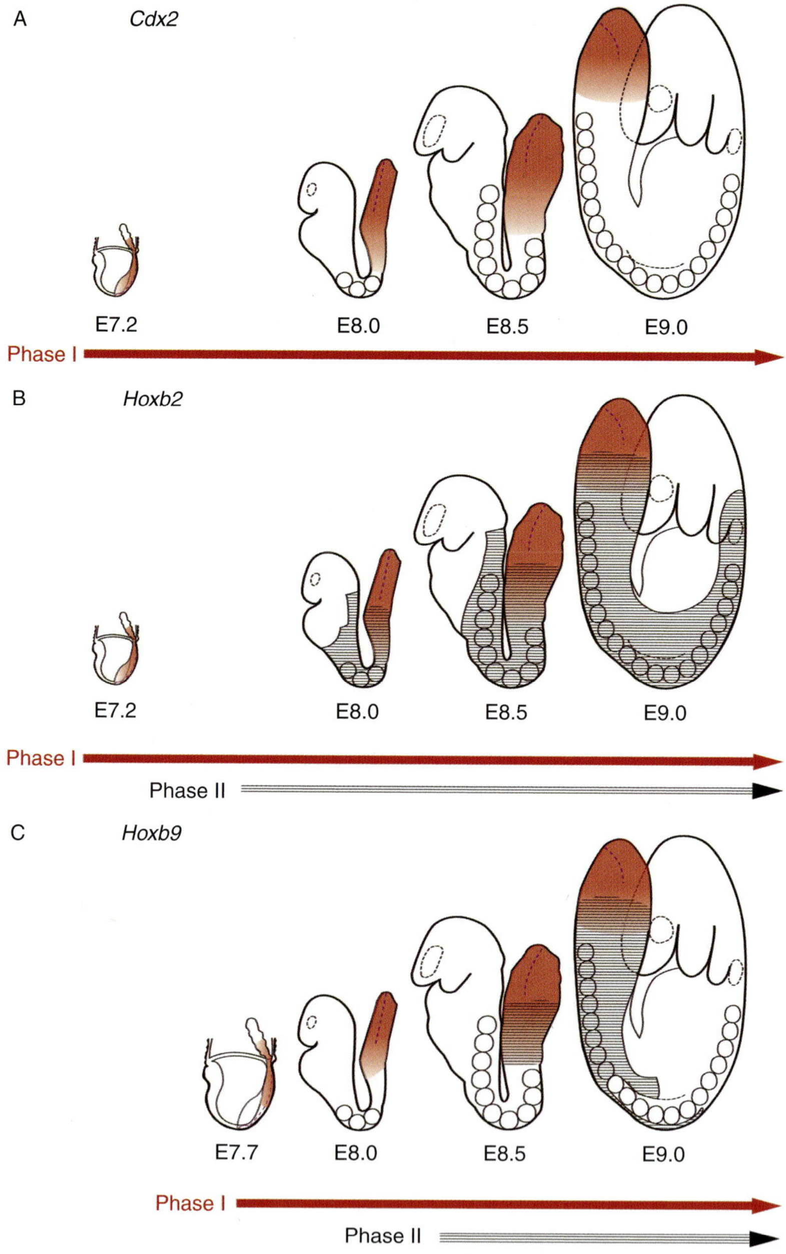

Teddy Young and Jacqueline Deschamps, Figure 8.2 Please refer to the legend in the text.

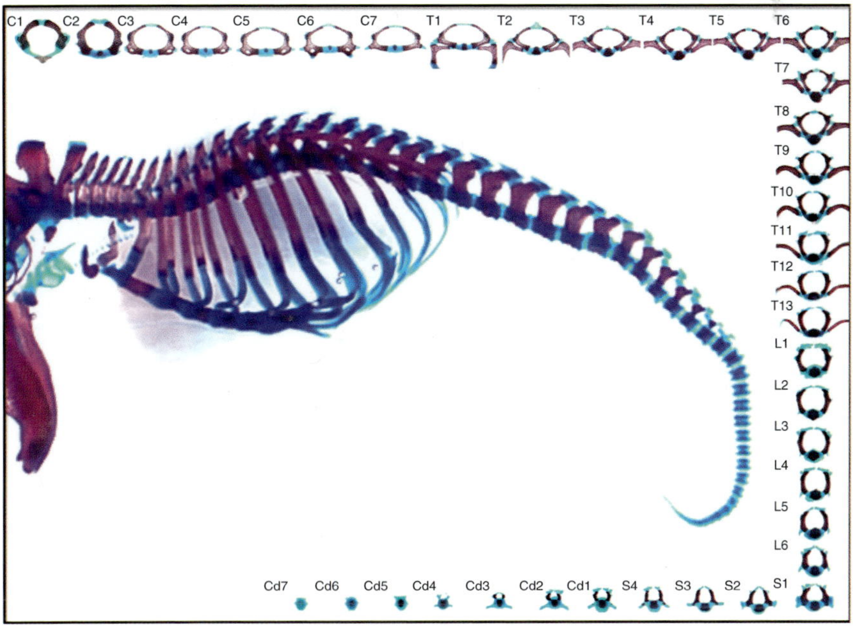

Deneen M. Wellik, Figure 9.2 Please refer to the legend in the text.

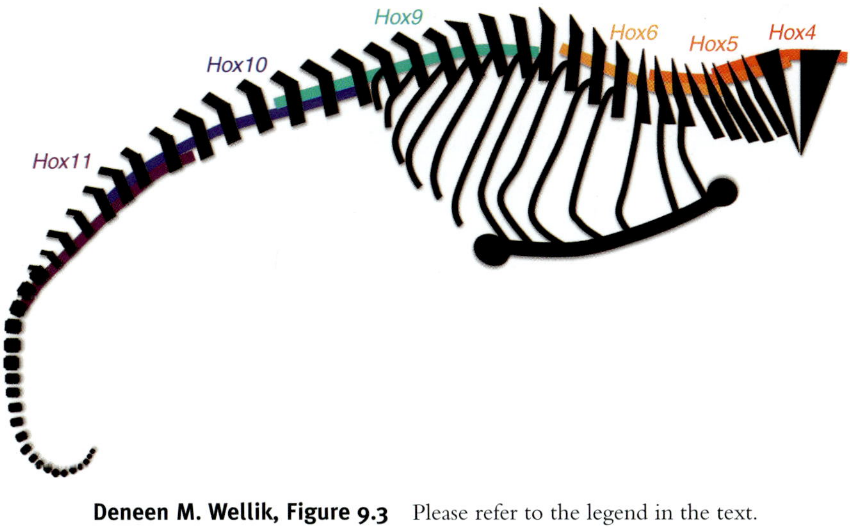

Deneen M. Wellik, Figure 9.3 Please refer to the legend in the text.

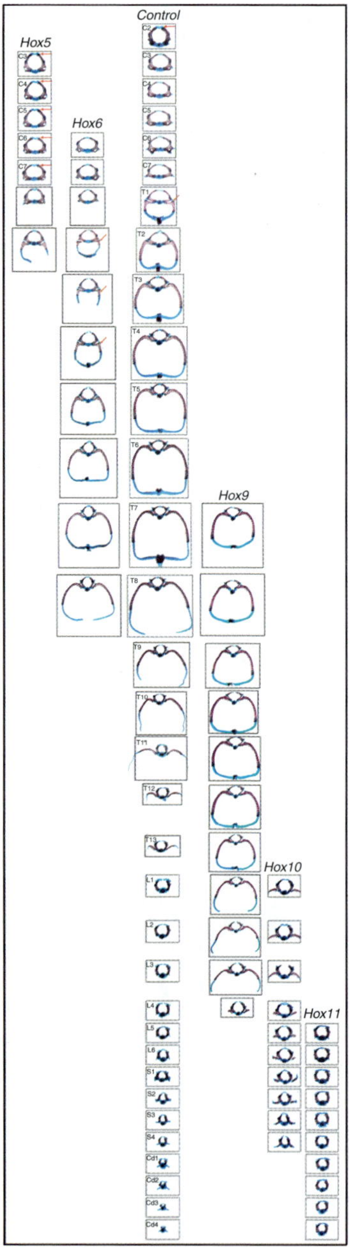

Deneen M. Wellik, Figure 9.4 Please refer to the legend in the text.